Renal Physiology and Hydrosaline Metabolism

Pedro A. Gallardo • Carlos P. Vio

Renal Physiology and Hydrosaline Metabolism

EDICIONES UC

Pedro A. Gallardo 🆔
Unit of Basic Science, Faculty of Medicine
Finis Terrae University
Santiago, RM - Santiago, Chile

Carlos P. Vio 🆔
Faculty of Medicine and Science
San Sebastian University
Santiago, Chile

The translation was done with the help of artificial intelligence (machine translation by the service DeepL. com). A subsequent human revision was done primarily in terms of content.

Translation from the Spanish language edition: Fisiología Renal y Metabolismo Hidrosalino. Segunda edición. By Pedro A. Gallardo and Carlos P. Vio. © Ediciones Universidad Católica de Chile, 2018. Original Publication ISBN 9789561422605. All rights reserved.

Jointly published with Ediciones Universidad Católica de Chile, Santiago. Original Spanish edition published by Ediciones Universidad Católica de Chile, Santiago, 2018.

ISBN 978-3-031-10258-5   ISBN 978-3-031-10256-1   (eBook)
https://doi.org/10.1007/978-3-031-10256-1

This Springer imprint is published by the registered company Springer Nature Switzerland AG
The registered company address is: Gewerbestrasse 11, 6330 Cham, Switzerland

Preface to the Second Edition

As the first edition, this second edition of *Renal Physiology and Hydrosaline Metabolism* is dedicated to medical and health sciences, and biology students as well as professional in these areas. This updated and revised edition in English is a translation of the second edition in Spanish, published by Ediciones UC. This title was selected by Springer Nature and Ediciones Universidad Católica to be translated and published in English. The authors are indebted to the publishers for this recognition and opportunity to reach a broader spectrum of readers.

Important advances in the field of renal physiology had occurred since the publishing of the first edition. In this second edition, we incorporated those advances in knowledge that we felt were important for students and professionals. Several chapters were rewritten to incorporate new relevant issues. For example, in Chap. 2, we included information on the molecular structure of the slit diaphragm junction and its relevance for the function of the glomerular filtration barrier. In Chap. 3, we included a discussion concerning the role of adenosine as a paracrine signal in tubuloglomerular feedback. We also included recent insights about the regulation of the function of the distal convoluted tubule and how this knowledge impacts the understanding of the regulation of potassium balance and regulation of effective circulating volume. We also included the anatomical base for the connecting tubule glomerular feedback (*CTGF*) in Chaps. 2 and 11. The importance of dietary potassium on sodium excretion, blood pressure regulation and prevention of cardiovascular diseases and hypertension, and public health regulations is highlighted in Chaps. 11 and 12. Chapter 13 is a new chapter concerning genetic alterations in tubular transport of Na^+, Cl^-, K^+, and water. We felt that this chapter was important for two reasons. First, it highlights the impact of molecular biology in understanding renal physiology. Second, from the pedagogical standpoint, the chapter underlines the relevance of specific proteins in the function of a given tubular segment. This is accomplished through the analysis of the impact of specific protein mutations in the function of the tubular segment and how this affects the function of downriver segments and whole electrolyte and water balance.

At the beginning of each chapter, the authors included a list of learning objectives that guide the reader about the goals of the chapter. At the end of each chapter, there is a list of review questions; the answer to each question requires the application of specific contents in the chapter. All the answers to review questions are contained in Chap. 14.

Chapter 6 is devoted to biologists and to all students interested in osmoregulation in non- mammalian vertebrates. The authors felt that biologists should have a broader knowledge about osmoregulation than the study of the mammalian kidney.

We want to thank the overwhelming number of comments from medicine students from Pontificia Universidad Católica and Universidad Finis Terrae who encouraged us to write this second and updated edition, and to the Universidad San Sebastián.

We are deeply indebted to Mr. Carlos Céspedes for his invaluable help in reviewing the manuscript and figures and to Ms. Fernanda Carter for her valuable help. We also thank the artwork originally made by Fabiola Solari and enriched by Felipe Serrano.

Santiago, Chile   Pedro A. Gallardo
                                                                             Carlos P. Vio

Acknowledgments

We would like to thank Mr. Carlos Céspedes for his rigorous and professional review of the chapters and figures and Felipe Serrano (www.illustrative-science. com/) for vectorizing our draft figures.

This textbook *Renal Physiology and Hydrosaline Metabolism* has been funded in part by ANID (Agencia Nacional de Investigación y Desarrollo, Chile) to the Basal Centre of Excellence in Aging and Regeneration CARE (Grant PIA CONICYT ACE210009), and by a donation from SQM to the Basal Center CARE from the Pontificia Universidad Catolica de Chile.

Contents

About the Authors

Carlos P. Vio, obtained his degree in medicine from Universidad Austral de Chile and the University of Chile. He pursued his postdoctoral studies in endocrinology through a Fogarty CTR-NIH grant in New York, USA. Dr. Vio is Full Professor of Physiology in the Faculty of Medicine and Science, Universidad San Sebastián, Santiago, Chile, and Full Professor of Physiology at Pontificia Universidad Catolica de Chile. He has a large experience teaching in Physiology and Pathophysiology at medical students, health sciences, and PhD programs in Physiology, Cell Biology and Biomedicine, and Medical Sciences. He is author of more than 100 research papers and books in physiology, hypertension, and chronic renal diseases. Dr. Vio is a member of the Chilean Academy of Medicine, Chilean Society of Hypertension, Chilean Society of Physiological Sciences, American Society of Nephrology, American Society of Physiology, InterAmerican Society of Hypertension, and American Heart Association, among others. He was chairman of physiology and vice president of research and doctoral studies at Pontificia Universidad Catolica de Chile (2000–2010) and currently is vice president of research and doctoral studies at Universidad San Sebastian, Chile.

Pedro A. Gallardo, MSc, PhD, obtained his degree in biological sciences at Universidad de Talca. He obtained his Master of Biological Sciences (Physiology) from Universidad de Chile and Doctor of Biological Sciences (Physiological Sciences) from Pontificia Universidad Católica de Chile. He holds a Fogarty postdoctoral position at NHLBI-NIH, Bethesda, USA, and is a graduate in pedagogy in teaching in health sciences. For 10 years, Dr. Gallardo was in charge of the physiology course for medicine students. He is an associate professor at the School of Medicine, Faculty of Medicine, Universidad Finis Terrae.

General Kidney Functions

1

Learning Objectives
- To understand the role of the kidney in the homeostasis of body fluids through the regulation of several physiological variables.

The kidney is a key organ in the regulation of the composition and volume of the extracellular fluid. Through this general function, the kidney contributes to the maintenance of homeostasis of body fluids. This huge regulatory task can be divided in the regulation of several physiological variables that are briefly reviewed below.

1.1 Excretion of Nitrogenous Metabolites

Several nitrogenous compounds like creatinine, urea, and uric acid are excreted by the kidney. The body is constantly producing these nitrogenous compounds; therefore, these are metabolic end products that must be excreted to keep their plasma concentrations constant. More broadly, the renal excretory function also includes the excretion of several organic endogenous (e.g., biliary salts) and exogenous compounds like drugs (e.g., antibiotics, diuretics, analgesics).

1.2 Conservation of Organic Solutes That Are Valuable to the Body

The importance of this process becomes evident if one considers that the total volume of extracellular fluid is filtered by the kidneys more than 10 times in a day (180 l/day approximately). If valuable solutes such as glucose and amino acids were not reabsorbed, their plasma concentration would decline rapidly. A similar situation applies to all electrolytes. The conservation of valuable solutes occurs through a wide variety of transepithelial transport mechanisms that take place along the renal tubule and collectively known as tubular reabsorption and secretion.

© Springer Nature Switzerland AG 2022
P. A. Gallardo, C. P. Vio, *Renal Physiology and Hydrosaline Metabolism*,
https://doi.org/10.1007/978-3-031-10256-1_1

1.3   Regulation of Plasma Osmolality

The kidney is able to regulate the concentration of osmotically active solutes in the extracellular fluid. Plasma osmolality is regulated through changes in renal water excretion, under the influence of antidiuretic hormone (ADH). The maintenance of a normal extracellular osmolality is essential for cell volume homeostasis. This is evident in the brain, where changes in osmolality produce alterations in neuronal volume, which can generate important disorders. For example, an increase in the volume of brain neurons (cerebral edema) results in symptoms and signs (headache, drowsiness, coma) and other alterations in the function of the nervous system, which can even lead to death.

1.4   Regulation of the Effective Circulating Volume

Regulation of effective circulating volume occurs through control of renal excretion of Na^+ under the influence of the sympathetic system, renin-angiotensin-aldosterone system, kallikrein-kinin, atriopeptin, and nitric oxide. The body maintains a balance between the intake and excretion of Na^+. Fine adjustments in renal excretion of Na^+ are directly related to maintaining effective circulating volume and blood pressure. Alterations in renal sodium handling lead to alterations in circulating volume and blood pressure.

1.5   Regulation of the Acid-Base Balance

Through the control of renal acid excretion, the kidney plays a key role in acid-base balance, reabsorbing bicarbonate from the glomerular filtrate. The kidney also regenerates an amount of bicarbonate equal to that consumed in neutralizing protons from acids derived primarily from protein and nucleic acid metabolism. The role of the kidneys in acid-base balance is integrated with the function of the liver and the lungs. The pulmonary ventilation eliminates CO_2 derived from aerobic metabolism, and hepatic production of glutamine by the liver is essential for bicarbonate regeneration in the kidneys.

1.6   Regulation of Potassium Balance

Potassium balance is maintained through two processes: internal and external balance. Internal balance maintains plasma potassium concentration during the postprandial period and other circumstances like exercise. This process allows the regulation of plasma potassium concentration within a narrow range which is essential to the function of excitable cells. Hypokalemia and hyperkalemia affect the resting membrane potential of all cells. Urinary potassium excretion roughly equals daily potassium ingestion.

1.7   Regulation of Calcium and Phosphate Balance

The kidney plays an important role in maintaining calcium and phosphate balance and bone metabolism. Parathyroid hormone (PTH) and calcitriol regulate calcium and phosphate renal handling.

1.8   Endocrine Function

The kidney is an endocrine organ that produces and secretes many hormones and vasoactive components. Renal cortical interstitial cells synthesize the peptide hormone erythropoietin that stimulates erythropoiesis. The proximal tubule is the site where the final step in calcitriol synthesis takes place; calcitriol participates in calcium and phosphate homeostasis. Although renin is not a hormone, this endopeptidase, secreted by granular cells of the juxtaglomerular apparatus, is the rate-limiting step in the formation of angiotensin I, and angiotensin I-converting enzyme generates angiotensin II, both very important in the renin-angiotensin system. Kallikrein is an enzyme secreted by the connecting cells of the connecting tubules; the enzyme participates in bradykinin formation, a peptide hormone that regulates several aspects of renal function.

1.9   Extrarenal Organs in the Maintenance of the Hydrosaline Balance

The mammalian kidney is, to a superlative degree, the main excretory organ in the maintenance of the hydrosaline balance. However, the above statement does not apply to the kidneys of other vertebrates, such as birds, reptiles, and fish, which are generally unable of excreting excess salt. In non-mammalian vertebrates, the extrarenal excretion of salts constitutes an effective mechanism to maintain the hydrosaline balance. The extrarenal excretion occurs in the fish gills and in reptilian and avian salt-secreting glands. The architecture of these glands bears no resemblance to that of a kidney, but at the cellular and molecular level, there is a convergence in the ultrastructure of the constituent epithelial cells and in the transporters and channels that carry out the secretion. On the other hand, the amphibian skin and bladder are important organs in hydrosaline metabolism in amphibians.

Given the wide range of physiological variables regulated by the kidneys, it is not surprising that a progressive failure in renal function results in a wide variety of clinical manifestations. Hence, a reduction in the renal clearance capacity increases plasma concentration of several nitrogenous metabolites. This gives rise to the uremic syndrome which includes, for example, anemia associated with decreased erythropoietin synthesis, bone disease triggered by the imbalance between calcium and phosphate, and alteration in the amount of water, Na^+, and K^+ of the body, characteristics of kidney failure. This picture, taken to its maximum clinical expression, is incompatible with life, unless some artificial method of replacement, such as hemodialysis or peritoneum dialysis, is used—at least some of the kidney functions described above. These therapeutic procedures were developed in the second half of the twentieth century from ingenious observations of basic renal physiology.

In this text the regulation of the different physiological variables mentioned above will be described separately, bearing in mind that in the organ all of them are regulated simultaneously and in an integrated manner.

1.10   Nomenclature of Solute Carriers, Water, and Ion Channels

The genes that code for carriers and channels and their respective protein products have a consensus acronym that is used internationally. The acronym refers to the molecule(s) transported or to permeating ionic species in the case of an ion channel. The nomenclature of the most important transport proteins that are present in the renal tubule and vascular elements is given below. In most cases, there is more than one family member or isoform in the renal tubule or associated vascular elements.

The common names, international acronyms, and locations of the most important proteins are listed in Table 1.1.

Table 1.1 Nomenclature used in this book for solute transporters, aquaporins, and ion channels

Transporter or channel	Acronym	Location
Water channels or aquaporins (AQP-x)	AQP-1	Proximal tubule, thin descending limb of Henle, vasa recta
	AQP-2, AQP-3, AQP-4	Connecting tubule, collecting duct
Na^+/glucose cotransporters	SGLT1 ($2Na^+$:1 glucose) SGLT2 ($1Na^+$:1 glucose)	Proximal tubule
Na^+-phosphate cotransporter	NPT2 ($2Na^+$:$1H_2PO_4$)	Proximal tubule
Cation-Cl^- cotransporters	Na^+-$K^+$$2Cl^-$ (NKCC2)	Thick ascending limb of Henle
	Na^+-K^+-$2Cl^-$ (NKCC1)	Salt-secreting glands
	Na^+-Cl^- (NCC)	Distal convoluted tubule
Na^+-Ca^{++} exchanger	NCX	Distal convoluted and connecting tubules
Na^+-H^+ exchanger	NHE3	Proximal tubule and thick ascending limb of Henle
Na^+-bicarbonate cotransporter	NBC	Proximal tubule

(continued)

Table 1.1 (continued)

Transporter or channel	Acronym	Location
Cl$^-$ anion exchangers	Cl$^-$-organic ion (formate)	Proximal tubule
	Cl$^-$-HCO$_3^-$ (AE1 basolateral) Cl$^-$-HCO$_3^-$ (pendrin, apical)	Connecting tubule and collecting duct
Epithelial K$^+$ channel	ROMK	Thick ascending limb of Henle, connecting tubule, and cortical collecting duct
Epithelial Na$^+$ channel	ENaC	Connecting tubule and cortical collecting duct
Cl channels$^-$ (ClCx)	CLC5	Proximal tubule
	CLC-K	Thick ascending limb, distal convoluted tubule
Epithelial Ca^{++} channel	ECaC (TRPV5)	Distal convoluted and connecting tubules
Urea transporters	UT1, UT3	Medullary collecting duct
	UT2	Thin descending limb of Henle
	UTB	Erythrocytes, descending vasa recta
Transporters without acronyms		
Na$^+$-pump	Na$^+$, K$^+$-ATPase	Basolateral membrane
Proton pumps	H$^+$-ATPase	Apical membrane, the entire tubule
	H$^+$, K$^+$-ATPase	Apical membrane connecting tubule and collecting duct
Ca^{++} pump	Ca^{++}-H$^+$-ATPase	Basolateral membrane, distal convoluted and connecting tubule

1.11 Conclusions

The kidney has the crucial role in maintaining the volume and composition of the extracellular fluid. Through this function, the kidney plays a pivotal role in fluid homeostasis. This task is accomplished through the regulation of several physiological variables of the extracellular fluid.

Bibliography

Gottschalk CW (2000) A history of renal physiology. In: Seldin D, Giebisch G (eds) The kidney: physiology and pathophysiology, vol I, 3rd edn. Lippincott Williams and Wilkins, Philadelphia

Hediger MA, Mount DB, Rolfs A, Romero MF (2004) The molecular basis of solute transport. In: Brenner BM (ed) The kidney, vol I, 7th edn. WB Saunders, Philadelphia, pp 261–308

Functional Anatomy of the Kidney

2

Learning Objectives

- To understand the macroscopic anatomy of the kidney.
- To understand the nephron as structural and functional unit of the kidney.
- To describe the ultrastructure of the glomerulus, its cell types, and the glomerular filtration barrier.
- To describe the structure of the renal tubule and the structure-function relationships in the different tubular segments.
- To understand the structure of the juxtaglomerular apparatus.
- To understand the organization of blood flow.

2.1 General Structure of the Kidney

Each adult human kidney weighs between 115 and 170 g. Blood vessels and the ureter enter and exit each kidney through the renal hilum. The macroscopic structure of the kidney can be seen in Fig. 2.1. Each kidney is wrapped by the renal capsule. The renal parenchyma consists of a surface area or cortex and a deeper area or medulla. The shape of the medulla resembles a pyramid, where the apex is oriented towards the pelvic space and the base towards the corticomedullary border. The human kidney is multipapillary and consists of several lobes, each formed by the cortex and medulla that end in the renal papilla. Each of the papillae drains urine into the pelvic space. The kidneys of smaller mammals, such as rodents, rabbits, and others, are unipapillar, and all urine passes through a single papilla into the pelvic space.

The organization and components of the cortex and medulla can be seen in Fig. 2.2. The renal cortex is made up of cortical labyrinths and medullary rays that, despite their name, are considered part of the cortex (Fig. 2.2). Cortical labyrinths have renal corpuscles and tubules that follow a convoluted path like proximal and distal convoluted tubules and connecting tubules (Fig. 2.2). The medullary rays contain straight tubules, such as straight proximal tubules, thick ascending loops of Henle, and cortical collecting ducts. The renal medulla consists

© Springer Nature Switzerland AG 2022
P. A. Gallardo, C. P. Vio, *Renal Physiology and Hydrosaline Metabolism*,
https://doi.org/10.1007/978-3-031-10256-1_2

Fig. 2.1 Macroscopic features of the human kidney anatomy. The capsule is a dense connective tissue covering the organ. The outer zone is the cortex, and the inner part is the renal medulla. The shape of the medulla is pyramidal with the base oriented to the cortex and the apex is oriented to the renal pelvis. The blood vessels and ureter enter and exit the kidney through the renal hilum

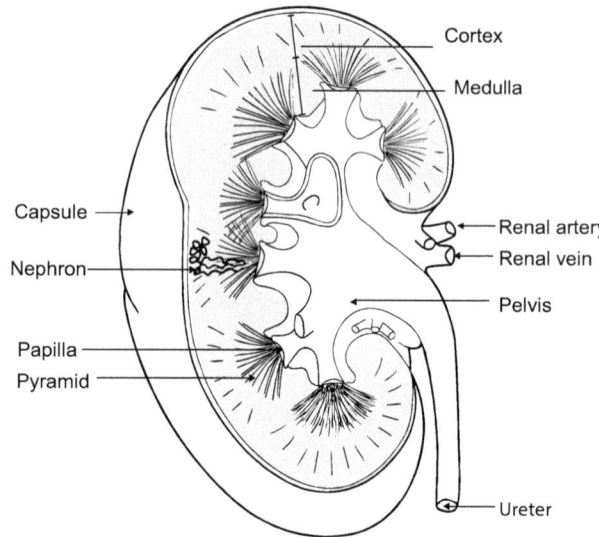

Fig. 2.2 Diagram of the renal architecture. The drawing shows the renal cortex containing a medullary ray between two cortical labyrinths. The renal medulla consists of an outer medulla further divided in an outer and inner stripe. The latter is in contact with the inner medulla. In the left, a juxtamedullary nephron is depicted; in the right a cortical nephron is shown. *1.* renal corpuscle; *2.* proximal convoluted tubule; *3.* proximal straight tubule; loop of Henle (*4, 5, 6*); *4.* thin descending limb; *5.* thin ascending limb (only in juxtamedullary nephrons); *6.* thick ascending limb; *7.* macula densa; *8.* distal convoluted tubule; *9.* connecting tubule; *10.* cortical collecting duct; *11* and *12.* outer and inner medullary collecting duct, respectively

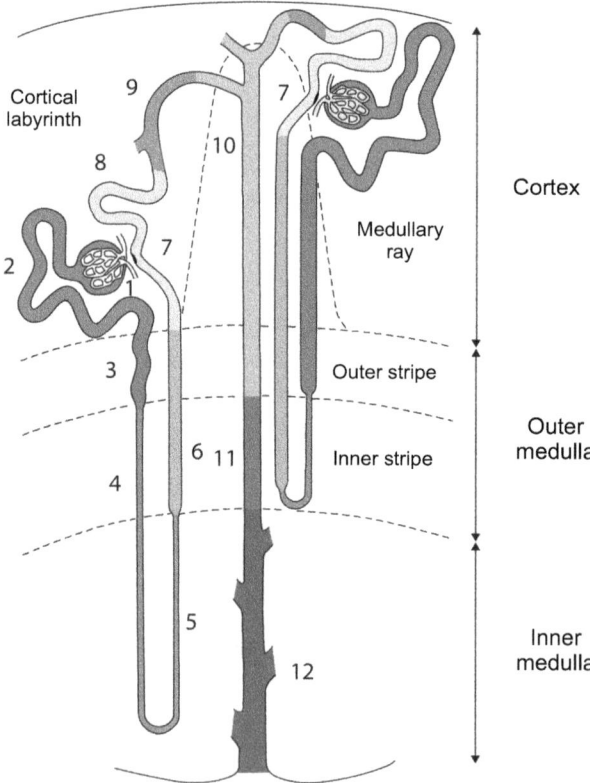

of an outer and an inner stripe. The outer medulla is subdivided into an outer stripe that is in contact with the cortex and an inner stripe that is in contact with the inner medulla. Figure 2.2 shows the spatial arrangement of the nephrons in the kidney.

The location of the renal corpuscles in the cortex allows the classification of the nephrons into two types. The cortical nephrons have their corpuscles in the superficial regions of the cortex; they also have a short loop of Henle that penetrates the outer medulla and bends towards the cortex at the boundary with the inner medulla. The second type is the juxtamedullary nephrons; they have their corpuscles deeper in the cortical labyrinths, near the corticomedullary boundary. This nephron type has long loops of Henle that penetrate in variable degree into the inner medulla before bending towards the cortex. Human kidneys have a higher number of cortical nephrons over juxtamedullary nephrons.

A relevant feature of the architecture of both types of nephrons is that the loop of Henle is arranged in a countercurrent pattern. This determines that the tubular flow also follows the countercurrent pattern: the tubular fluid descends through the thin descending limbs and ascends through the ascending limbs. The tubular fluid descends again through the collecting ducts. This organization, in addition to the functional characteristics of these segments, is a determining factor in the renal mechanism for concentrating and diluting urine, a function performed with greater efficiency by nephrons with long loops of Henle.

2.2 Nephron Structure

The nephron is the structural and functional unit of the kidney; each is composed of a renal corpuscle and tubule (Fig. 2.2). Each human kidney has approximately 1 to 1.3 million of nephrons. In the human kidney, there is a predominance of cortical nephrons over the juxtamedullary ones.

2.2.1 Renal Corpuscle

The term renal corpuscle or Malpighian corpuscle is the anatomical expression that defines the spherical structure located at the beginning of each nephron. This structure was described in 1666 by the Italian anatomist Marcello Malpighi (1628–1694). In renal physiology, the term corpuscle has little use and is commonly referred as the "glomerulus," a term that will be used from here on. In the glomeruli occurs the ultrafiltration of plasma, which is the first process in urine formation. The average diameter for human glomeruli is 200 μm. The structure of the glomeruli provides a large surface area available for filtration, which in the case of the human kidney can reach approximately 0.136 mm^2 per glomerulus. The glomeruli are always located in the cortical labyrinths (Fig. 2.2). In a glomerulus (Fig. 2.3) two poles can be distinguished. The vascular pole is a small indentation where the afferent arteriole enters, and efferent arteriole exits each glomerulus. The urinary

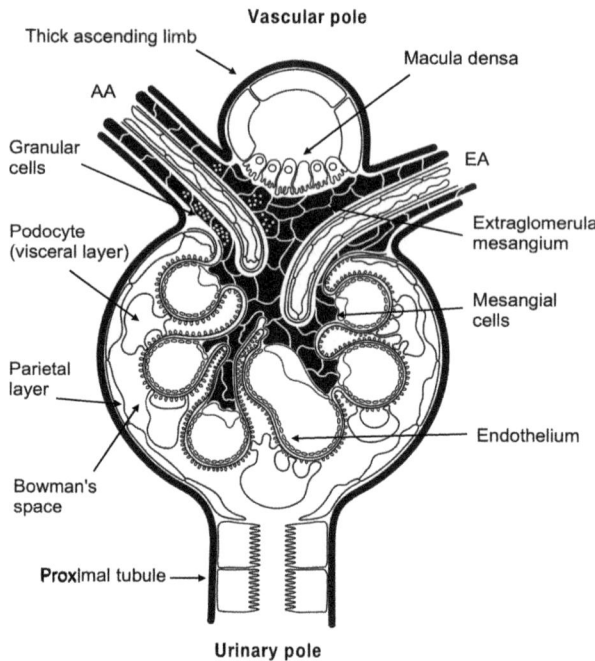

Fig. 2.3 Structure of the renal corpuscle. It is formed by Bowman's capsule and the glomerular tuft. Bowman's capsule is composed of a parietal layer that invaginates at the vascular pole forming the visceral layer formed by podocytes. Both layers are separated by the urinary space (Bowman's space). The urinary pole is the site where the parietal layer originates the proximal tubule. The glomerular tuft consists of the glomerular capillaries and the mesangium composed of mesangial cells and the mesangial matrix. The blood supply to glomerular capillaries derives from the afferent arteriole that enters the corpuscle at the vascular pole. Blood is drained from the capillaries by the efferent arteriole. The juxtaglomerular apparatus is formed at the vascular pole; it is composed of the granular cells in the wall of the afferent arteriole, the extraglomerular mesangial cells, and the macula densa cells of the thick ascending limb of the same nephron

pole is located opposite to the vascular pole and is the site where the glomerulus is connected with the renal tubule.

The English anatomist William Bowman (1816–1892) made the first histological observations of the structure of the glomerulus; he established the structural and functional relationship between the glomerulus and the renal tubule. Each glomerulus is formed by the Bowman capsule and the glomerular capillaries, which are inside the capsule. Bowman's capsule consists of two layers of epithelial cells: the visceral and parietal layers, separated by the urinary space. The visceral layer is formed by epithelial cells called podocytes. The podocytes are very complex and differentiated cells are unable to undergo mitosis. Thus, podocyte damage might imply irreversible glomerular damage. They have a very prominent cell body that protrudes into the urinary space. The cell body gives rise to primary processes that extend and branch several times towards the glomerular capillaries (Fig. 2.4a), giving rise to "fine foot processes" or pedicels. These foot processes are anchored

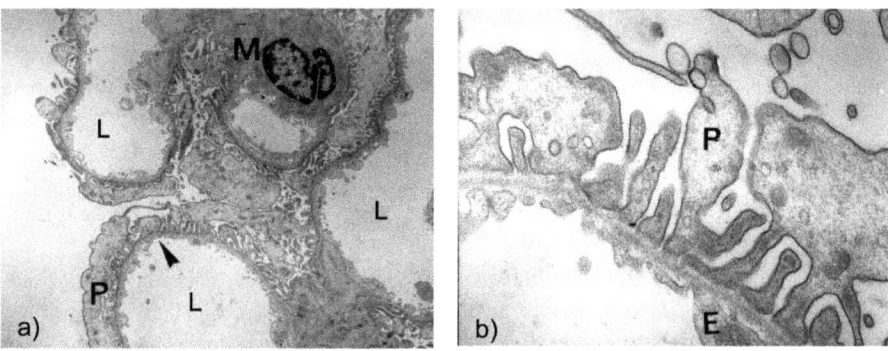

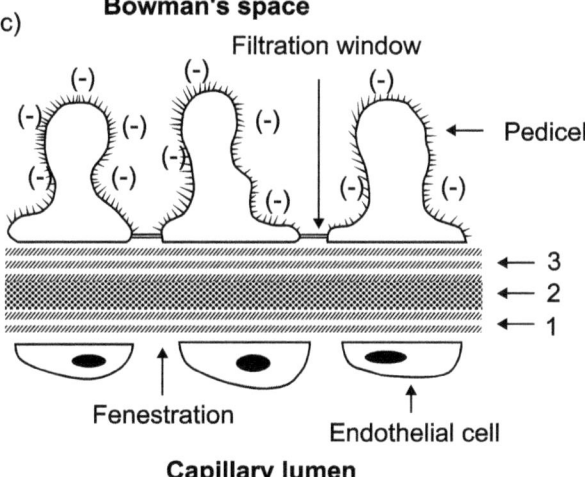

Fig. 2.4 Ultrastructure of the glomerular filtration barrier. (**a**) Electron micrograph (3.000×) depicting a podocyte (P) and lumen of a glomerular capillary (L). (**b**) Details of the glomerular filtration barrier pointed by the arrowhead in (**a**). The electron micrograph (20.000×) shows the capillary lumen and endothelial cells (E), the basal lamina, and the foot process of podocytes (P); most of the podocyte plasma membrane is oriented to the urinary space. (**c**) Scheme of the glomerular filtration barrier. Glomerular endothelium is fenestrated; the basal lamina is formed by a lamina rara interna (*1*) in contact with the basolateral membrane of endothelial cells and lamina densa (*2*) and a lamina rara externa (*3*) in contact with the basolateral membrane of podocyte foot processes. Adjacent foot processes are bound by the filtration slit junction formed by podocyte transmembrane proteins that appear as a diaphragm in transmission electron micrographs. Transmission electron micrographs were kindly provided by Dr. C. Bosco (Department of Morphology, ICBM, University of Chile)

to the basal lamina. Interdigitations of adjacent pedicels in the surface of glomerular capillaries leave a narrow space bridged with a thin diaphragm. The narrow spaces are called filtration slits (Fig. 2.4b); through these spaces occurs the drainage of the ultrafiltrate into the urinary space. The structural and functional integrity of podocytes is crucial for proper glomerular filtration. As mentioned above, podocytes are epithelial cells; the apical membrane is oriented to the urinary space and the basolateral membrane is in contact with the basal lamina. The limit between the apical and basolateral domains is the slit diaphragm junction. This cell junction is established between adjacent feet processes that are attached to the basal lamina. Podocyte injury is the prelude to the damage of the filtration barrier and the generation of proteinuria. Between the pedicels and the endothelium of the glomerular capillaries is the basal lamina, secreted by the podocytes and endothelial cells (Fig. 2.4b, c), which is described later.

The parietal layer of Bowman's capsule is a simple squamous epithelium, which is continuous with the renal tubule at the urinary pole. Between both layers is the urinary or Bowman's space, which receives the glomerular ultrafiltrate.

The glomerular capillaries derive from the branching of the afferent arteriole, through which blood enters each glomerulus. The glomerular capillaries form a tuft and coalesce inside the glomerulus to form the efferent arterioles, through which blood leaves each glomerulus. The endothelium of the glomerular capillaries is fenestrated and contains large pores, of approximately 700 Å (Fig. 2.4b, c), allowing the free passage of ions and of organic solutes such as glucose, urea, and creatinine and also of some plasma proteins. Blood cells cannot pass through the pores of the endothelium. The apical membrane of the endothelial cells is covered with a glycocalyx and a rich endothelial surface layer composed of secreted negatively charged proteoglycans and glycosaminoglycans. The basolateral membrane of endothelial cells expresses receptors for vascular endothelial growth factor, secreted by the podocytes and for integrins that anchor the basolateral membrane to the basal lamina.

The basal lamina (300 nm thick in human) is interposed between the pedicels of the podocytes and the capillary endothelium and consists of three layers: the inner rare lamina that is in contact with the capillary endothelium, the dense lamina, and the outer rare lamina, in contact with the pedicels (Fig. 2.4c). The mature basal lamina corresponds to extracellular matrix and is made up of several proteins: type IV collagen as the main collagen fiber, type II laminin (laminin 521), and heparin sulfate proteoglycans like agrin and perlecan. The type IV collagen fibers form a net-like structure to which the abovementioned glycoproteins are attached. The monomers that make up the collagen network are made up of a helix consisting of three chains. There are six isoforms of collagen IV chains (COL4A1 to COL4A6); in humans, the most common combination for the triple helix is COL4A1-COL4A2-COL4A1. In each monomer (Fig. 2.5), the three collagen chains are oriented in the same direction; therefore, there are an amino terminal end called the 7S domain, a triple central helix, and a globular terminal carboxyl end called the non-collagenic domain or NC1. The NC1 domain is important in the binding of collagen monomers and dimer formation; the 7S domain is important in the formation of tetramers and binding of other proteins, giving rise to a collagen structure to which acid glycoproteins and proteoglycans bind, which determine the size of the pores present in the basal lamina and their anionic character.

The podocytes are polarized cells. Their apical domain is in contact with the urinary space and the basal membrane is in contact with the lamina rara externa of

Fig. 2.5 Collagen IV structure. (**a**) Collagen IV monomer. (**b**) Collagen IV protomer formed by $\alpha3$, $\alpha4$, and $\alpha5$ chains. 7S domain is critical for tetramer formation and protein interaction in the basal lamina; NC1 domain is important for dimer formation

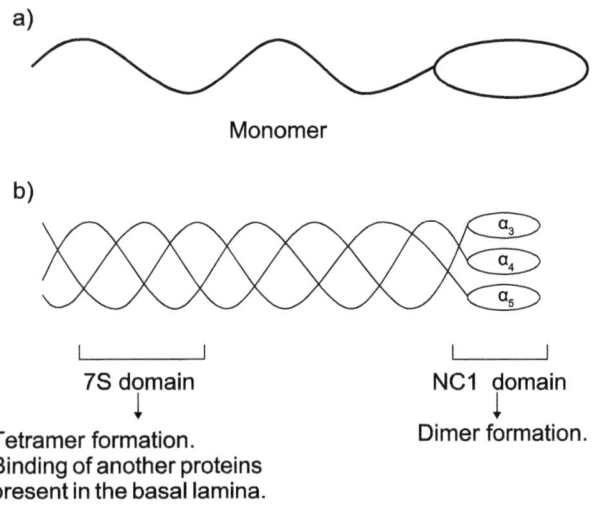

the glomerular basement membrane. The limit between both membrane domains is a cell junction located within the slit diaphragm which is part of the filtration slits between two adjacent pedicels.

The filtration slits (30–40 nm width) are the pathway through which the ultrafiltrate passes into the urinary space. These are narrow spaces between adjacent pedicels covered by a thin diaphragm. From a histological standpoint, the filtration slit is a cell junction similar to an adherens junctions made up of multiple proteins, such as nephrin, NEPH1, Fat, and podocin. Nephrin is a single-pass transmembrane protein with a big extracellular domain that establishes homophilic interactions with the same domain from adjacent pedicels. The interaction between these domains forms the zipper-like structure in electron micrographs of the filtration slit. The protein is essential in the development and maintenance of the slit junction. Mutations with loss of function are the base for the Finnish type nephrotic syndrome. NEPH1 is a transmembrane protein whose exact role in the slit junction remains to be elucidated. Fat1 is a protein member of the cadherin family; its extracellular domain establishes homophilic interactions that contribute to the zipper-like image of the slit diaphragm. Podocin is a transmembrane protein also necessary for the integrity of the slit junction. Podocin mutations are also associated with nephrotic syndrome.

The windows in the slit unction have smaller pores (40 × 140 Å) and are covered by a basal film that forms a diaphragm. The structural and functional integrity of this junction is a key determinant of the quality of glomerular ultrafiltrate.

The structural integrity of the glomerular basement membrane is crucial for the formation of a protein-free glomerular ultrafiltrate. Several glomerular diseases are good examples of how loss-of-function mutations in specific proteins result in a glomerular pathology. One example is Alport syndrome, linked to the X chromosome. The syndrome is caused by mutations in the gene encoding for the α-5 chain of collagen IV. Several mutations had been identified; all translate into an altered basal lamina, whose renal manifestations are hematuria and proteinuria. Another example is the Goodpasture disease. In this pathology, the NC1 domain of the α-3 chains of collagen IV is the amino acid sequence that acts as an antigen. The antibodies against the NC1 domain of the chains α-3 that are generated determine an inflammatory reaction that in the kidney produces hematuria and renal failure. The abovementioned diseases affect the type IV collagen and hence the glomerular basement membrane. The filtration slit can also be affected by protein mutations. One example is the Finnish type of nephrotic syndrome, caused by mutations in the nephrin gene. This pathology is characterized by increased permeability of the glomerular barrier to proteins and urinary excretion of proteins.

The assembly formed by the pedicels, the filtration slits, the basal lamina, and the fenestrated endothelium is known as the glomerular filtration barrier (Fig. 2.4c). The normal structure of the glomerular filtration barrier—in terms of structure and components—is essential for the formation of a normal glomerular filtrate in composition and volume. Removal of glycosaminoglycans by enzymatic digestion increases the permeability of the glomerular filtration barrier to proteins such as ferritin and bovine albumin, suggesting that the negative charges present in this barrier play an important role in the permeation-selectivity function.

Another glomerular cell type is the mesangial cell. This cell type plays a crucial role in the maintenance of the glomerular capillary tuft. The mesangial cell secretes the extracellular matrix or mesangial matrix interposed between the basolateral membrane of the glomerular capillaries and the mesangial cell (Fig. 2.6). Intraglomerular mesangial cells perform a wide range of functions: they secrete extracellular matrix, have phagocytic activity, and secrete prostaglandins and cytokines, which are important messengers in inflammation. Mesangial cells exhibit contractile activity in the presence of hormones such as angiotensin II and

Fig. 2.6 Anatomical relationship between the main cell types in the glomerular tuft. Fine processes from podocyte body contact the lamina rara externa of the basal lamina. The rest of the podocyte body is oriented to the urinary space. The glomerular endothelium is fenestrated and its basolateral membrane contacts the lamina rara interna of the basal lamina. The endothelial cell nucleus and most of the cytoplasm are located opposite to the podocyte and close to the mesangial cells

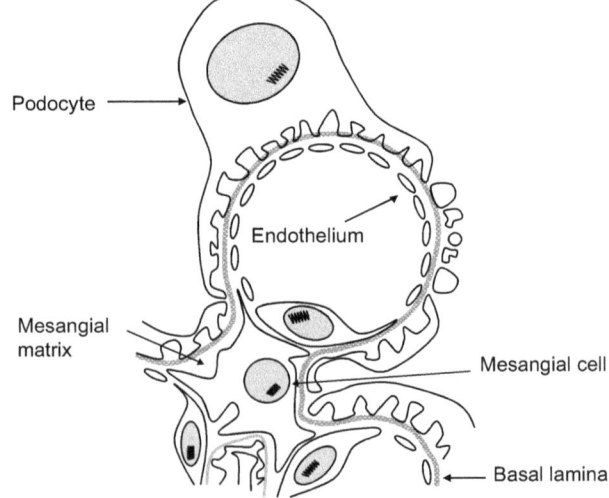

vasopressin, activity that modifies the area available for glomerular filtration and thus affects the speed of glomerular filtration. Extraglomerular mesangial cells are part of the juxtaglomerular apparatus.

2.2.1.1 Juxtaglomerular Apparatus (JGA)

The juxtaglomerular apparatus is a tubular-vascular structure that is formed at glomerular vascular pole. It is located at the point where the cortical segment of the thick ascending loop of Henle passes through the angle formed by the afferent and efferent arterioles of the parent glomerulus. The JGA (Fig. 2.3) is composed of:

(a) Macula densa cells of the cortical segment of Henle's thick loop.
(b) Extraglomerular mesangial cells.
(c) Granular cells from the wall of the afferent arteriole.

The macula densa consists of epithelial cells from the cortical segment of the thick ascending loop of Henle, which are oriented towards the glomerular vascular pole. These tubular cells differ morphologically from the adjacent cells of the thick ascending limb. They are taller and have a very large nucleus. The apical membrane has short studded microvilli. The lateral membrane has few interdigitations and the basal membrane has few basal infoldings. The lateral intercellular space has a dilated appearance. Among other transporters, the apical membrane expresses the NKCC2 cotransporter; the basolateral membrane expresses the Na^+, K^+-ATPase. The basolateral membrane of macula densa cells is oriented to the extraglomerular mesangial cells, which fills the space delimited by the angle formed by the afferent and efferent arterioles in the vascular pole. A third component are the granular cells in the wall of the afferent arteriole. These are modified smooth muscle cells that are located towards the more terminal portion of the arterioles. Granular cells synthesize and store in granules the enzyme renin (Fig. 2.7a). At the electron microscope, granular cells exhibit a collection of electron-dense granules that store renin. Renin activity is the rate-limiting step in the renin-angiotensin system. Extraglomerular mesangial cells, renin-secreting cells, and smooth muscle cells of the afferent arteriole are functionally coupled by gap junctions formed mainly by connexin 43, whose expression is crucial in the regulation of glomerular filtration rate and renal blood flow.

Renin secretion is controlled by the sympathetic system, angiotensin II, and adenosine, through A1 receptors. Norepinephrine secreted by the sympathetic nerve endings and epinephrine secreted by the adrenal medulla stimulate renin secretion through activation of β_1 adrenergic receptors. These receptors are coupled to Gs heterotrimeric protein, stimulating adenylyl cyclase activity and enhancing cytosolic cAMP levels, which in turn activates protein kinase A (PKA), responsible for the activation of the secretory mechanism. Angiotensin II is the hormone derived from renin activity and inhibits renin secretion. This inhibitory effect is mediated by angiotensin II binding to AT_1 receptors coupled to Gq heterotrimeric protein. AT_1 receptor activation leads to an increase in cytosolic calcium and activation of protein

a)

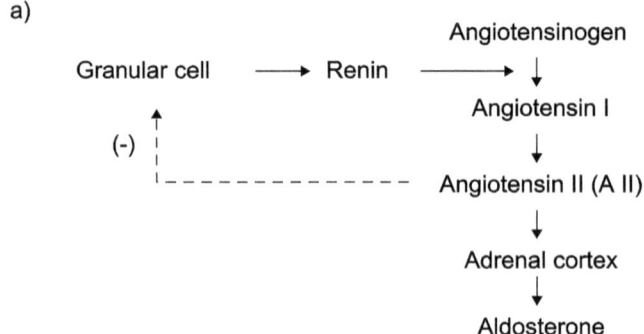

b)

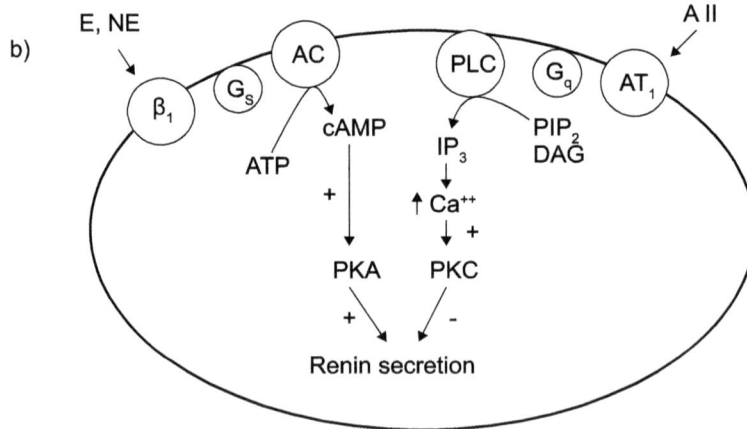

Fig. 2.7 Granular cells and renin-angiotensin-aldosterone system. (**a**) Negative feedback between angiotensin II and renin secretion from granular cells. Angiotensin II inhibits renin secretion from granular cells. (**b**) Main signal transduction cascades in granular cells; these cells receive sympathetic innervation. Norepinephrine and epinephrine stimulate renin release through the activation of β1-adrenergic receptors and cAMP-protein kinase A cascade. As a part of a negative feedback, angiotensin II binding to AT1 receptor results in inhibition of renin release. This effect is mediated by an increase in cytosolic Ca^{++} concentration and protein kinase C activation

kinase C. Hence, unlike other secretory mechanisms, renin secretion is inhibited by agents that increase cytosolic calcium. The angiotensin II inhibitory effect constitutes a negative feedback for the renin-angiotensin system activity.

Adenosine is derived from ADP metabolism in macula densa cells and exits the cells through pannexin release channels located at the basolateral membrane. Adenosine levels at the juxtaglomerular apparatus are directly correlated with NaCl transport through the apical membrane. An increase in apical NaCl transport is correlated with an increase in adenosine concentration in the JGA. Adenosine inhibits renin secretion through a Gi protein-dependent mechanism (Fig. 2.7b).

Hence, any extracellular signal that activates the cyclic cAMP-PKA signal cascade will activate renin secretion. Activated PKA not only stimulates renin secretion but is also responsible for basal transcription of renin gene through CBREB/ATF transcription factor.

2.2.2 Renal Tubule

The renal tubule is formed by a single layer of epithelial cells. The apical membrane is oriented to the tubular compartment, and the basolateral membrane is oriented towards the interstitial compartment that contains blood capillaries derived from efferent arterioles. The tubular epithelium functions as a selective barrier to the movement of solutes and water between the tubular and the interstitial compartment. Transepithelial transport is responsible for modifying the volume and composition of the glomerular ultrafiltrate.

The tubular cells are connected to each other by the junctional complex. Within the junctional complex, the tight junction (*zonula occludens*) forms the boundary between the apical and the basolateral membrane domain. The lateral membranes of adjacent cells delineate the lateral intercellular space. The lateral membranes of adjacent cells interdigitate. This creates a lateral intercellular space with a complicated geometry formed by tortuous channels that are filled with extracellular fluid. The basal membrane of the cells rests on the basal lamina (Fig. 2.8). The basal membrane can exhibit basal infoldings of variable depths. The apical domain can exhibit specializations like microvilli. The most abundant and conspicuous microvilli are found in the proximal tubule. In electron micrographs, such an abundance of microvilli is collectively called "brush border." The basolateral membrane also develops specializations such as the lateral membrane interdigitations described above and basal membrane infoldings. Both apical and basolateral membrane specializations increase the membrane surface available for transepithelial transport.

Transport activity across the tubular epithelium and along the nephron accounts for the modification of the volume and composition of the ultrafiltrate. The transepithelial transport is manifested in two processes: tubular reabsorption and secretion. Reabsorption is the movement of molecules from the tubular lumen to the blood capillaries. Tubular reabsorption is a transcellular process: transport of molecules occurs first across the apical and then the basolateral membrane. Paracellular reabsorption occurs through the tight junction. Tubular secretion is the movement of molecules from blood capillaries into the tubular lumen. The process is transcellular. Both processes occur along the nephron.

Figure 2.9 shows an outline of the nephron and the different segments that form the renal tubule. The denomination of the tubular segments follows a microanatomical criterion or a combination of anatomical and functional criteria. The nomenclature used in this text is the standardized one in renal physiology and essentially follows anatomical and functional criteria. The tubular segments are:

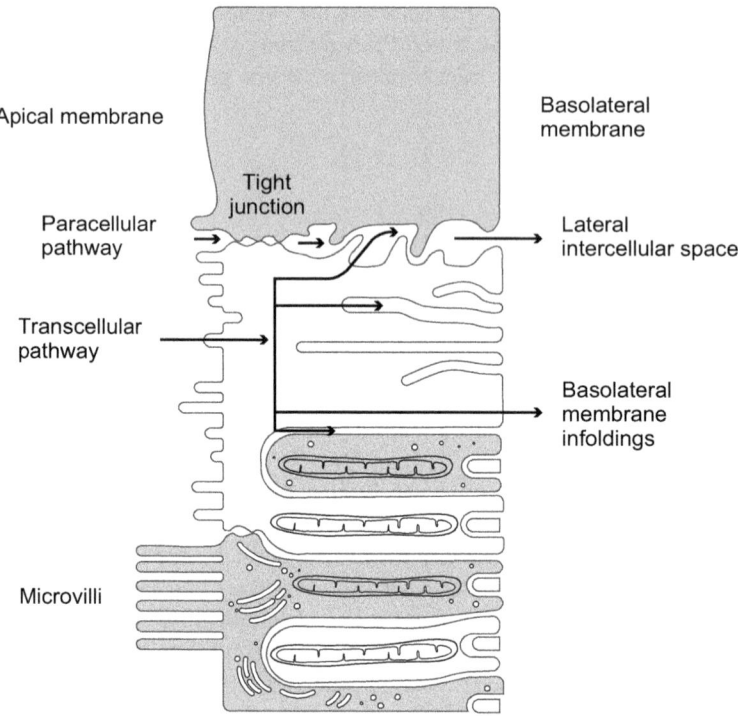

Fig. 2.8 Morphofunctional features of the tubular epithelium. The tubular epithelium is a simple cuboidal epithelium. The apical membrane faces the tubular fluid and may exhibit long and abundant microvilli forming the brush border as in the proximal tubule. In other segments, microvilli are scarce and shorter. All tubular cells except intercalated cells exhibit a primary cilium (not shown). The tight junction forms the boundary between apical and basolateral domain. The lateral membranes of adjacent cells exhibit interdigitations rendering a tortuous lateral intercellular space. The basal membrane may exhibit deep infoldings forming membrane pockets that contain elongated and flat mitochondria. The basolateral membrane contains the Na$^+$, K$^+$-ATPase, main primary active transport mechanism in the tubular epithelium (Adapted from Christensen EI, Wagner CA, Kaissling B. Uriniferous tubule: structural and functional organization. *Compr Physiol* 2:805–861, 2012)

proximal tubule, loop of Henle, distal nephron formed by the distal convoluted tubule, connecting tubule, and cortical collecting duct. From the embryological point of view, the collecting duct is not part of the nephron since it does not originate in the metanephric mesenchyme, but in the ureteral yolk. However, the collecting duct functions in a coordinated way with the rest of the tubular segments; hence, it is considered as part of the functional nephron. As mentioned above, each tubular segment is made up of epithelial cells with particular characteristics closely related to the function of the segment in question. Figure 2.9 illustrates the epithelial cells that are typical of each segment.

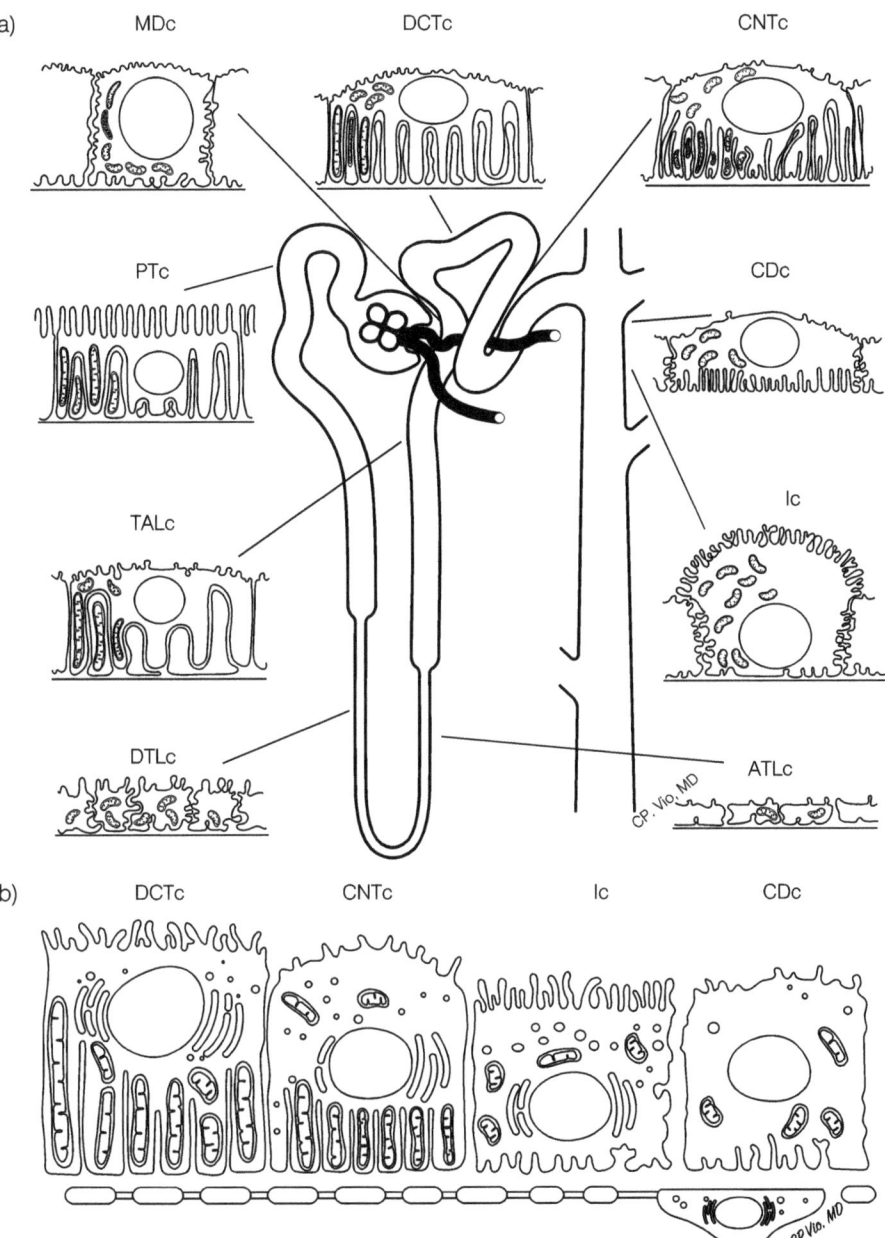

Fig. 2.9 Tubular epithelial cells. Proximal tubule cells (PTc) are characterized by a prominent brush border in the apical membrane and basal membrane infoldings containing abundant mitochondria. The thin descending epithelium (DTLc) is formed by cells that are shorter; the apical microvilli are scarce and shorter. Lateral membranes have deep interdigitations. The ascending thin limbs (ATLc) are formed by a cuboidal epithelium with few microvilli and lateral interdigitations The thick ascending limb cells (TALc) have few and short microvilli; the basal membrane displays

2.3   Proximal Tubule

The proximal tubule has the greater capacity for reabsorption of organic solutes, electrolytes, and water among all tubular segments. The proximal epithelium exhibits morphological and functional adaptations that enable high rates of tubular reabsorption of solutes and water.

The proximal tubule consists of a convoluted part (proximal convoluted tubule, PCT), which originates in the urinary pole, and a straight part (proximal straight tubule, PST) that extends into the outer stripe of the outer medulla. According to the ultrastructural characteristics of proximal tubular cells (height and abundance of microvilli, abundance of lateral interdigitations and basal infoldings, abundance of mitochondria and lysosomal apparatus), this segment was divided into three subsegments: S_1, S_2, and S_3. The subsegment S_1 and the first part of S_2 are included within the convoluted part, while the rest of S_2 and S_3 are part of the proximal straight tubule.

The apical membrane of proximal tubule cells exhibits abundant microvilli, which greatly increases the surface for reabsorption. From S_1 to S_3, the height and abundance of microvilli decreases significantly. The apical membrane is characterized by the expression of sodium-dependent transporters such as SGLT1, SGLT2, NHE3, and NPT2 and AQP-1 water channel. The cytoplasm contains abundant elongated mitochondria, which decrease in number from S_1 to S_2. The cytosol has a well-developed vacuolar-lysosomal system, especially in the S_1 and S_2. The proximal tubule is a leaky epithelium, characterized by a low transepithelial resistance due to the scarce number of protein strands forming the tight junction; it also exhibits a high osmotic permeability compared to other tubular segments.

The lateral aspect of the basolateral membrane exhibits interdigitations that form the boundary of the lateral intercellular space. The basal aspect of the basolateral membrane exhibits several infoldings; this membrane is oriented to the basal lamina. The Na^+, K^+-ATPase is localized to this membrane domain.

2.4   Loop of Henle

The loop of Henle is formed by three subsegments (Fig. 2.2): thin descending limb, thin ascending limb, and thick ascending limb. The length of the thin descending limbs of juxtamedullary nephrons (long-looped nephrons) penetrates to variable

←

Fig. 2.9 (continued) deep infoldings forming narrow sacs containing flat and elongated mitochondria. Macula densa cells (MDc) are oriented to the vascular pole; they are taller than the typical thick ascending limb cells. Distal convoluted tubule cells (DCTc) appear shorter after the macula densa; they have the most prominent and abundant basal infoldings containing elongated mitochondria. Basal infoldings are less prominent in the connecting tubule cells. The connecting tubule exhibits cellular heterogeneity with more than one cell type: connecting cells (CNTc) and intercalated cells (Ic). The cortical collecting duct exhibits principal cells (CDc) as the major cell type and intercalated cells

degrees into the inner medulla where they turn into a hairpin towards the cortex. In contrast, the thin descending limbs of cortical nephrons (short-looped nephrons) penetrate to the limit between the outer and inner medulla. In any case, the thin descending limbs turn into a hairpin giving rise to the thin ascending limb which is very short or absent in cortical nephrons but is present in the juxtamedullary nephrons, extending to the limit between the inner and outer medulla. The thick ascending loop of Henle (TAL) is present in both types of nephrons. It is composed of a medullary and a cortical portion; the latter penetrates the cortex and passes through the angle formed by the afferent and efferent arterioles at the vascular pole of the parent glomerulus in the juxtaglomerular apparatus. At this point, the cells oriented to the extraglomerular mesangium forms a plaque known as the macula densa. The thick ascending limb ends immediately shortly after the macula densa, giving rise to the distal convoluted tubule, first segment of the distal nephron.

The thin descending loops are formed by three different types of epithelial cells, which are very notorious in rodent loops from desert environments. In general, the apical membrane exhibits short and scarce microvilli. The lateral membrane may exhibit interdigitations; the basal membrane shows shallow infoldings without mitochondria. The morphological features of the thin descending limb suggest that this segment does not perform active transport. This segment exhibits osmotic water permeability due to apical and basolateral expression of AQP-1; this segment is also capable of urea secretion mediated by facilitated urea transporter 2 (UT2) on apical and basolateral membranes.

The thick ascending limb is formed by cubic cells. The apical membrane has short microvilli and a single primary cilium, probably involved in tubular flow detection. The apical membrane hosts several Na^+-dependent transporters like NKCC2 (a feature of this segment) and NHE3 as well as ion channels like ROMK (renal outer medullary potassium channel). The lateral aspect of the basolateral membrane exhibits extensive interdigitations and the basal aspect has abundant narrow infoldings containing flat, elongated mitochondria. Electron micrographs reveal that some basal infoldings penetrate deeply into the cytosol, very near to the apical membrane. The proximity between mitochondria and the localization of the Na^+, K^+-ATPase in these basal infoldings are characteristics of epithelia involved in active ion transport. Besides the Na^+, K^+-ATPase, the basolateral membrane hosts the CLC-K chloride channel and a potassium channel. On the other hand, the apical and basolateral membrane of these cells lack aquaporins, which explains the water impermeability of this segment.

2.5   Distal Nephron

The term distal nephron groups three tubular segments located after the macula densa and comprises the distal convoluted tubule (DCT), the connecting tubule (CNT), and the cortical collecting duct (CCD). The distal nephron is the site of fine regulation of Na^+ and water reabsorption and K^+ secretion. All these processes are under hormonal regulation.

2.5.1 Distal Convoluted Tubule

The DCT is located in the cortical labyrinths and begins shortly after the macula densa. The distal convoluted cell is a cuboidal cell; the apical membrane has short microvilli and a single primary cilium. The apical membrane expresses the NCC cotransporter as a mechanism for Na^+ reabsorption. As in the thick ascending limb, the apical and basolateral membranes of the DCT cell are devoid of aquaporins; hence, this segment is impermeable to water. The basal aspect of the basolateral membrane has abundant and prominent narrow infoldings that contain elongated and flat mitochondria. The lateral membranes of adjacent cells are interdigitated. Such an abundance of basal infoldings greatly increases the basolateral membrane area. Besides, the DCT has the highest Na^+, K^+-ATPase activity. DCT cells are sensitive to angiotensin II through AT_1 receptors.

2.5.2 Connecting Tubule

The distal convoluted tubule empties into the connecting tubule. In some species, the transition between DCT and connecting tubule is gradual and in others is abrupt. The connecting tubules of juxtamedullary nephrons merge into one connecting tubule that ascends to the superficial cortex forming an arcade. In cortical nephrons, a single connecting tubule flows directly into the cortical collecting duct. The connecting tubule is the first segment exhibiting cellular heterogeneity, that is, there is more than one type of cell with notoriously different morphology and function (Fig. 2.9). The most abundant and characteristic cell type is the connecting cells (CNTc). This cell type is involved in Na^+, Ca^{++}, and water reabsorption as well in K^+ secretion. The apical membrane has few microvilli and expresses the epithelial sodium channel (ENaC), the epithelial calcium channel (ECaC), and the apical K^+ channel (ROMK). The basolateral membrane has abundant infoldings; as in the DCT, they contain elongated mitochondria. CNT cells are sensitive to aldosterone and vasopressin. Recent studies had demonstrated that the late portion of the DCT and CNT are the first responders in the homeostatic response to a low-Na^+ diet or a high-K^+ diet. Studies carried out in mice with selective knock-out of ENaC in the cortical collecting duct demonstrated that under low-sodium diet, the increased reabsortive function of the connecting tubule is sufficient to maintain the Na + balance. They are the only renal cells that synthesize and secrete the proteolytic enzyme kallikrein; this endopeptidase cleaves kininogen to generate the hormone bradykinin, which participates in the homeostasis of Na^+, renal hemodynamics, and vascular tone. Another cell type present in this segment is the intercalated cell, which will be described in the next segment.

2.5.3 Cortical Collecting Duct (CCD)

The cortical collecting duct is located in the medullary rays. The cortical collecting duct is made up of three cell types: the most abundant cell is the principal cell (Pc); the other cell types are type A intercalated cell and type B intercalated cell. Both types of intercalated cells are also present in the CNT. The apical membrane of Pc has few studded microvilli and a single primary cilium involved in fluid flow sensing. Pc cell apical membrane expresses ENaC for Na^+ reabsorption and ROMK for K^+ secretion. Na^+ reabsorption and K^+ secretion are under the control of aldosterone. Pc is also sensitive to vasopressin; they express AQP-2 in the apical membrane and in subapical tubulovesicles. In comparison to DCT and CNT cells, Pc cell cytoplasm has few mitochondria, most of them with a round shape instead of the elongated and flat mitochondria observed in DCT and CNT cells. The basal membrane has abundant short infoldings that do not contain mitochondria; it contains the Na^+, K^+-ATPase and AQP-3. The intercalated cells (Ic) participate in acid-base transport. Type A (α) intercalated cell is involved in acid secretion. The apical membrane of Ic type A cells (Ic A) has abundant microprojections and the apical cytoplasm has abundant tubulovesicles and round-shaped mitochondria. Carbonic anhydrase is present in the cytoplasm; its activity generates H^+ and HCO_3^-. The apical microprojections and the tubulovesicles near the apical membrane contain H^+-ATPase. This membrane also displays the H^+, K^+-ATPase. The presence of H^+ secreting pumps in the apical membrane is correlated with the ability to secrete H^+ to the lumen, which is mainly carried out by the H^+-ATPase. The basolateral membrane contains few basal infoldings which are devoid of mitochondria; this membrane contains the anion exchanger AE1 that exports HCO_3^- to the extracellular fluid. Type B (Ic B) intercalated cells have a round shape; they have mitochondria closely associated with the basolateral membrane. Type B intercalated cells secrete HCO_3^- via luminal pendrin exchanger; H^+-ATPase is localized to the basolateral membrane, which explains the close association of this organelle to this membrane domain. Hence, besides function, there is a clear morphological difference between both cell types in the localization of mitochondria.

2.5.4 Medullary Collecting Duct (MCD)

The histological features of the principal cell present in the CCD gradually change towards the medullary collecting duct. The main changes are the decrease in abundance and depth of basal membrane infoldings. There is also a reduction on the size and number of mitochondria, which are scattered in the cytosol and not located in the basal membrane infoldings. The complete collecting duct is permeable to water in the presence of vasopressin. AQP-2 is localized to the apical membrane and apical tubulovesicles throughout the entire collecting duct. The basolateral membrane expresses AQP-3 throughout the entire duct; besides, the inner medullary collecting duct has AQP-4 in the basolateral membrane.

The inner medullary collecting duct end is at the tip of the papilla and, through a sieve or perforated area, empties the urine into the pelvic space.

In conclusion, the different tubular segments are characterized by presenting morphological characteristics that account for some of their functional properties. On the other hand, the different segments express different apical and basolateral transporters, which give it typical functional properties. Table 2.1 summarizes some properties of tubule segments.

Table 2.1 Major functional features of the tubular segments. Acronyms for each transport protein are indicated in parenthesis

Segment	Subsegments	Cellular types	Marker	Characteristic function
Proximal tubule	S_1 (convoluted)	S_1	Cotransporter Na^+/phosphate (NPT2) Cotransporters Na^+/glucose (SGLT1, SGLT2)	Massive reabsorption of organic solutes, Na^+, K^+, Cl^-, HCO_3^-, and water
	S_2 (convoluted/ straight)	S_2		
	S_3 (straight)	S_3		
Thick ascending limb	Medullary and cortical portion	TAL	Apical cotransporter Na^+/ K^+/$2Cl^-$ (NKCC2) Tamm-Horsfall protein (apical)	Na^+ reabsorption, K^+, Cl^-, and HCO_3^- Impermeable to water
Distal nephron	Distal convoluted tubule	DCTc	Apical cotransporter Na^+/ Cl^- (NCC)	Na^+ reabsorption, Cl^-, Ca^{++}. Impermeable to water
	Connecting tubule	CNTc	Kallikrein	Na^+ reabsorption, Cl^-, Ca^{++}. Secretion of K^+. water reabsorption
		Ic-A	Apical H^+-ATPase, H^+, K^+-ATPase.	Secretion of H^+
		Ic-B	Cl^-/HCO_3^- apical	Secretion of HCO_3^-
	Cortical collecting duct	CCDc	Epithelial channel of Na^+ (ENaC)	Similar to connecting cells
		Ic	A B	Secretion of H^+ Secretion of HCO_3^-

2.6 Renal Circulation

The renal circulation serves several purposes:

(a) As a part of the systemic circulation, it provides cells with nutrients and oxygen and removes CO_2.
(b) It provides the glomeruli with an adequate plasma flow to support the glomerular filtration.
(c) It removes solutes and water that have been reabsorbed along the renal tubule and supplies the tubular epithelium with solutes for secretion.
(d) It participates in the mechanism of concentration and urinary dilution, through the medullary blood circulation in the vasa recta capillaries.
(e) It helps in blood pressure regulation.

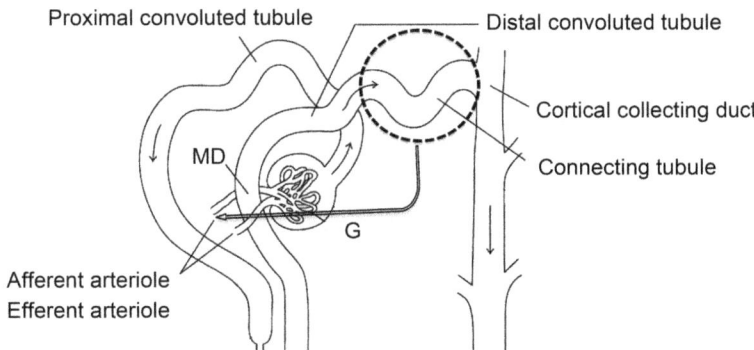

Proximal convoluted tubule

Distal convoluted tubule

MD

Cortical collecting duct

Connecting tubule

G

Afferent arteriole
Efferent arteriole

Fig. 2.10 Tubulo-vascular relationship between the connecting tubule and afferent arteriole. This segment (dotted circle) contacts the afferent arteriole of the parent nephron, giving rise to a second loop and a new structure composed of the juxtaglomerular apparatus and the connecting tubule. This structure is the anatomical base for the connecting tubule glomerular feedback (CTGF), involved in the tubuloglomerular feedback (*G* glomerulus, *MD* macula densa)

The renal blood flow in a human adult is 1200 mL/min, which is nearly one fourth of the resting cardiac output, although the kidney's weight is only 0.5% of the total body weight. This high flow correlates with the plasma clearing function of the kidneys. The general organization of the renal circulation is shown in Fig. 2.10.

Each kidney receives blood through the renal artery derived from the abdominal aorta. The renal artery branches off, giving rise to the radial cortical arteries, which in turn give rise to the arciform arteries, which run parallel to the corticomedullary boundary, and interlobular arteries—which run along the cortex from the corticomedullary to more superficial portions of the renal cortex. The cortical renal circulation is organized into two capillary beds. The first one is formed by the glomerular capillaries which originate from afferent arterioles derived from interlobular arteries. Glomerular capillaries merge forming the efferent arteriole, which drains the blood from each glomerulus. The efferent arterioles from cortical and mid-cortical nephrons give rise to the second capillary bed which is formed by the peritubular capillaries. These fenestrated capillaries absorb fluid reabsorbed by the renal tubules mainly present in the cortical labyrinths. The efferent arterioles from juxtamedullary nephrons give rise to the capillaries of the vasa recta, composed of a descending and ascending limb (Figs. 2.11 and 2.12). The descending vasa recta carry blood to deeper portions of the outer and inner medulla; they form a hairpin and form the ascending vasa recta that carry blood from the medulla to the cortex and deliver to the venous system. The arrangement of this capillary system plays a key role in the urinary mechanism of concentration and dilution.

Therefore, in the cortical renal circulation there are two capillary beds arranged in series: glomerular capillaries, interposed between two resistance vessels, such as afferent and efferent arterioles, and peritubular capillaries derived from cortical and mid-cortical nephron efferent arterioles (Figs. 2.11 and 2.12). As will be seen in the next chapter, the interposition of the glomerular capillaries between two resistance

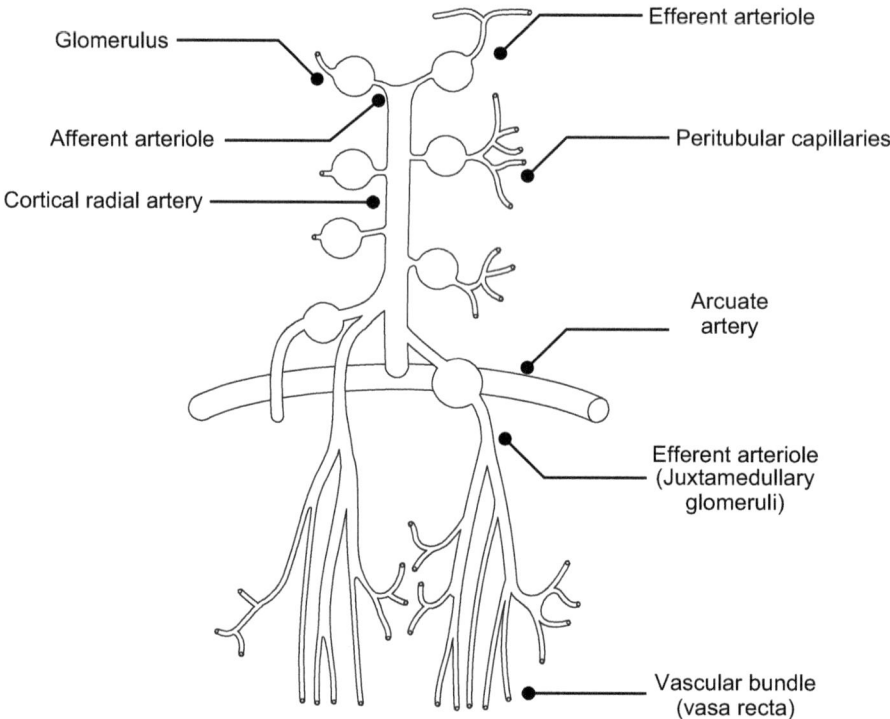

Fig. 2.11 Renal circulation. Spatial organization of the pre- and postglomerular arterial circula-tion. The efferent arterioles of cortical nephrons originate in the peritubular capillaries; these fenestrated capillaries reabsorb fluid in the renal cortex. The efferent arterioles from juxtamedullary nephrons give rise to vasa recta capillaries that accompany medullary tubular segments. Vasa recta capillaries are important in the urine-concentrating mechanism

vessels, such as the afferent and efferent arterioles, has an important effect on the hydrostatic pressure in these capillaries. The venous system runs parallel to the arterial system, and the vessels are called the equivalent of arteries.

2.7   Conclusions

The gross anatomy of the mammalian kidney is divided into two zones: cortex and medulla. The cortex is further divided into cortical labyrinths and medullary rays. The renal medulla is further divided into an outer and inner medulla.

The nephron is the structural and functional unit of the kidney. It consists of a renal corpuscle and the renal tubule. The renal corpuscle contains Bowman's capsule and the glomerular capillary tuft. The glomerular filtration barrier is a specialized structure for ultrafiltration of plasma. The renal tubule consists of several segments with different morphofunctional properties allowing tubular reabsorption and secretion.

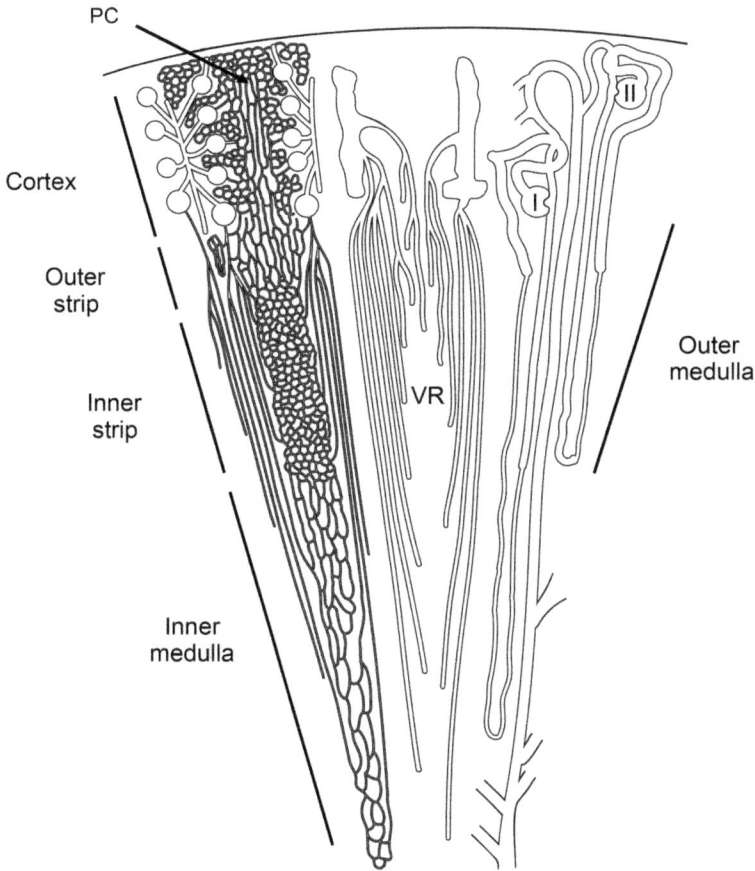

Fig. 2.12 Renal microcirculation. Spatial organization of the postglomerular microcirculation formed by peritubular capillaries (PC) in the renal cortex and vasa recta (VR) in the medulla in relation to the architecture of juxtamedullary (*I*) and cortical nephrons (*II*). Peritubular capillaries accompany tubules of the renal cortex. Vasa recta capillaries accompany tubular segments that project to deeper portions of the medulla and then return to the cortex

Long-looped nephrons (juxtamedullary) have their renal corpuscles in the corticomedullary limit. Their renal tubule penetrates deep into the inner medulla. Short-looped nephrons (cortical) have their corpuscles in the cortex; their renal tubule does not penetrate the inner medulla.

The renal cortex has two vascular beds. The first is the system formed by afferent arterioles, glomerular capillaries, and efferent arterioles. This vascular bed is histologically suited for plasma ultrafiltration. The second vascular bed is formed by peritubular capillaries that irrigate the renal cortex; these fenestrated capillaries are ideally suited for fluid reabsorption. They drain into postcapillary venules.

The renal medulla has one vascular bed. This is formed by vasa recta capillaries derived from efferent arterioles of long-looped nephrons. These capillaries penetrate into the medullary interstitium, making a hairpin at different levels in the medulla and returning to the cortex. They participate in the urinary-concentrating mechanism.

Review Questions

1. What are the main differences between a cortical or short-looped nephron and a juxtamedullary or long-looped nephron?
2. What are the components of the glomerular filtration barrier?
3. What are the components and how is the juxtaglomerular apparatus organized?
4. A histological section of rat kidney was incubated with an unknown antibody. The immunostaining procedure revealed abundant immunoreactivity in the apical and basolateral membrane of proximal tubules. Which cellular protein could be responsible for immunoreactivity?
5. Which are the main features of the proximal tubule that allow the reabsorption of huge amounts of solutes and fluid?

Bibliography

Fenton RA, Prætorius J (2016) Anatomy of the kidney. In: Brenner BM, Rector FC (eds) The kidney, vol I, 10th edn. Elsevier, Philadelphia, pp 42–82

Garg P (2018) A review of podocyte biology. Am J Nephrol 47(Suppl 1):3–13

Kriz W, Bankir L (1988) A standard nomenclature for structures of the kidney. Kidney Int 33:1–7

Kriz W, Kaissling B (2013) Structural organization of the mammalian kidney. In: Seldin D, Giebisch G (eds) The kidney: physiology and pathophysiology, vol I, 5th edn. Academic Press, San Diego, pp 595–691

Pollak MR, Quaggin SE, Hoenig MP, Dworkin LD (2014) The glomerulus: the sphere of influence. Clin J Am Soc Nephrol 9:1461–1469

Rubera I, Loffing J, Palmer LG, Frindt G, Fowler-Jaeger N, Sauter D, Carrol T, McMahon A, Hummler E, Rossier BC (2003) Collecting duct-specific gene inactivation of α-ENaC in the mouse kidney does not impair sodium and potassium balance. J Clin Invest 112(4):554–565

Glomerular Filtration and Renal Blood Flow

3

Learning Objectives

- To understand the processes of glomerular filtration as the first process involved in urine formation.
- To describe the physical determinants of the glomerular filtration rate.
- To describe the functions of renal blood flow and the mechanisms of autoregulation.
- To understand the application of the clearance to determine the glomerular filtration rate.

The first step in urine formation is the production of an ultrafiltrate of plasma through the process glomerular filtration. Daily, in a healthy adult, the kidneys filter about 180 L of water, which corresponds to a glomerular filtration rate (GFR) of 100–125 mL/min. Knowledge of the chemical composition of the ultrafiltrate and the ultrastructure of the glomerular filtration barrier is a key point to understand the glomerular ultrafiltration.

3.1 Composition of Glomerular Filtrate

The first data about the chemical composition of the glomerular ultrafiltrate were obtained in 1924 in frog kidney. Only in 1970, the micropuncture of superficial glomeruli from rat kidney (Munich-Wistar strain) allowed the collection of fluid samples from Bowman's space. The chemical analysis of ultrafiltrate samples leads to the conclusion that the glomerular ultrafiltrate is an almost protein-free fluid with identical concentration of other solutes like electrolytes and organic solutes like glucose, urea, and creatinine.

To understand the factors that influence the filtration of a molecule, it is necessary to consider the properties of the molecules (molecular weight, hydrated molecular radius, net electrical charge, etc.) and the ultrastructure of the glomerular filtration barrier, reviewed in the previous chapter. Some of the barrier characteristics can be

© Springer Nature Switzerland AG 2022
P. A. Gallardo, C. P. Vio, *Renal Physiology and Hydrosaline Metabolism*,
https://doi.org/10.1007/978-3-031-10256-1_3

Table 3.1 Relationship between the filterability of molecules and parameters, such as molecular weight and hydrated molecular radius. Filterability refers to the ratio between the filtration of a molecule under study and that of inulin, which filters freely

Molecule	Molecular weight (Da)	Molecular radius hydrated (Å)	Filterability
Water	18	1	1
Na$^+$	23	1.4	1
Urea	60	1.6	1
Glucose	180	3.6	1
Inulin	5500	14.8	1
Myoglobin	17,000	19.5	0.75
Hemoglobin	68,000	32.5	0.03
Albumin	69,000	35.5	<0.01

deduced from the data in Table 3.1, which shows the relationship between the filterability of a molecule and its hydrated molecular radius.

As shown in Table 3.1, the higher the molecular weight and hydrated radius of a substance, the lower its filterability. For molecules with a hydrated radius of less than 18 Å, their filtration is 100%, which corresponds to a filterability = 1. This means that cations, anions, and neutral molecules with a hydrated radius lower than 18 Å are freely filtered. Above this radius, filterability decreases. From these data, it can be deduced that the glomerular filtration barrier discriminates solutes by size. Hence, the glomerular filtration barrier imposes a mechanical restriction to the passage of solutes. The origin of the restriction relies in the size of the pores of the glomerular filtration barrier: those present in the endothelium of the glomerular capillaries are large (700 Å) and only limit the passage of blood cells, but the pores present in the basal lamina and filtration windows are smaller and can restrict the passage of molecules according to their size. A second factor that determines the filterability of a molecule is its net electrical charge. Figure 3.1 shows the results of experiments where the filterability of dextrans (polysaccharides) of different hydrated molecular radius and net electrical charge (cationic, anionic, and neutral) was measured. For dextrans with a molecular radius less than 18 Å, the filterability is independent of the net electrical charge and therefore these dextrans are freely filtered at 100% or have a filterability of 1. Above this molecular radius value, the net electric charge affects the filterability of the molecule, delaying the filtration of anionic dextrans compared to neutral and cationic ones. In other words, starting from a critical molecular radius (18 Å), the net charge of the molecule determines, together with the molecular radius, the filterability of the molecule.

Albumin is an abundant plasma protein with many functions. Filtration of albumin and its appearance in urine is a hallmark in many glomerular diseases. As shown in Table 3.1, albumin hydrated radius is 35.5 Å. At the physiological pH of 7.4, albumin has a negative net electrical charge. An anionic dextran with the same molecular radius to that of albumin has filtration near *zero*, compared to a cationic dextran with the same molecular weight. Therefore, two properties of albumin, its molecular hydrated radius together with the negative net electrical charge, are crucial

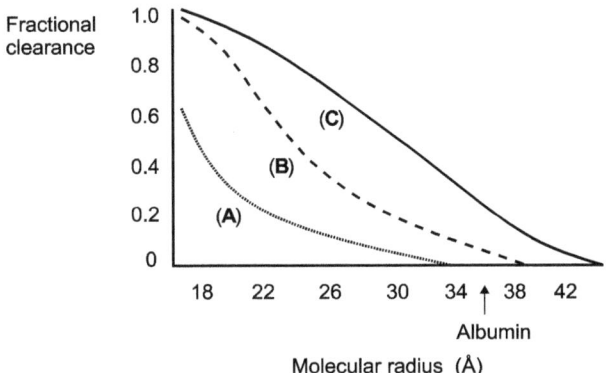

Fig. 3.1 Fractional clearance of dextrans. The fractional clearance of anionic (*A*), neutral (*B*), and cationic (*C*) dextrans was assayed in relation to their effective molecular radius. The fractional clearance is a comparison between the filtration of a specific molecule and the filtration of a molecule that is completely filtered. For a molecular ratio greater than 18 Å, the filterability of a cationic dextran is greater than for a neutral or anionic dextran. The arrow indicates the albumin effective radius (35.5 Å); the protein filtration is greatly reduced because of its size and net negative electrical charge at physiological pH

for its very low glomerular filtration. The pore size and the negative charges of the basal lamina and filtration slits are two properties of the glomerular filtration barrier that play a key role in the mechanical as well as electrical restriction to the passage of plasma proteins like albumin. Although the exact mechanisms by which the glomerular filtration barrier functions as a selectivity barrier are not completely understood, it is clear that negative charges play an important role. The absence of negative charges by digestion of glycosaminoglycans results in proteinuria. Cell and molecular biology studies of some glomerular diseases highlighted the importance of some proteins. In Alport syndrome, the mutations in collagen IV chains result in heavy proteinuria including albuminuria. Nephrin is a structural and signaling transmembrane protein expressed at the cell junction formed in the filtration slits. More than 50 mutations of this protein had been identified causing the Finnish type of the congenital nephrotic syndrome. All identified mutations cause heavy proteinuria that can start during intrauterine life.

3.2 Dynamics of Glomerular Filtration

The forces involved in the filtration of plasma from the glomerular capillaries into Bowman's space are the same Starling forces that drive ultrafiltration in any capillary bed: capillary hydrostatic pressure and interstitial oncotic pressure are the forces that favor filtration. Capillary oncotic pressure and interstitial hydrostatic pressure oppose filtration.

However, the process is qualitatively and quantitatively different in glomerular capillaries compared to systemic capillaries, due to their characteristics:

Fig. 3.2 Nefrovascular units.
In the cortex there are two
nefrovascular units. The first
is composed of the glomerular
capillaries, interposed
between two resistance
vessels like the afferent (*AA*)
and efferent arterioles (*EA*).
The second unit is composed
of the peritubular capillaries
originated from efferent
arterioles of cortical nephrons

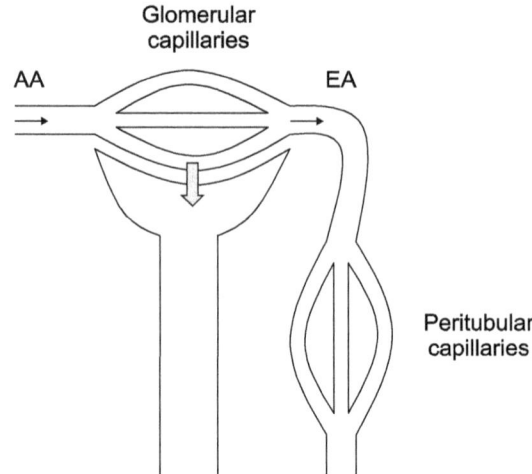

1. They have a high hydraulic permeability (Lp), about 100 times more than muscle capillaries. This is correlated with the finding that glomerular capillaries are fenestrated, with pores of an average size of 700 Å.
2. They have a large surface area (S) for filtration, calculated between 100 m^2/100 g of the kidney. Around 20–50% of the endothelial surface is occupied by fenestrations.
3. The glomerular capillaries are interposed between an afferent and an efferent arteriole (Fig. 3.2). This important feature determines two important facts: first, the capillary hydrostatic pressure is higher than that in systemic capillaries and second the hydrostatic pressure is maintained along the capillary length.

In a systemic capillary, fluid filtration decreases along the length of the capillary, due to the drop in capillary hydrostatic pressure, while the capillary oncotic pressure remains relatively constant. In glomerular capillaries, the hydrostatic pressure is higher and changes little along the capillary from the afferent to the efferent side and is the main force driving plasma filtration. The fact that plasma proteins have low filtration has two important consequences. First, as filtration occurs, there is an increase in capillary oncotic pressure along the length of the capillary. Second, since plasma proteins do not filter, the oncotic pressure in the urinary space is close to *zero*. Since ultrafiltrate is constantly draining from Bowman's space to the proximal tubule, the hydrostatic pressure in Bowman's capsule remains relatively constant.

Therefore, in the glomerular capillaries, filtration ceases or decreases according to the increase in capillary oncotic pressure, due to the proteins that do not filter. The determinants of the glomerular filtration are arranged in Eq. 3.1:

$$GFR = Lp \cdot S[(PHgc - PHbs) - (\pi gc - \pi bs)] \qquad (3.1)$$

where:

GFR = glomerular filtration rate
Lp = hydraulic permeability
S = surface available for filtration
PH = hydrostatic pressure; gc: glomerular capillary; bs: Bowman's space
π = oncotic pressure; gc: glomerular capillary; bs: Bowman's space

Equation 3.1 is also known as the Starling equation and is the general expression for fluid filtration in capillaries. Ultrafiltration pressure (UFP) is the pressure difference that determines plasma filtration and is expressed in Eq. 3.2:

$$UFP = [(PH_{cg} - PH_{cb}) - (\pi_{cg} - \pi_{cb})] \tag{3.2}$$

Since plasma proteins are almost not filtered, oncotic pressure in Bowman's space is close to *zero* and Eq. 3.2 can be rewritten as:

$$UFP = [(PH_{cg} - PH_{cb}) - (\pi_{cg})] \tag{3.3}$$

where:

UFP = ultrafiltration pressure difference

The variables of Eq. 3.3 have been estimated from measurements in rats, rabbits, dogs, and primates.

At present, the values in Table 3.2 are the closest to that of primates. In some species, the ultrafiltration pressure decreases to *zero* towards the efferent end of the glomerular capillary, reaching a filtration equilibrium. This condition is defined as a balance between the hydrostatic pressure and the oncotic pressure in the glomerular capillaries. Hence, under filtration equilibrium, the net ultrafiltration pressure is *zero*. In primates and probably in humans, the filtration equilibrium condition is not reached. In this setting, ultrafiltration pressure decreases along the capillary length (Fig. 3.3).

The dynamics of glomerular filtration is affected by several factors. One of these is renal plasma flow (Fig. 3.3). Changes in renal plasma flow affect the glomerular capillary oncotic pressure and therefore the ultrafiltration pressure. A high renal plasma flow determines a lower elevation of the oncotic pressure along the length of

Table 3.2 Estimated and approximate hydrostatic (PH) and oncotic pressure (π) values for the human kidney. The negative sign means that this value is against filtration

Variable	Afferent arteriole (mmHg)	Efferent arteriole (mmHg)
PHcg	60	58
PHcb	−15	−15
πcg	−28	−35
πcb	0	0
UFP	17	8

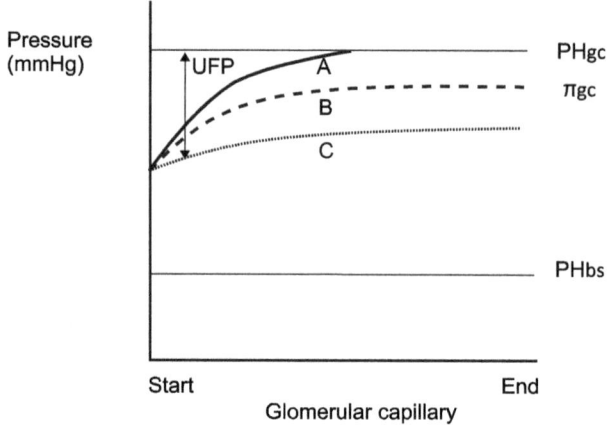

Fig. 3.3 Physical factors involved in the ultrafiltration of plasma along the length of the glomerular capillary. The hydrostatic pressure in glomerular capillaries (PHgc) is almost constant because the capillaries are interposed between two arterioles. The hydrostatic pressure in Bowman's space (PHbs) is constant as long as the ultrafiltrate is drained into the proximal tubule. The oncotic pressure in the glomerular capillaries (πgc) increases due to fluid filtration but not protein filtration. The lines *A*, *B*, and *C* represent different degrees of renal plasma flow. The magnitude of the renal plasma flow determines the ultrafiltration pressure (UFP) through the oncotic pressure in the glomerular capillaries

the glomerular capillary as filtration occurs and, therefore, the ultrafiltration pressure experiences a lower drop throughout the glomerular capillary. With reduced renal plasma flow, filtration produces a more rapid rise in oncotic pressure in the glomerular capillaries. Hence, the ultrafiltration pressure falls more prematurely along the capillary and filtration decreases reaching equilibrium filtration.

A second factor that influences the dynamics of glomerular filtration is the contractile tone of the smooth muscle of the afferent and efferent arterioles (Fig. 3.4). The contraction of the afferent arteriole causes a decrease in the hydrostatic pressure in the glomerular capillaries and, consequently, a reduction in the ultrafiltration pressure. Conversely, a decrease in the contractile tone of the afferent artery causes an increase in plasma flow, raising hydrostatic pressure and GFR. Therefore, changes in afferent arteriolar tone cause changes in the same direction in renal plasma flow and GFR. However, this is not the case with changes in efferent arteriolar tone. An increase in efferent arteriolar smooth muscle tone increases capillary hydrostatic pressure and GFR. This increase in GFR is counteracted by a rise in the oncotic pressure in the glomerular capillary. The increase in efferent arteriole resistance also diminishes renal blood flow and hence renal plasma flow. Therefore, increasing efferent arteriole has a dual effect on GFR: first, the increase is due to an increase in capillary hydrostatic pressure and then a decrease is due to the increase in oncotic capillary pressure caused by the increase in filtration.

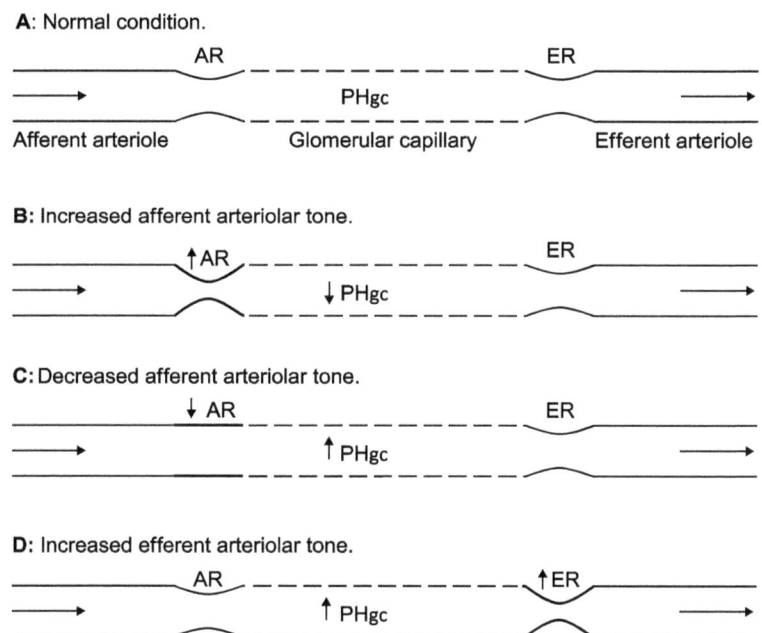

Fig. 3.4 Effect of changes in afferent arteriolar resistance (*AR*) and efferent arteriolar resistance (*ER*) on the glomerular capillary hydrostatic pressure (PHgc). (**a**) Normal condition, (**b**) increase of AR, (**c**) decrease of AR, and (**d**) increase of ER. See text for detailed explanations

3.3   Renal Blood Flow

The renal blood flow (RBF) from both kidneys in a normal adult resting subject is 1200 mL/min, corresponding to 25% of the resting heart rate. The two kidneys (300 g, both kidneys) of a 70 kg adult represent 0.4% of body weight and receive a much higher blood flow than well-irrigated organs such as the heart, liver, and brain (Table 3.3).

The high renal blood flow is due to the filtration function of the kidney, which in turn is closely related to the other functions the kidney performs in the context of maintaining homeostasis of body fluids. 90–95% of the RBF is distributed to the cortex where the glomeruli are, and 5–10% to non-filtering structures such as the medulla, pelvis, hilum, and renal capsule. The high oxygen consumption of the kidneys is related to high rates of active transport, mainly related to primary active transport carried out by the Na^+, K^+-ATPase in tubular cells of different nephron segments.

The renal blood flow has several functions:

1. Indirectly, it determines the glomerular filtration rate.
2. It modifies the proximal tubular reabsorption of solutes and water.

Table 3.3 Comparison of blood flow and oxygen consumption of the kidneys and other organs with high blood flow

Organ	Mass (g)	Blood flow (mL/min/100 g)	Oxygen consumption (μmol/100 g/min)
Kidneys	300	400	267
Heart	300	84	431
Brain	1400	54	147
Skin	3600	13	15
Skeletal muscle at rest	31,000	3	7

3. It participates in the mechanism of urinary concentration and dilution (medullary blood flow through the capillaries of the vasa recta capillaries).
4. It delivers oxygen, nutrients, and hormones to the cells.
5. It participates in tubular secretion by providing molecules (ions, endogenous, and exogenous organic compounds) to the tubular cells for their tubular secretion.

As in any organ, renal blood flow is determined by pressure gradient (ΔP) and vascular resistance (R) and is described in Eq. 3.4:

$$\text{RBF} = \frac{\Delta P}{R} = \frac{P \text{ renal artery} - P \text{ renal vein}}{\text{Renal vascular resistance}} \tag{3.4}$$

According to Eq. 3.4, an increase in pressure or a fall in resistance translates into an increase in renal blood flow, and, conversely, a fall in pressure or a rise in resistance should determine a fall in flow. However, these changes are limited by the self-regulating ability of the kidney's blood flow.

3.3.1 Autoregulation of Renal Blood Flow

The kidney is able to maintain a constant blood flow in the face of changes in blood pressure. The autoregulation of renal blood flow is intrinsic to the kidney, since it occurs in the absence of any neural (sympathetic) and hormonal influences. Since glomerular filtration rate is dependent on renal blood flow, the autoregulation of flow is crucial for the regulation of glomerular filtration rate (Fig. 3.5). In turn, the regulation of glomerular filtration rate is a critical process for maintaining a constant filtering load of solutes that will be delivered to the tubular segments. In a more specific way, autoregulation of blood flow prevents a possible overload or NaCl to segments of the distal nephron where the fine regulation of NaCl reabsorption occurs.

Autoregulation of renal blood flow operates within a range of mean blood pressure between 80 and 180 mmHg. Two mechanisms have been proposed to explain autoregulation: a pressure-dependent or myogenic mechanism and a flow-dependent mechanism or tubuloglomerular feedback.

Fig. 3.5 Autoregulation of renal blood flow (RBF) and glomerular filtration rate (GFR). (**a**) Renal blood flow as a function of mean arterial blood pressure. For a range between 80 and 180 mmHg, renal blood flow is constant; (**b**) GFR as a function of mean arterial blood pressure; glomerular filtration exhibits the same behavior as renal blood flow

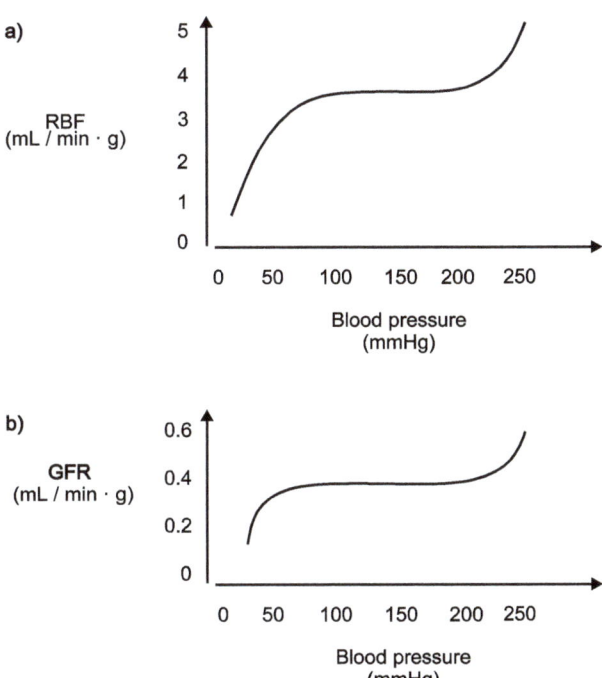

The myogenic mechanism operates in the preglomerular arteries, particularly in the afferent arteriole (Fig. 3.6a). Arteriolar vascular smooth muscle has the property of contracting when transmural pressure increases. In the afferent arteriole, an increase in blood pressure elevates the transmural pressure and stretches the smooth muscle cells of the arteriolar wall. In response to stretching, smooth muscle cells contract with the consequent increase in resistance and drop in flow (Fig. 3.6b). An important fact is that the autoregulatory response can be completely inhibited with calcium channel blockers. The molecular basis of the myogenic mechanism is related to the activation of stretch-sensitive cation-selective channels expressed in the sarcolemma of smooth muscle cells. Membrane stretching activates these channels, depolarizing the membrane potential, and triggers the opening of voltage-dependent calcium channels. Calcium entry increases cytosolic ionic calcium and activates the contractile machinery of arteriolar wall smooth muscle cells. Therefore, increasing the resistance in the afferent arterioles (and other preglomerular vessels) reduces renal blood flow, renal plasma flow, and glomerular filtration rate. In other words, changes in afferent resistance induce changes in these variables in the same direction.

The flow-dependent mechanism operates as a tubuloglomerular feedback; this results from an integration between tubular function and the resistance of preglomerular vessels. This mechanism involves a feedback from the macula densa cells of the juxtaglomerular apparatus to the afferent arteriole (Fig. 3.7a). The variable detected is NaCl concentration in the tubular fluid delivered to the

Fig. 3.6 Myogenic
mechanism. (**a**) An increase in
the internal pressure
(Ip) produces an increase in
the transmural pressure
(Ip-Op) and arteriolar wall
tension. (**b**) Vascular
resistance as a function of the
mean arterial pressure. An
increase in mean arterial
pressure increases afferent
arteriole but not efferent
arteriolar resistance

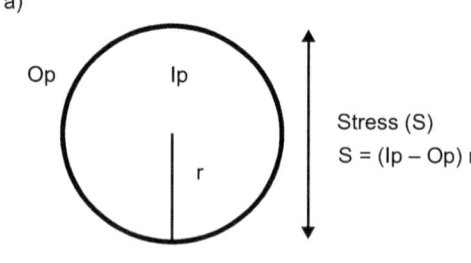

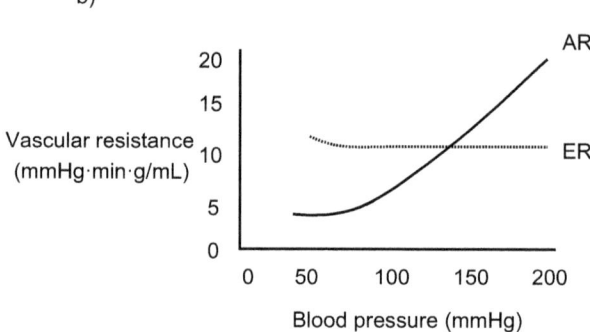

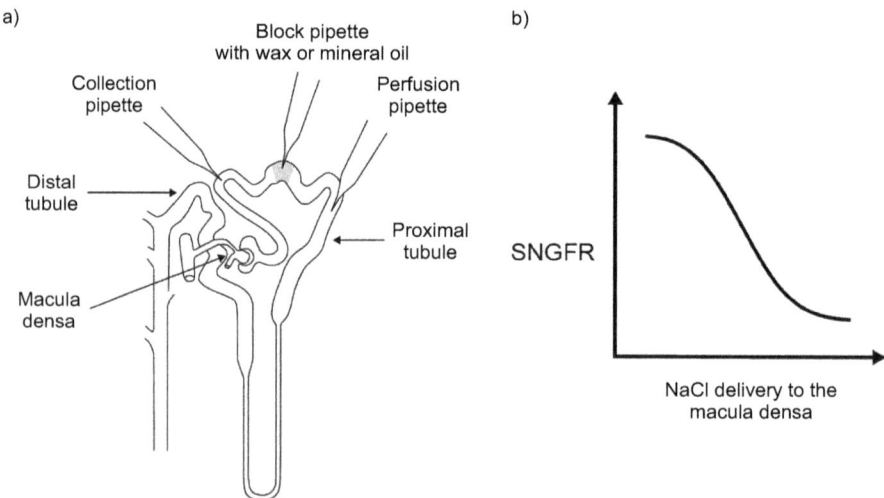

Fig. 3.7 Mechanism of tubuloglomerular feedback. (**a**) Drawing of a nephron microperfusion experiment to demonstrate the tubuloglomerular feedback. The collection pipette allows measurement of single-nephron glomerular filtration rate (SNGFR); a wax drop is installed to block fluid movement. The perfusion pipette delivers fluid at a known rate to the proximal tubule. (**b**) Single-nephron glomerular filtration rate as a function of the rate of NaCl delivery to the macula densa. As NaCl delivered to the macula densa increases, the SNGFR decreases

macula densa. An increase in NaCl concentration in fluid delivered to the macula densa leads to an increase in the resistance of the afferent arteriole. Inversely, a decrease in the NaCl concentration in the tubular fluid reduces the resistance in the afferent arteriole. The tubuloglomerular feedback has been studied in single-nephron tubular microperfusion experiments measuring its GFR (SNGFR, single-nephron glomerular filtration rate) (Fig. 3.7b). It was demonstrated that an increase in tubular flow results in a reduction in capillary hydrostatic pressure and plasma flow to the parent glomerulus. Both effects occur only when afferent arteriolar tone increases. Also, it was demonstrated that the tubuloglomerular feedback response can be abolished if the tubular lumen is perfused with a NaCl solution containing furosemide. This drug blocks the NKCC2 cotransporter, which is expressed in the apical membrane of the thick ascending limb cells. Hence, NKCC2 activity is an important component in the mechanism by which the macula densa cells sense the NaCl concentration delivered in the tubular fluid.

The physiological role of the tubuloglomerular feedback mechanism is to keep constant the filtered load of solutes, especially NaCl, delivered to distal tubular segments. This is of vital importance, as it prevents large amounts of NaCl from reaching segments with low reabsorption capacity, such as those of the distal nephron, where fine regulation of NaCl transport occurs to maintain the sodium balance.

The tubuloglomerular feedback raises several questions regarding the nature of the signal that finally modifies afferent arteriolar resistance. Multiple lines of evidence point to adenosine as the primary signal arriving from macula densa cells. The role of adenosine is inserted in the following sequence of events (Figs. 3.8 and 3.9):

(a) Macula densa. An increase in the luminal concentration of NaCl increases the activity of apical NKCC2 cotransporter. This leads to an increase in basolateral Na^+, K^+-ATPase activity, which generates ADP. This nucleotide can exit through pannexins expressed at the basolateral membrane. In the extracellular space, ATP and ADP can be degraded to adenosine by a membrane-bound $5'$-nucleotidase expressed in extraglomerular mesangial cells.

(b) Adenosine binds to A1 receptors expressed on extraglomerular mesangial cells, granular cells, and smooth muscle cells of the afferent arteriolar wall. A1 receptors are coupled to Gq protein and cause an increase in the concentration of cytosolic Ca^{++}.

(c) Extraglomerular mesangial cells, granular cells, and smooth muscle cells are functionally coupled by gap junctions formed mainly by connexin 43. This allows the Ca^{++} wave to expand between coupled cells. The increase in cytosolic Ca^{++} activates arteriolar smooth muscle cell contraction.

(d) In granular cells of the afferent arteriole, the increase in cytosolic Ca^{++} ion inhibits renin secretion and synthesis. Therefore, an increase in NaCl delivered to the macula densa translates into increased adenosine production. In turn, adenosine functions as a brake to the increase in GFR and NaCl filtered load.

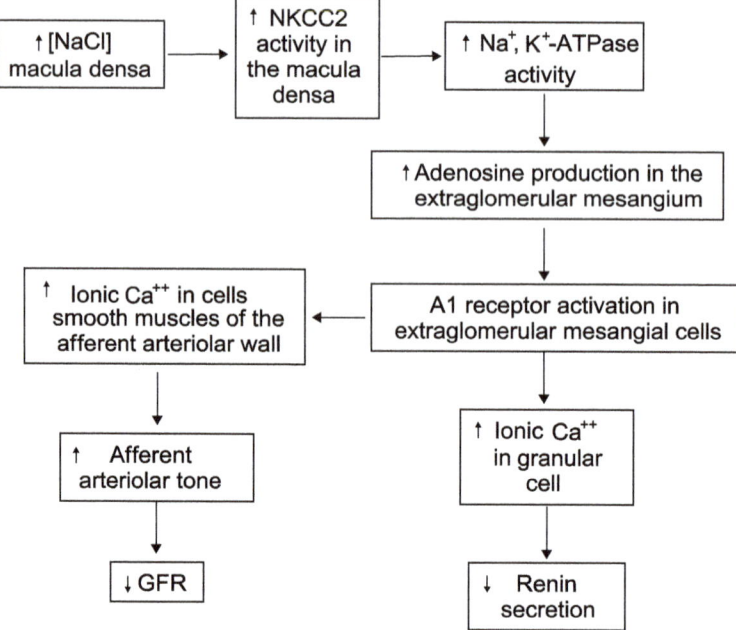

Fig. 3.8 Sequence of events involved in the mechanism of tubuloglomerular feedback. The initial signal is the NaCl delivery to the macula densa followed by NaCl transport by NKCC2 cotransporter. The increased NaCl transport across the apical membrane increases Na⁺, K⁺-ATPase activity and adenosine production. A1 adenosine receptors increase afferent arteriolar resistance, thus reducing glomerular filtration rate

3.3.2   Neurohumoral Regulation of Renal Blood Flow

Several chemical messengers are involved in the regulation of renal blood flow and thus of glomerular filtration. Among them are angiotensin II, norepinephrine, prostaglandins, nitric oxide, endothelin, atriopeptin, bradykinin, etc.

Angiotensin II (AII) This vasoconstrictor hormone is synthesized systemically and intrarenally. Its actions are mediated through AT_1 receptors coupled to phospholipase C through Gq protein, which increases cytosolic Ca^{++}. AII contracts afferent arterioles and other preglomerular vessels, decreasing renal blood flow. Besides, the vasoconstrictor effect on preglomerular vessels is rapidly counteracted by the production of vasodilators, stimulated by the same angiotensin II. The most powerful vasoconstrictor effect of angiotensin II is exerted on efferent arteriole. This action has two important effects. The first is a reduction in renal blood flow and the second is a moderate increase in glomerular filtration rate. The latter results from the increase in glomerular capillary hydrostatic pressure. AII reduces glomerular filtration rate by decreasing the area available for filtration. This effect is caused by the contraction of glomerular mesangial cells.

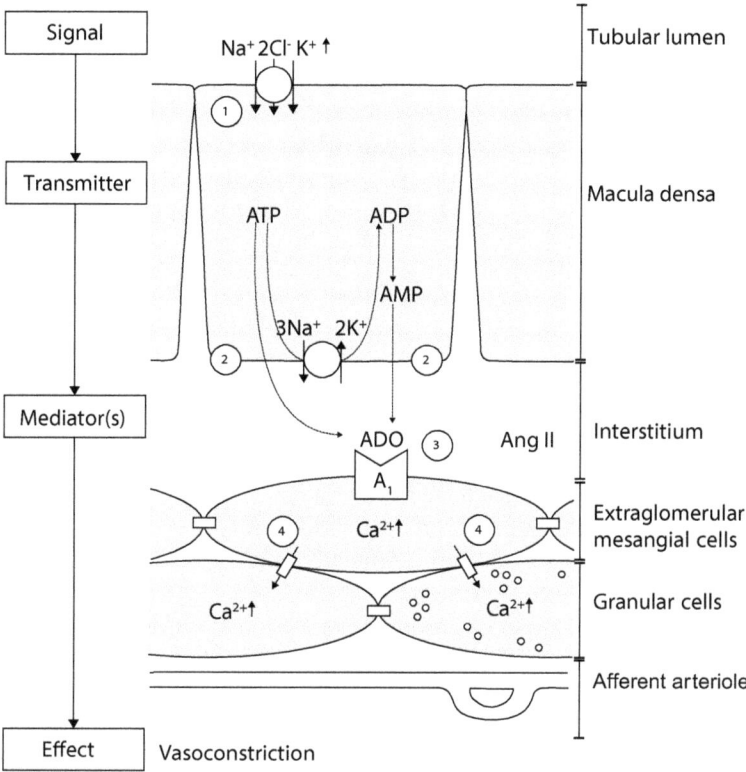

Fig. 3.9 Sequence of cellular and molecular events involved in the tubuloglomerular feedback. The initial signal is an increase in NaCl delivery to macula densa cells, activating NaCl transport and adenosine (ADO) production. Paracrine adenosine binds to A1 receptors in extraglomerular mesangial cells. These cells are functionally coupled with smooth muscle cells and granular cells in the afferent arteriole through gap junctions. The subsequent increase in cytosolic Ca^{++} contracts smooth muscle cells and inhibits renin secretion. The net effect is a decrease in renin secretion and glomerular filtration rate. (Adapted from Vallon V. Tubuloglomerular feedback and the control of glomerular filtration rate. *Physiology* 18:169–174, 2003)

Angiotensin II is part of the renin-angiotensin-aldosterone system and it plays an important physiological role in the regulation of the effective circulating volume; general aspects of its physiology will be discussed in the respective chapter.

Norepinephrine   Both glomerular arterioles receive sympathetic innervation. When the effective circulating volume is normal, the influence of the sympathetic tone on the renal function is low, but it is important in conditions of volume depletion. Norepinephrine, secreted by postganglionic sympathetic fibers, binds to α1 adrenergic receptors coupled via phospholipase C via Gq protein, leading to vascular smooth muscle contraction.

Nitric oxide (NO) Nitric oxide (endothelium-derived relaxing factor) is synthesized from L-arginine by nitric oxide synthase (NOS). NO is a powerful vasodilator with an important physiological role in renal hemodynamics. The absence of NO causes a marked increase in renal vascular resistance, which implies that under basal conditions it provides a vasodilator tone that counteracts the effects of vasoconstrictor agents, such as AII. NO-induced changes in renal hemodynamics not only affect glomerular filtration rate but also tubular transport, especially of Na^+ and water. Several agents such as acetylcholine, bradykinin, and histamine stimulate NO synthesis and so do physical factors such as increased shear stress stimulating NO synthesis.

Bradykinin This vasodilator peptide is part of the kallikrein-kinin system. Kallikrein is a proteolytic enzyme synthesized in the kidney exclusively in the connecting cells of the renal tubule. The peptide bradykinin formed by kallikrein is degraded by the angiotensin I-converting enzyme (ACE) that forms AII from angiotensin I. Thus, this vasodilator system is closely linked to a vasoconstrictor system formed by renin and angiotensin II. Bradykinin exerts its vasodilatory effect through receptors coupled to phospholipase C and stimulates the synthesis of vasodilatory prostaglandins and NO.

Endothelins These are peptides that have a powerful vasoconstrictor effect. Endothelin increases renal vascular resistance and affects tubular transport, cell proliferation, and extracellular matrix synthesis.

3.4   Measurement of Glomerular Filtration Rate

The measurement of glomerular filtration rate is very relevant from the clinical point of view, since it is the best index of the global renal function in both healthy and diseased subjects, and it can be used to estimate the magnitude of the functional renal mass in a subject. Many relevant decisions for patients, such as drug doses or entry into a dialysis program, are made using estimates of glomerular filtration rates. The measurement of the glomerular filtration velocity is done using the concept of renal clearance.

3.4.1 Concept of Renal Clearance

The term clearance refers to the concept of cleaning or purifying the plasma of a given substance. In terms of renal physiology, this concept emphasizes the ability of the kidney to clear a substance from the plasma and thus the renal excretory function. To apply this concept, the kidney is considered as a black box where substance can only enter through the renal artery. However, the kidney has two exits for a substance: the renal vein and the ureter (Fig. 2.1). If the substance appears only in the urine, it means that that particular substance is being excreted from the body. The appearance of the substance in the renal vein means that it has been returned to the body. The fate of a specific substance depends on its particular renal handling. Recall that the renal clearance does not give any information concerning the mechanisms by which a substance is handled by the kidney.

The general clearance for a substance "x" (C_x) is expressed in Eq. 3.5:

$$C_x = \frac{V \cdot [x]u}{[x]pl} \tag{3.5}$$

where:

C_x (mL/min) = clearance of the molecule "x"
V (mL/min) = urinary output
$[x]$ (mg/mL) = concentration of "x" molecule in arterial plasma (pl) and urine (u)

The renal clearance of some substances can be used to estimate GFR. That particular substance has to meet several requirements:

(a) Freely filtered in the glomerular filtration barrier.
(b) Should not be secreted, reabsorbed or stored in the tubular cells. Therefore, all the filtered amount should be excreted in the urine.
(c) The substance should not be produced or metabolized by the cells of the body.
(d) The substance should not be toxic and should not affect the renal function. It circulates freely in the plasma, without binding to plasma proteins or any other plasma molecule.

It should be mentioned that it is possible to measure the renal clearance of many molecules; however, their clearances might not be an estimation of GFR. Inulin (not insulin) is a one of the few molecules that fulfill all the requirements listed above. It is a plant origin polymer made of fructose units with a molecular mass of 5200. Inulin is freely filtered in the glomerular filtration barrier and is not reabsorbed nor secreted by the renal tubule (Fig. 3.10). Given its characteristics, Eq. 3.6 represents the renal handling of inulin:

$$\text{Filtered load of inulin} = \text{Excreted load of inulin} \tag{3.6}$$

Equation 3.6 means that all of the amounts filtered in all the functional glomeruli should appear in the final urine. Equation 3.5 can be rewritten as follows:

$$\text{GFR} \cdot [\text{Inulin}]pl = \dot{V} \cdot [\text{Inulin}]u \tag{3.7}$$

where:

GFR (mL/min) = glomerular filtration rate
[Inulin] (mg/mL) = inulin concentration on arterial plasma (pl) and urine (u)V
   (mL/min) = urine output

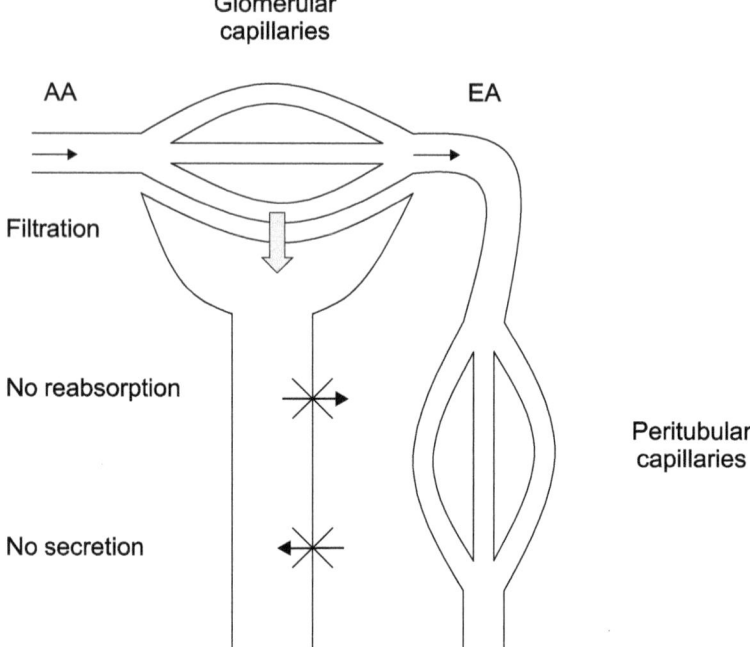

Filtered load of inulin = GFR [Inulin] plasma

Inulin excreted load = V [Inulin] urine

GFR [Inulin] plasma = V [Inulin] urine

GFR = Cin = V [Inulin] urine / [Inulin] plasma

Fig. 3.10 Renal handling of inulin. This polymer is freely filtered across the glomerular barrier and is not reabsorbed or secreted by the renal tubule. *AA*: afferent arteriole; *EA*: efferent arteriole. In this case, the inulin filtered load equals the excreted load of inulin. Therefore, inulin clearance equals the glomerular filtration rate

The left side in Eq. 3.7 represents the filtered load of inulin and the right side is the excreted load of inulin. Since the amount of filtered inulin is not reabsorbed and there is no inulin secretion, then the filtered load equals the excreted load.

If the term GFR in Eq. 3.7 is cleared:

$$\text{GFR} = \frac{V\,[\text{Inulin}]\text{u}}{[\text{Inulin}]\text{pl}} \tag{3.8}$$

In Eq. 3.8, GFR has the units of mL/min, as in Eq. 3.5. Inspection of both equations shows that they are the same. Hence, the inulin clearance is an estimation of GFR. Recall that the inulin clearance involves a method carried out in the whole organism or in perfused kidneys under experimental settings. Inulin clearance is not an estimate of GFR in a particular nephron. Therefore, the inulin clearance is an estimate of the GFR of all functional glomeruli of both kidneys.

In clinical practice, the inulin clearance is of limited use, because inulin must be infused into the patient until stable concentrations are reached in the blood and urine. The whole procedure requires taking several blood and urine samples, in which inulin concentration should be measured. In addition, the method for determining inulin in biological fluids might not be routinely available and it is not implemented in many laboratories.

Instead, the clearance of an endogenous substance such as creatinine is used to estimate GFR. Creatinine is a nitrogenous organic compound derived from the metabolism of skeletal muscle creatine. Its production and subsequent release from the muscle compartment into the plasma is relatively constant. Therefore, its plasma concentration is relatively stable and varies less than 10% per day in normal subjects. In the plasma compartment, creatinine circulates freely. However, renal handling of creatinine makes it an imperfect marker. Creatinine is a small organic cation freely filtered in the glomerular barrier. In contrast to inulin, it is secreted to the tubular lumen by the organic cation secretory system in the proximal tubule. Tubular secretion of creatine induces an overestimation of creatinine clearance since the excreted load will always be higher than the filtered load. The exact amount of this overestimation has been estimated as 10–15%. However, as plasma creatinine measurements have been improved, this number might be higher. Therefore, the creatinine excreted load exceeds the filtered load by about 10–15%. This is why the clearance of creatinine will be 10–15% higher than inulin. The creatinine clearance is expressed in Eq. 3.9:

$$Ccr = \frac{V[\text{Creatinina}]u}{[\text{Creatinina}]pl} \tag{3.9}$$

The clearance of creatinine is approximately 95 ± 20 mL/min in women and 120 ± 25 mL/min in men, per 1.73 m^2 of surface.

To calculate creatinine clearance the patient has to provide a 24-hour urine collection to measure the diuresis and urinary creatinine concentration. A venous blood sample is also needed to determine plasma creatinine.

Despite its widespread use in the clinic, two factors may affect the accuracy of the clearance of creatinine as an estimator of GFR:

(a) 24-h urine collection by the patient might not be very accurate. This fact induces error in the calculation of creatinine urinary excretion.
(b) Increased tubular creatinine secretion as kidney function decreases with age or kidney disease.

Plasma creatinine concentration can also be used as an estimator of GFR. This is based on the following principle: if creatinine excretion occurs mainly by filtration and its daily production by muscle mass is relatively constant, then a steady state is maintained between production and excretion, which can be expressed in the following equation:

$$\text{Creatinine production} = \text{Creatinine excretion} \qquad (3.10)$$

Rewriting Eq. 3.10:

$$\text{Constant} = \text{GFR [Creatinina]plasma} \qquad (3.11)$$

Therefore, in the presence of relatively constant production, an increase in plasma creatinine necessarily implies an equivalent reduction in GFR. In other words, the relationship between plasma creatinine and GFR is inverse.

However, the predictive value of plasma creatinine as an indicator of GFR is limited. In patients with kidney failure, the number of functional glomeruli can be reduced considerably; however, the clearance of creatinine stays within normal levels. This happens because creatinine excretion is maintained at the expense of increased proximal tubular creatinine secretion. This increase results in an overestimation of the GFR.

The plasma creatinine concentration can also be used to calculate the clearance of creatinine. In normal subjects there is a good equivalence between the clearance measured and calculated creatinine clearance. This does not occur in subjects with impaired kidney function. The relationship considers body weight, age, sex, and muscle mass. For an adult male in creatinine steady state, the expression defined by Cockcroft corresponds to Eq. 3.12:

$$Ccr = \frac{[140 - \text{age(years)}] \cdot \text{Body weight (kg)}}{([\text{Creatinine}]\text{pl} \left[\frac{\text{mg}}{\text{dL}}\right]) \cdot 72} \qquad (3.12)$$

In women, the result of Eq. 3.7 must be multiplied by 0.85. This accounts for the fact that the percentage of body weight corresponding to muscle mass is lower than in men.

Normal plasma creatinine value ranges are 0.8 to 1.3 mg/100 mL for men and 0.6 to 1.0 mg/100 mL for women.

Today, the Cockcroft formula (Eq. 3.12) is no longer valid. Three main limitations to this formula are described. First, it lacks precision for values above 60 mL/min. Second, it overestimates GFR because of the body weight factor in the numerator. The latter is especially valid in obese or edematous patients. Third, the formula was developed based on the old method of determining serum creatinine.

Currently, the glomerular filtration rate is estimated using the MDRD formula, derived from a diet modification study in kidney disease. The formula considers variables such as age, gender, race, and standardized serum creatinine concentration. The MDRD formula for estimating GFR is expressed as follows:

$$\text{GFR}\left(\frac{\frac{\text{mL}}{\text{min}}}{1.73\text{m}^2}\right) = 175 \cdot [\text{Creatinine}]^{-1.154} \cdot \text{Age}^{-0.203} \cdot 0.742 \text{ (for women)} \qquad (3.13)$$

where [Creatinine] is the standardized serum creatinine concentration, expressed in mg/dL. The full result should be multiplied by 1210 for African-Americans.

Another endogenous molecule, whose clearance has been used as an indicator of kidney function, is urea. Traditionally, plasma urea concentration has been measured as blood urea nitrogen (BUN, normal range: 8–12 mg/100 mL). Like creatinine, urea filters freely. There is also an inverse relationship between the GFR and its plasma concentration. However, many factors affect urea renal handling: protein intake and catabolism, state of water balance, and effective circulating volume. About 50% of the urea filtered load is passively reabsorbed in the proximal tubule. This percentage may increase in states of hypovolemia, due to the higher reabsorption of sodium and water. These factors make the clearance of urea as an unreliable estimate of GFR.

The BUN/plasma creatinine ratio has clinical application. Under normal conditions it is about 10, and a reduction in the VFG would raise the BUN and plasma creatinine, so the ratio would remain constant. An increase in the ratio is indicative of increased urea production in the liver or a state of hypovolemia.

3.4.2 Measuring Renal Plasma Flow (RPF)

Resting renal blood flow is 1200 mL/min, representing 25% of cardiac output at rest. The renal plasma flow is about 700 mL/min, of which 125–100 mL/min is filtered. The fraction of the renal plasma flow that filters into the glomeruli is known as the filtration fraction (FF) and is approximately 15–20%.

RPF can be measured using the clearance of para-aminohippuric acid (PAH). This exogenous molecule is excreted by the combination of filtration and proximal secretion (Fig. 3.11). With plasma PAH concentrations below 10 mg/100 mL, virtually all of the PAH entering the kidney through the renal artery will be excreted in the urine by two mechanisms (Fig. 3.11). This is because the transport system involved in the proximal secretion of PAH is not saturated and therefore the PAH molecules delivered to the peritubular capillaries will be secreted into the tubular lumen.

$$RPF[PAH]pl = \dot{V}[PAH]u \tag{3.14}$$

where:

RPF (mL/min) = renal plasma flow
[PAH] (mg/mL) = PAH concentration in arterial plasma (pl) and urine (u)
V = urine flow (mL/min)

If the term RPF is cleared from Eq. 3.14:

$$RPF = CPAH = \frac{V \cdot [PAH]u}{[PAH]pl} \tag{3.15}$$

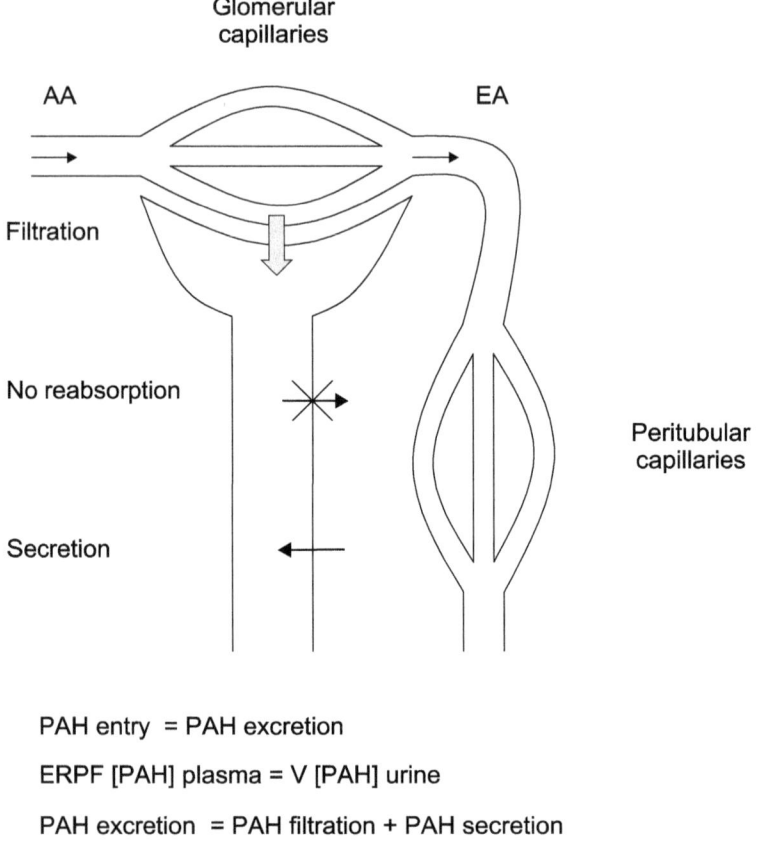

PAH entry = PAH excretion

ERPF [PAH] plasma = V [PAH] urine

PAH excretion = PAH filtration + PAH secretion

ERPF = CPAH = V [PAH] urine / [PAH] plasma

Fig. 3.11 Renal handling of para-aminohippuric acid (PAH) and PAH clearance as indicators of the effective renal plasma flow (ERPF) *AA*: afferent arteriole; *EA*: efferent arteriole. PAH is freely filtered and also secreted in the proximal tubule. Therefore, PAH excretion is a function of filtration and secretion. With plasma PAH concentrations below 10 mg/dL, all PAH that enters the afferent arteriole is excreted by filtration and tubular secretion

The form of Eq. 3.15 is identical to the general clearance formula (Eq. 3.5). Hence, the PAH clearance is an estimate of the renal plasma flow. Recall that the full extraction of PAH from blood comprises different parts of the nephron that are in the cortical labyrinths like glomeruli (PAH filtration), proximal tubules, and peritubular capillaries (both involved in PAH secretion). Therefore, the renal plasma flow determined by the PAH clearance actually represents the effective renal plasma flow (ERPF), which corresponds to the plasma flow to the renal cortex. PAH clearance cannot represent total RPF because approximately 10% of renal blood flow does not pass through the glomeruli and irrigates other structures. Thus, the PAH contained in this fraction of blood will not be extracted (neither by filtration nor secretion), returning to circulation through the renal vein. In other words, the renal extraction of PAH is not 100%, but 90%.

It is necessary to keep in mind that the clearance of PAH only represents the ERPF when the arterial plasma concentration of PAH is below 10 mg/100 mL. Values of PAH concentration above this limit will saturate the proximal tubule secretory mechanism, which saturates at 80 mg/min.

Knowing renal plasma flow and hematocrit (Hct), renal blood flow can be calculated according to Eq. 3.16:

$$RBF = \frac{RPF}{(1 - Hct)} \tag{3.16}$$

Usually, the clearance of PAH is about 700 mL/min, and, considering a hematocrit of 0.40, the renal blood flow is about 1200 mL/min.

In summary, renal blood flow at rest corresponds to 25% of cardiac output (1200 mL/min), of which 700 mL/min corresponds to effective renal plasma flow, determined by clearance of PAH. Of this, 100–120 mL/min filters into the glomeruli of both kidneys, which constitutes the glomerular filtration rate, determined through the clearance of inulin or endogenous indicators such as creatinine. In other words, a filtration fraction of 15–20% of the effective renal plasma flow is filtered in the glomeruli at a rate of 100–120 mL/min.

3.5 Conclusions

The glomerular ultrafiltrate composition is identical to plasma, except for the absence of plasma proteins like albumin. Normal plasma ultrafiltration occurs at high rates and requires an intact glomerular filtration barrier formed by fenestrated glomerular endothelium, basal lamina, and the filtration slit junction.

Glomerular filtration rate depends on renal plasma flow and the Starling forces along the glomerular capillary.

Glomerular filtration rate can be estimated through the clearance technique. The inulin clearance is the gold standard marker. Its clearance represents the glomerular filtration rate. However, it has little applicability in clinical practice. Creatinine is an endogenous nitrogenated molecule derived mainly from muscular mass. Its renal handling makes its clearance a good marker of the state of the functional renal mass.

Renal blood flow and glomerular filtration rate exhibit autoregulation. This intrinsic property is based on the myogenic and tubuloglomerular feedback mechanisms. Both control the afferent arteriolar resistance.

PAH clearance, under certain restrictions, is an estimate of the effective renal plasma flow and can be used to estimate the renal blood flow if the hematocrit is known.

Review Questions

1. The data in the table below corresponds to a study of clearance of inulin in a healthy subject. Calculate the inulin clearance (C inulin).

Period	Urinary flow (mL/min)	$[Inulin]_{plasma}$ (mg/mL)	$[Inulin]_{urine}$ (mg/mL)	C_{inulin} (mL/min)
Protocol A: Increasing plasma inulin concentration				
1	1.2	0.9	90	
2	1.3	1.5	136	
3	1.0	2.3	282	
4	1.4	3.8	336	
5	1.2	5.7	570	
Protocol B: Constant plasma inulin concentration and water overload				
6	1.3	0.5	46	
7	2.2	0.6	34	
8	3.1	0.4	16	
9	6.0	0.5	10	
10	6.6	0.5	9.2	

    (a) Graph the inulin clearance as a function of plasma inulin concentration and explain the relationship.

    (b) What properties of inulin explain the relationship you deduced in the previous question?

    (c) Graph the relationship between the excreted load of inulin and the filtered inulin load. How do you explain this relationship? What is the slope?

2. At the beginning of the proximal tubule the ratio of inulin concentration in tubular fluid/plasma (TFin/Pin) is 1. At the end of the proximal tubule, the ratio TFin/Pin is 3. How do you explain this change in TFin/Pin?

3. Glucose is freely filtered in the glomerular filtration barrier. Why does the glucose clearance do not serve as an estimate of the glomerular filtration rate?

4. The following data from creatinine clearance correspond to a subject of 30 years, 80 kg of weight, and 1.73 m² of body surface area.

Period	Creatinine plasma mg/dL	Creatinine urinary mg/dL	Urinary flow mL/d	Clearance creatinine mL/min	Clearance estimated[a]	1/ creatinine
Start study	1.7	146	1280			
1 year later	2.1	112	1600			
2 years later	3.0	180	1000			

[a]Cockcroft's formula

(a) If the subject's muscle mass has not undergone major alterations, how can you explain what is happening with the clearance of creatinine as a function of time?

(b) Graph the ratio 1/creatinine in plasma as a function of time. If after 3 years the plasma creatinine was 8 instead of 3.5 mg/dL, what do you think has happened?

Bibliography

Grahammer F, Schell C, Huber TB (2013) The podocyte slit diaphragm-from a thin grey line to a complex signalling hub. Nat Rev Nephrol 9:587–598

Greka A, Mundel P (2012) Cell biology and pathology of podocytes. Annu Rev Physiol 74:299–323

Ma R, Pluznick JL, Sansom SC (2005) Ion channels in mesangial cells: function, malfunction, or fiction. Physiology 20:102–111

Munger KA, Kost CK, Brenner BM, Maddox DA (2016) The renal circulations and glomerular ultrafiltration. In: Brenner BM, Rector FC (eds) The kidney, vol I, 10th edn. Elsevier, Philadelphia, pp 83–110

Navar LG, Inscho EW, Majid SA, Imig JD, Harrison-Bernard LM, Mitchell KD (1996) Paracrine regulation of the renal microcirculation. Physiol Rev 76:425–536

Pollak MR, Quaggin SE, Hoenig MP, Dworkin LD (2014) The glomerulus: the sphere of influence. Clin J Am Soc Nephrol 9:1461–1469

Schnermann J, Castrop H (2013) Function of the juxtaglomerular apparatus: control of glomerular hemodynamics and renin secretion. In: Seldin D, Giebisch G (eds) The kidney: physiology and pathophysiology, vol I, 5th edn. Academic Press, San Diego, pp 595–691

Tryggvason K, Wartiovaara J (2005) How does the kidney filter plasma? Physiology 20:96–101

Vallon V, Mühlbauer B, Osswald H (2006) Adenosine and kidney function. Physiol Rev 86:901–940

Transport of NaCl, Organic Solutes, and Water in the Renal Tubule

4

Learning Objectives

- To relate the functional structure of the renal tubule to the general principles of transepithelial transport of solutes and water.
- To describe the processes of tubular reabsorption and tubular secretion participating in urine formation.
- To understand the physiological role of the electrochemical Na^+ gradient and Na^+, K^+-ATPase in the tubular function.
- To describe the function of the tubular segments.
- To describe the neurohormonal mechanisms in the regulation of tubular function.

4.1 Renal Function and Tubular Reabsorption

The magnitude of renal function in humans is exemplified in Table 4.1. Solutes like sodium, chloride, and bicarbonate are reabsorbed above 99%. With normal blood glucose levels, glucose is reabsorbed to 100%. Water is also reabsorbed above 99%. Solutes like urea are only reabsorbed nearly 60%, while 93% of the filtered potassium load is reabsorbed. The lesson from this table is that under normal circumstances, the kidneys handle each molecule independent of others.

The transport activity of the epithelium lining the renal tubule modifies the ultrafiltrate both in composition and volume through two processes: tubular reabsorption and tubular secretion. Both processes occur throughout the renal tubule and they are crucial in urine formation. The foundations for these processes are in transepithelial transport of solutes and water, which will be briefly discussed in this chapter.

© Springer Nature Switzerland AG 2022
P. A. Gallardo, C. P. Vio, *Renal Physiology and Hydrosaline Metabolism*,
https://doi.org/10.1007/978-3-031-10256-1_4

Table 4.1 Magnitude of normal renal function in a healthy adult of 70 kg with a renal blood flow of 1200 mL/min and a renal plasma flow of 600 mL/min. Glomerular filtration rate is 120 mL/min and filtration fraction is approximately 20%

Compound	[Plasma] (mM)	Filtered/24 h		Excreted/24 h		Reabsorption (%)
		mmol	gram	mmol	gram	
Sodium	140	25,200	577	100	2.9	>99
Chloride	105	18,900	670	103	3.7	>99
Bicarbonate	25	4500	274	2	0.1	>99
Potassium	4	720	28	50	19.5	>93
Glucose	5.5	1000	180	0	0	100
Urea	5	900	54	360	22	60
Water	–	–	180 L	–	1–1.5 L	>99

4.2   General Concepts of Transepithelial Transport

The renal tubule is a simple cuboidal epithelium. Epithelial cells are joined together by tight junctions, formed by several types of claudins and occludins. The tight junction forms the seal and boundary between the apical and basolateral membrane domains. The luminal or apical membrane is in contact with the tubular fluid. The basolateral domain can be regarded as composed of two aspects: lateral and basal. The lateral aspect of the basolateral membrane of two neighbor cells forms the boundary of the lateral intercellular channel. The basal aspect of the basolateral membrane is in contact with the basal lamina through different anchoring junctions like focal points and hemidesmosomes. The basolateral membrane domain contains the Na^+, K^+-ATPase, the most important mechanism of primary active transport in the renal tubule (Fig. 4.1).

The epithelia of the renal tubule exhibit morphological and functional polarity. The latter is because the transport proteins expressed in the apical domain are different than those in the basolateral domain, which allows vectorial transport of solutes and water. Vectorial transport allows the flux of solutes and water to occur in a defined direction. In tubular reabsorption the molecules move from tubular fluid into the blood. In tubular secretion, flux of molecules takes place in the opposite direction. A typical example of tubular reabsorption is the transport of Na^+. Na^+ enters the cell through Na^+-dependent transporters or Na^+ epithelial channels expressed in the apical membrane. Na^+ is pumped out of the cell through the Na^+, K^+-ATPase or sodium pump, expressed only in the basolateral membrane (Fig. 4.1). In pathologies such as acute tubular necrosis caused by ischemia, the polarity of the proximal tubular epithelium is lost and a disorganization of the tight junction occurs. In this condition, the Na^+, K^+-ATPase appears in the apical membrane and the net result is the loss of vectorial transport of Na^+ and other solutes. The net result is the increase in the delivery of sodium and other solutes to more distal nephron segments and an increase in Na^+ urinary excretion.

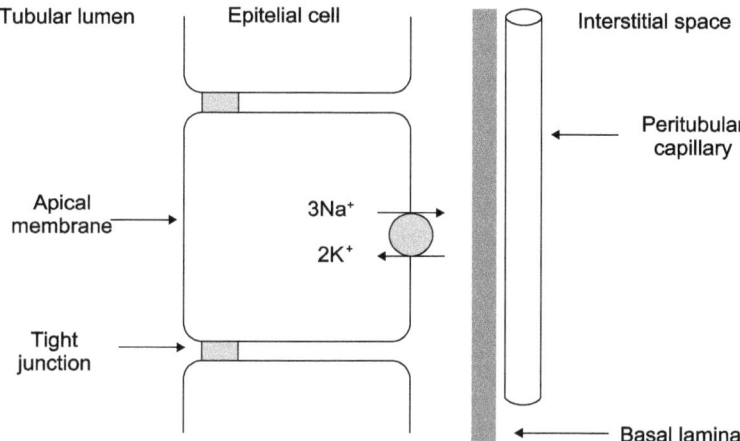

Fig. 4.1 Basic components of the tubular epithelium. The apical membrane and basolateral membrane are oriented to the tubular fluid and extracellular fluid, respectively. The tight junction is the limit between the apical and basolateral membrane domains. The basal lamina and peritubular capillaries are located beneath the basolateral membrane. The basolateral membrane contains the Na^+, K^+-ATPase that plays a key role in the tubular transport of solutes and water

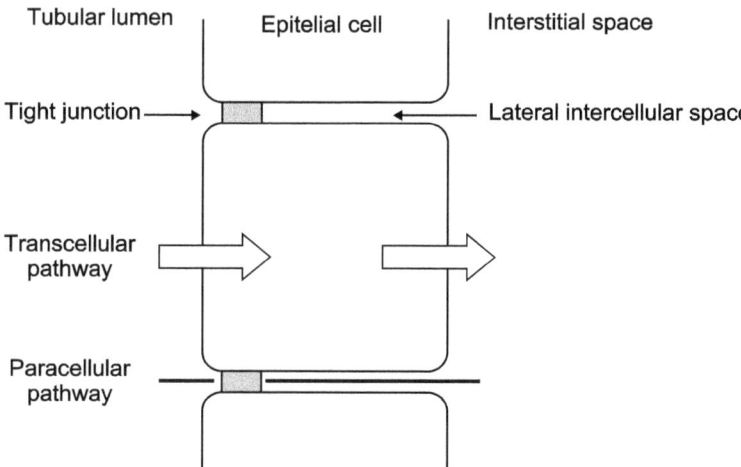

Fig. 4.2 Transepithelial transport routes. In the transcellular pathway the molecules have to cross the apical and basolateral membranes. In the paracellular pathway, the molecules move across the tight junctions. The permeability of the paracellular pathway depends on gradients generated by the transcellular transport and the claudins and occludins expressed in a specific tight junction

In an epithelium there are two potential pathways for vectorial transport: transcellular and paracellular (Fig. 4.2). Transcellular transport is a two-step process: the solute must pass through two membranes: the apical and basolateral membrane. In paracellular transport, the solute moves through the tight junction and thus between rather than across the epithelial cells. The magnitude and characteristics of the paracellular pathway depend on the type and abundance of members of the claudin and occludin protein family forming the tight junction. In general, tubular reabsorption is a transcellular and a paracellular process, while tubular secretion is always a transcellular process.

From the electrophysiological standpoint, epithelia have been classified into two categories. The first type of epithelium is the so-called leaky or low-resistance transepithelial epithelium (Rte). In the kidney, this kind of epithelia is best represented by the proximal tubule. Histologically, the tight junction is formed by few strands and discontinuous protein strands that form a belt at the limit between the apical and basolateral membrane. This small tight junction formed by claudin 2 in human proximal tubule offers little electrical resistance. Through this tight junction, solutes (ions, small organic solutes like urea) can flow following an electrochemical gradient. This means that the movement of a cation like Na^+ can be easily followed by an anion. Since there is little separation of electrical charges across the epithelium, the transepithelial potential difference (ΔVte) is small. In addition, leaky epithelia have a high hydraulic or osmotic permeability. In the case of the proximal tubule this is associated with high constitutive expression of aquaporin-1 in the apical and basolateral membrane. The net result is that the flow of water is tightly coupled to the solute flow. Table 4.2 shows the electrical properties of the proximal tubule.

The second type is the "tight" or compact epithelium, characterized by high transepithelial electrical resistance (Table 4.2). It is associated with more complex tight junctions, formed by several protein strands that present ramifications and anastomosis. Examples of this epithelial type are the segments of the distal nephron (connecting tubule and cortical collecting duct). Transport in these epithelia is primarily transcellular. Therefore, a significant ΔVte lumen negative is generated. In tight epithelia, hydraulic permeability is normally low and considerable transepithelial osmolality gradients are produced: this is due to the fact that solute transport is accompanied by a low rate of water movement. However, this hydraulic permeability can increase considerably in the presence of hormones like vasopressin. This peptide hormone controls the abundance of aquaporin-2 in the apical membrane of connecting and principal cells. Tight epithelia in non-mammalian vertebrates like the amphibian urinary bladder were the first substrate for the research in the vasopressin-stimulated osmotic water transport.

Table 4.2 Electrophysiological and hydraulic permeability properties of tubular segments of the mammalian nephron. The $\Delta mOsm$ corresponds to the difference in osmolality between the tubular fluid in the segment and the interstitial medium

Segment of the nephron	Rte $\Omega \cdot cm^2$	ΔVte (mV)	Δ mOsm (mOsm/kg H_2O)	Tight junction
Proximal tubule	5	−3 to +3	4	Weak
Thick ascending limb of Henle	34	+10	200–400	Intermediate
Collecting duct	800	−25	200–800	Compact

The proximal tubule and collecting ducts represent the extremes in terms of their electrophysiological properties and water permeability. Table 4.2 also shows tubular segments with intermediate electrophysiological properties, such as the thick ascending limb of Henle, which has high ionic permeability and very low hydraulic permeability due to the absence of aquaporins in the apical and basolateral membrane. The tight junction is formed by claudin 16 that forms a cation-selective paracellular pathway.

4.3   Bioenergetics of Sodium Transport and the Na⁺, K⁺-ATPase

The electrochemical sodium gradient plays a key role in transepithelial transport along the renal tubule. The concentration of sodium in the tubular fluid in the proximal tubule is the same as in the plasma (145 mM) and in the cell is 15 mM, generating a chemical or concentration gradient that favors the movement of Na^+ into the cell.

$$\text{Chemical potential Na}^+\text{gradient} = \Delta\mu Na^+ = RTLn\frac{[Na^+]i}{[Na^+]e} \qquad (4.1)$$

where:

R = universal gas constant (8.32 Joule/mol. °K)
T (K) = absolute temperature
$[Na^+]_i$ and $[Na^+]_e$ = sodium concentration in the intracellular and extracellular compartments

Equation (4.1) represents the free energy or useful energy to do work, which is stored in the transmembrane Na^+ concentration gradient.

The cells have a membrane potential that is negative inside the cell (−60 mVolts, approximately), which constitutes an electrical force that also favors the entry of sodium into the cell. The effect of the resting membrane potential is considered in Eq. (4.2) of the electrical gradient:

$$\text{Electrical potential Na}^+ \text{ gradient} = zF\Delta Vm \qquad (4.2)$$

z = valence of the Na ion⁺ (z = +1)
F = Faraday constant 96,500 Coul/mol. Volt
ΔVm (Volts) = membrane potential difference ($V_i - V_e$)

The sum of the chemical and electrical potential gradients (Eqs. 4.2 and 4.3, respectively) constitutes the electrochemical potential gradient:

$$\text{Electrochemical Na}^+ \text{ gradient} = RTLn\frac{[Na^+]i}{[Na^+]e} + zF\Delta Vm \qquad (4.3)$$

Equation (4.3) contains the expressions for the two possible sources of free energy that can move Na^+ across the membrane: the chemical potential gradient and the electrical potential gradient. In the case of the electrochemical gradient of Na^+, the chemical and electrical gradients contribute to the total free energy accumulated and available to do work.

Considering a temperature of 37 °C, the concentrations of Na^+, and the resting membrane potential, the accumulated free energy in the Na^+ gradient is nearly −12,000 Joule/mol. The negative sign indicates that the process of moving a mole of sodium from the tubular fluid into the cell is spontaneous and occurs upon releasing energy.

The electrochemical sodium gradient is maintained by the operation of the Na^+, K^+-ATPase in the basolateral membrane. The Na^+ pump exchanges $3Na^+$ from cytosol to extracellular medium in exchange of 2 K^+ and hydrolyzing one ATP molecule.

Under physiological conditions, the Km of the Na^+, K^+-ATPase for cytosolic Na^+ is 10–20 mM and the concentration of Na^+ intracellular is approximately 17 mM. It follows that, under baseline conditions, the rate of Na^+, K^+-ATPase is 40–50% of its maximum rate. Therefore, an increase in cytosolic Na^+ activates the pump, and this allows the sodium that entered the cell through the Na^+ pump. Therefore, the Na^+, K^+-ATPase has the key function in maintaining a low intracellular Na^+ concentration and therefore the electrochemical sodium gradient.

The electrochemical Na^+ gradient that exists between the tubular lumen and the cytosol allows the entry of Na^+ to the cell through Na^+-dependent cotransporters, exchangers, and epithelial Na^+ channel. Cotransporters and exchangers mediate secondary active transport. The energy for its operation comes from the electrochemical sodium gradient, which in turn is maintained by the Na^+, K^+-ATPase (Fig. 4.3).

In the kidney there is a very close coupling between oxygen consumption and active sodium transport, with a directly proportional relationship between the two processes. Approximately 95% of the ATP generated comes from the oxidative metabolism and fuels the Na^+, K^+-ATPase. The tubular segment with the highest Na^+ pump activity is the distal convoluted tubule, followed by the thick ascending limb of Henle and the proximal tubule. Thus, the Na^+, K^+-ATPase is the most important primary active transport mechanism in the kidney, followed by other ATPases, such as H^+-ATPase, H^+, K^+-ATPase, and Ca^{++}-ATPase.

4.4   Tubular Processes in Urine Formation

As mentioned before, tubular reabsorption and secretion are the two processes that occur along the renal tubule that modifies the composition and volume of the ultrafiltrate. Both reabsorption and secretion indicate the direction of transport without reference to the transport mechanism. In tubular reabsorption, solutes move from the tubular fluid into the blood, through the transcellular and paracellular pathways. In tubular secretion, molecules move from the blood into the tubular fluid, often by the transcellular route (Fig. 4.4). The transepithelial transport that occurs

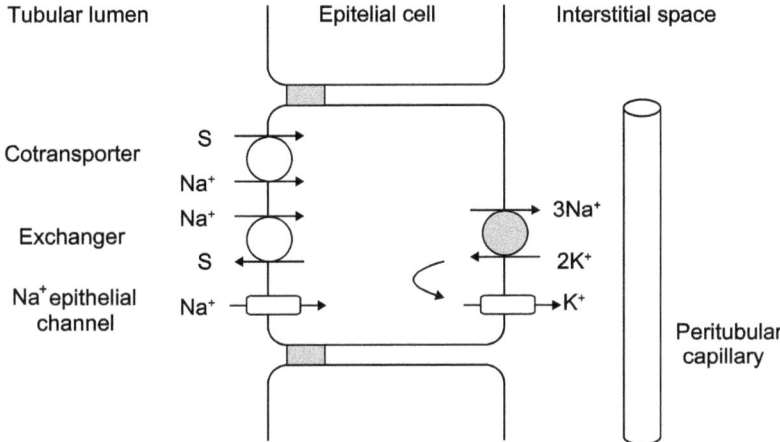

Fig. 4.3 Mechanisms of apical entry and basolateral exit across the tubular epithelium. Na^+ entry across the apical membrane is passive and powered by an electrochemical gradient. In general terms, Na^+ entry is mediated by Na^+-dependent cotransporters, Na^+-dependent exchangers, or Na^+ channels. Na^+ exit across the basolateral membrane is active and mediated by the Na^+, K^+-ATPase that maintains the Na^+ electrochemical across the plasma membrane

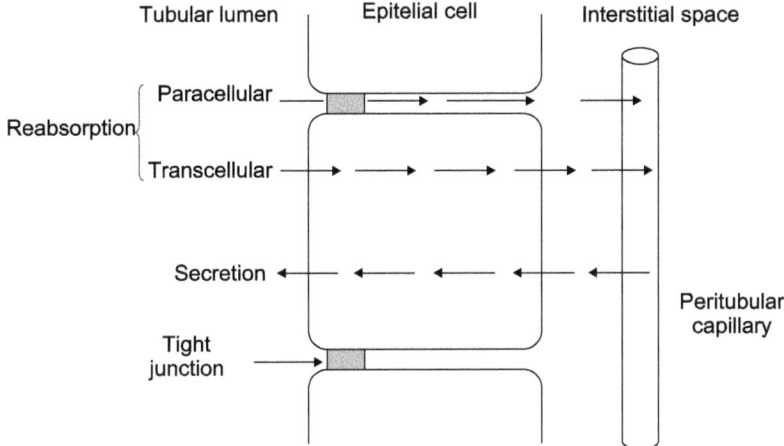

Fig. 4.4 Tubular processes contributing to urine formation. Two processes across the tubular epithelium are responsible for urine formation. Reabsorption is the movement of molecules from the tubular fluid to the peritubular capillaries. Transcellular reabsorption creates the gradient for paracellular reabsorption. Tubular secretion is the transcellular movement of solutes from peritubular capillaries to the lumen

along the nephron modifies the composition and volume of the glomerular filtrate. The urinary excretion of any solute will be the result of the processes of filtration, tubular reabsorption and secretion, expressed below in an equation:

Table 4.3 Daily filtration, reabsorption, and excretion of electrolytes, solutes, and water

Substance	Measure	Filtered load	Excreted load	Amount reabsorbed	% reabsorbed
Water	L/day	180	1.5	1785	99.2
Na^+	mEq/day	25,200	150	25,050	99.4
K^+	mEq/day	720	100	620	86.1
Ca^{++}	mEq/day	540	10	530	98.2
HCO_3^-	mEq/day	4320	0	4318	100
Cl^-	mEq/day	18,000	150	17,850	99.2
Glucose	g/day	180	0	180	100

$$\text{Urinary excretion} = \text{Filtration} + \text{Secretion} - \text{Reabsorption} \qquad (4.4)$$

Equation (4.4) is an expression of the principle of mass balance in the kidney. The entry of any solute into the kidney through the renal artery is balanced with its exit. The output of the solute will depend on the renal handling of each individual solute. For example, if a solute is completely reabsorbed, it should leave the kidney through the renal vein and not in the urine. If a solute is completely excreted, the entire amount that enters the kidney will appear in the urine.

Table 4.3 shows that, under physiological conditions, the kidney is capable of handling different solutes and water independently. The reabsorptive work carried out by the kidney is evident if it is considered that 25,200 mEq of Na^+ (equivalent to 1462 g NaCl) and 180 liters of water, of which only less than 1% is excreted; a similar situation occurs with bicarbonate and valuable organic solutes such as glucose. In the case of potassium, as will be seen below, the amount excreted is variable and depends essentially on the intake of K^+ on the diet. In general, it is valid to state that, within certain margins, the kidney can adjust the excretion of the various solutes and water independently, to satisfy the homeostatic needs of the organism.

The different transport mechanisms of electrolytes, organic solutes, and water that operate in each segment and actively contribute to urine formation will be discussed in general terms below.

4.4.1 Proximal Tubule

The proximal tubule reabsorbs large amounts of solutes and water: 67% of the filtered water and 67% of the filtered Na^+ and Cl^-. Nearly 25,200 mEq of Na^+ are filtered daily and about 16,884 mEq are reabsorbed in the proximal tubule. Two thirds of the reabsorbed sodium occur through the transcellular pathway; the remaining third is reabsorbed through the paracellular pathway. In addition, the proximal tubule reabsorbs 100% of the filtered load of glucose and amino acids and 80% of the filtered bicarbonate are reabsorbed.

In Fig. 4.5, the ratio of solute concentration in tubular fluid (TFx) to plasma concentration (Px) (TFx/Px) is plotted as a function of the length of the proximal tubule. The graph shows that for some solute the ratio remains 1, while for others it decreases or increases. This is a clear manifestation that the proximal tubule handles different solutes in an independent manner. Organic solutes such as glucose and amino acids have a TF/P ratio that starts at the unit and decreases rapidly to *zero* along the length of the proximal tubule. This means that under physiological conditions, the whole filtered load of glucose and amino acids is completely reabsorbed. The case of HCO_3^- is slightly different. Its TF/P ratio declines to 0.2, which means that 80% of the filtered load was reabsorbed. The TF/P ratio for osmolality of the tubular fluid remains close to 1. This is because the reabsorption of Na^+ and water is coupled and occurs under isoosmotic conditions. The TF/P ratio

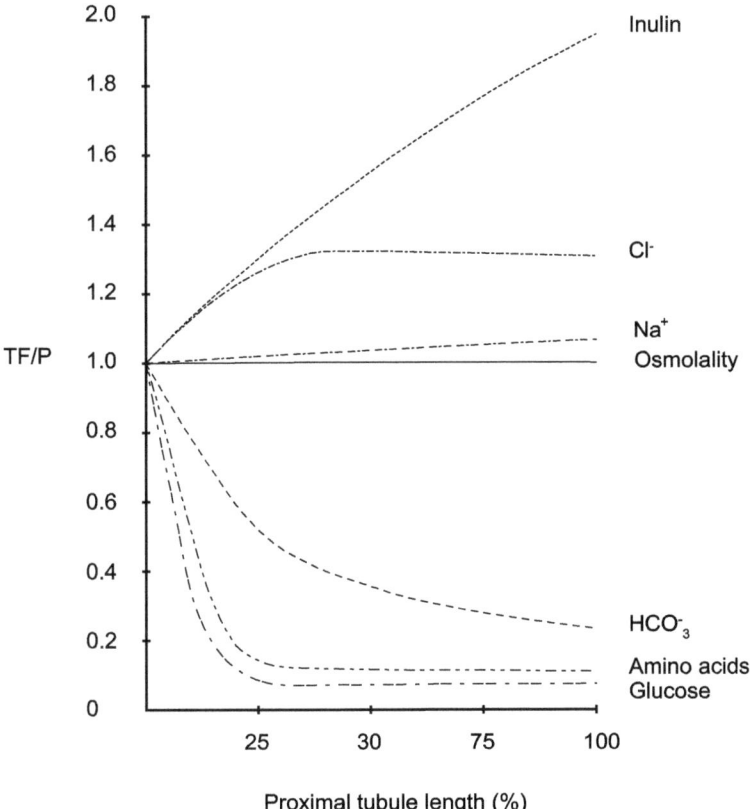

Fig. 4.5 Tubular fluid (TF) to plasma (P) concentration ratio for different solutes and osmolality along the length of the proximal tubule. Solutes like glucose, amino acids, and bicarbonate are reabsorbed in the first half of the proximal tubule, lowering their TF/P. Inulin is not reabsorbed or secreted, and an increase in its TF/P ratio implies water reabsorption. TF/P ratio for osmolality maintains near 1 because proximal water reabsorption occurs in isoosmotic conditions. The first part of the proximal tubule is less permeable to Cl^-, increasing its concentration the tubular fluid

for inulin reflects the tubular handling of this solute: it is neither reabsorbed nor secreted, and therefore an increase in its tubular concentration represents volume reabsorption.

Proximal reabsorption can be understood in two stages. In the first half of the proximal tubule, Na^+ reabsorption occurs with neutral organic solutes (glucose, amino acids, lactate), phosphate, and HCO_3^-. The reabsorption of Na^+ with neutral solutes generates a lumen-negative transepithelial potential difference (-3 mVolt). In the second phase, preferential chloride reabsorption occurs. The preferential reabsorption of Na^+ with organic solutes, bicarbonate, and phosphate causes volume reabsorption which is evidenced as an increase in the TF/P ratio for inulin. In turn, the reabsorption of water causes an increase in the luminal concentration of chloride, resulting in an increase in its TF/P ratio. The chloride gradient that develops favors its transcellular and paracellular reabsorption. The reabsorption of chloride in the second phase makes the transepithelial potential that was negative in the lumen positive.

4.5   Glucose Reabsorption

Under physiological conditions, all the filtered load of glucose is completely reabsorbed in the proximal tubule. According to Table 4.3, 180 grams of glucose is filtered daily, and the excreted load is *zero*.

The pioneer studies of renal glucose handling were carried out in whole animal by Shannon in 1938. The experiments demonstrated that glucose was filtered and reabsorbed by an unknown mechanism; glucose reabsorption also exhibited a transport maximum (Tmax). The results of an experiment like this are shown in Fig. 4.6. Glucose is a neutral organic solute and its hydrated radius is less than 18 Å and thus is freely filtered. For a constant glomerular filtration rate, the filtered load is directly proportional to glycemia. However, this relationship is not observed for the return of glucose through the renal vein, which represents the reabsorbed glucose. When glycemia is under the range of 180–200 mg/dL, all filtered glucose is reabsorbed and urinary glucose excretion is *zero*. With glycemia values higher than 200 mg/dL, glucose appears in the urine and reabsorption reaches a maximum value (transport maximum, Tmax), where it stabilizes and becomes independent of glycemia. The maximum glucose reabsorption rate (Tmax) is 375 mg/min. The transition between reabsorption directly proportional to glycemia and maximal reabsorption is gradual. This is explained by nephron heterogeneity, since nephrons differ in glomerular size and the length of the proximal tubule. The concentration of glucose at which it appears in the urine (glycosuria) is known as the renal threshold of glucose.

Since the normal range of glycemia is between 0.7 and 1 g/L, the kidney then has a high capacity to reabsorb glucose. From a clinical point of view, an elevation of glycemia above the glucose threshold value, as can occur in uncontrolled diabetes mellitus, results in the appearance of glycosuria. Recall that the proximal tubule is the only segment involved in glucose reabsorption. Therefore, those glucose

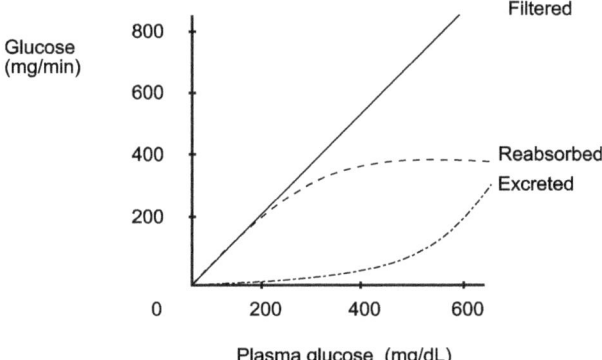

Fig. 4.6   Renal glucose handling. Filtered, reabsorbed, or excreted glucose is plotted as a function of plasma glucose concentration. Glucose is freely filtered, which explains the straight line between the filtered glucose and plasma glucose concentration. When the filtered load is about 375 mg/min, glucose reabsorption becomes constant. When glucose concentration is lower than 180–200 mg/dL, glucose excretion is 0. Glucose excretion is greater than 0 when glucose concentration exceeds the 180–200 mg/dL range

molecules escaping the proximal tubule into downriver segments will behave as effective osmoles retaining water in the tubular lumen. The direct consequence is a state of osmotic diuresis with a variable loss of water that alters the water balance. The elevation of plasma osmolality due to water loss and high glycemia triggers the feeling of thirst that diabetic patients complain.

Glucose reabsorption is a transcellular process that occurs only in the proximal tubule. Glucose is reabsorbed though Na^+-dependent secondary active transport in the apical membrane. The energy for the uphill transport of glucose is provided by the sodium electrochemical potential gradient maintained by Na^+, K^+-ATPase at the basolateral membrane. The mechanism is shown in Fig. 4.7. In the apical membrane glucose binds to Na^+-glucose cotransporters that are members of the SGLT family. Glucose binding to its site is enhanced by the binding of Na^+. Two members of the SGLT (SGLT1, SGLT2) family are expressed in the proximal tubule; they differ in their stoichiometry and affinity for glucose. SGLT2 is a low-affinity, high-capacity cotransporter expressed in the apical membrane of S_1 and part of S_2. This cotransporter is expressed only in the proximal tubule. It has a stoichiometry of $1Na^+$:1glucose. SGLT2 activity is responsible for the reabsorption of 97% of the filtered load of glucose. SGLT1 cotransporter is a high-affinity, low-capacity cotransporter expressed in the apical membrane of the last part of S_2 and S_3 and is responsible for the reabsorption of the remaining 3% of the filtered load of glucose. SGLT1 has stoichiometry of $2Na^+$:1 glucose and a high affinity for glucose. This cotransporter is also expressed in absorptive enterocytes of the small intestine. Both cotransporters are electrogenic, because in each cycle they carry out the transport of 1 or 2 Na^+ with a neutral molecule (glucose) into the cell and thus its activity tends to depolarize the cell. Basolateral glucose exit occurs through passive transporters of the GLUT family. In the basolateral cell membrane of the proximal

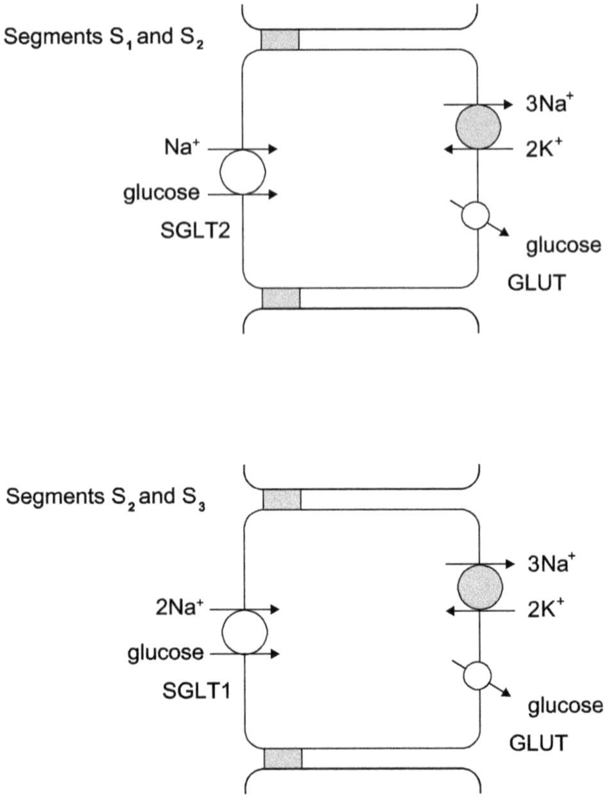

Fig. 4.7 Proximal glucose reabsorption. Two members of the SGLT family are responsible for glucose reabsorption along the proximal tubule. SGLT2 mediates most of the glucose reabsorption, whereas SGLT1 is responsible for the rest of the glucose reabsorption. Under physiological conditions, glucose concentration in the tubular fluid leaving the proximal tubule should be 0

tubule, the GLUT 1 and 2 members have been identified. In summary, the apical entry of glucose is secondary active and occurs against a chemical potential gradient and its basolateral exit is passive, in favor of a chemical gradient.

4.6   Amino Acids and Proteins

Amino acid reabsorption generally follows the same pattern of glucose reabsorption. There are several amino acid transporters that show specificity according to the chemical group present in the side chain of the amino acid structure. The basolateral exit is usually passive. Therefore, multiple Na^+-dependent cotransporters carry out amino acid reabsorption at the apical membrane. The situation with peptides and proteins is different. The small peptides are degraded by peptidases expressed in the brush border, which are similar to those of the small intestine, while the amino acids are reabsorbed by specific transporters. Larger proteins are endocited by receptor-mediated endocytosis mediated by the megalin-cubilin protein system expressed in the apical membrane. The endosomes fuse with lysosomes where the endocyted proteins are degraded, generating amino acids that exit through the basolateral membrane.

The case of albumin is particularly interesting, since it is important for maintaining the oncotic pressure of the plasma. Approximately 70,000 g of albumin pass through the glomeruli daily and the filtered load is about 8 g/day, representing 0.01%. However, the urinary excretion of albumin is about 30 mg/day and therefore about 99% of the filtered load of albumin is reabsorbed. Albumin reabsorption occurs in the proximal tubule through receptor-mediated endocytosis; the size of albumin does not allow for its paracellular reabsorption. In the apical membrane of the proximal tubule, albumin binds to the megalin-cubilin protein complex and the clathrin-coated vesicles are endocyted and fused with lysosomes. Fusion with lysosomes is important to cause a drop in the pH of the vesicle, releasing the albumin molecule from the receptor. The receptor proteins megalin and cubilin are recycled to the apical membrane. The albumin molecules are degraded in the lysosomes and the amino acids are absorbed through the basolateral membrane. Although there is no conclusive evidence, there is a possibility that exocytosis of whole albumin molecules occurs across the basolateral membrane. This transcellular transport mechanism in vesicles is known as transcytosis. The knowledge of the mechanisms of albumin reabsorption, as well as of the molecular entities involved, is key since it has been postulated that some forms of proteinuria are not only due to alterations in the permeability to albumin in the filtration barrier, but also to alterations in the mechanism of its reabsorption in the proximal tubule.

4.7 Bicarbonate Reabsorption

As shown in Fig. 4.5, at the end of the proximal tubule the TF/P concentration ratio for bicarbonate is 0.2. This indicates that the proximal tubule reabsorbs about 80% of the HCO_3^- filtered load; this occurs mostly in the S_1 and of S_2 subsegments. Proximal bicarbonate reabsorption depends mostly on H^+ secretion carried out mainly by the apical electroneutral NHE3 isoform of the Na^+/H^+ exchanger and the H^+-ATPase, apical and cytosolic carbonic anhydrase activity, and basolateral NBC (Na^+/HCO_3^- cotransporter). The mechanism of HCO_3^- reabsorption will be described extensively in Chap. 8.

4.8 Chloride Reabsorption

Chloride reabsorption in the proximal tubule occurs through the transcellular and paracellular pathway and is linked to Na^+ reabsorption (Fig. 4.8). In segments S_1 and first of S_2, Na^+ reabsorption occurs preferentially with neutral organic solutes like glucose, amino acids, and HCO_3^-. The Na^+ reabsorption with neutral organic solutes generates a lumen-negative transepithelial potential difference. Besides, solute reabsorption drives fluid reabsorption, which increases the TF/P ratio for Cl^-. The lumen-negative ΔVte and the increase in luminal concentration from 105 to 140 mEq/L are driving forces for chloride reabsorption in the second part of the proximal tubule (part of S_2 and S_3). As mentioned above, in the second part of

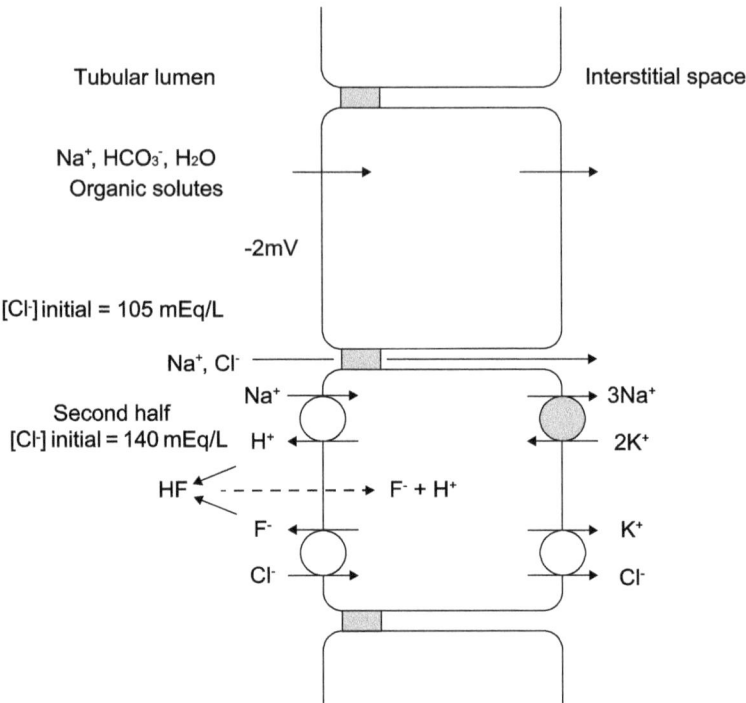

Fig. 4.8 Mechanisms of chloride reabsorption in the proximal tubule. Organic solute and bicarbonate reabsorption favors water reabsorption, increasing chloride concentration in the tubular fluid from 105 to 140 mEq/L. The lumen-negative transepithelial potential favors paracellular chloride reabsorption. In the second half, chloride is reabsorbed through the transcellular pathway. In the apical membrane, the parallel operation of the Na^+/H^+ exchanger and the Cl^-/formiate (F^-) exchanger renders NaCl reabsorption. The energy for the apical processes is the Na^+, K^+-ATPase. Basolateral chloride exit is mediated by a K^+/Cl^- cotransporter

the proximal tubule, chloride reabsorption is transcellular and paracellular. The transcellular mechanism results from the parallel operation in the apical membrane of NHE3 Na^+/H^+ exchanger and the Cl^-/base exchanger. H^+ secretion by NHE3 serves to acidify a base; the acid enters the cell through the apical membrane by non-ionic diffusion. Inside the cell, the acid dissociates generating the base which exits the cell through Cl^-/ base exchanger in the apical membrane. The most common acid/base pair involved in transcellular chloride reabsorption is formic acid with a pKa of 3.74. With a luminal pH of 6.8, the majority is in the form of formic acid, which is neutral and can diffuse through the membrane by non-ionic diffusion. In proximal tubular cells, cytosolic pH is about 7.1; this value favors formic acid dissociation into formate which can recycle to the lumen in exchange for chloride in the Cl^-/base exchanger. Therefore, the parallel operation of the apical NHE3 and Cl^-/formate exchangers results in NaCl reabsorption. Basolateral chloride exit occurs through chloride channels and a KCC cotransporter.

4.9   Secretion of Organic Anions and Cations

A number of compounds that are excreted in the urine have a clearance which is larger than inulin, which means there is tubular secretion process. The proximal tubule secretes several organic anions and cations, both endogenous and exogenous. The secretory processes are transcellular and occur via transporters. These transporters exhibit broad specificity and there is competition between various anions and between various cations to occupy sites on their transporters. Some of the endogenous anions secreted are bile salts, cAMP, prostaglandins, oxalate, and urate. Among the exogenous anions are diuretics (acetazolamide, furosemide, chlorothiazide), penicillin, salicylate, and p-aminohippurate (PAH).

Some endogenous cations secreted are represented by acetylcholine, catecholamines (dopamine, epinephrine), histamine, and serotonin. Among the exogenous ones are atropine, morphine, etc.

4.10   Coupling of Water Reabsorption with Solutes

As shown in Fig. 4.5, the tubular fluid/plasma ratio for osmolality remains close to unity. The reabsorption of water in the proximal tubule occurs under conditions very close to isoosmolality; this is because the proximal tubule is a "leaky" epithelium and also has a high osmotic permeability due to the high expression of AQP-1.

The proximal tubule has a great capacity to reabsorb solutes and water. The enormous capacity of water reabsorption is due to its high osmotic permeability. It is based on three components: first, an abundant constitutive expression of AQP-1 both at the apical and basolateral membrane domains; second, a paracellular hydraulic permeability; and third, the fact that the highest density of basolateral transporters is concentrated in the lateral aspect of the basolateral membrane, which forms the boundary of the lateral intercellular space between neighboring epithelial cells (Fig. 4.9). For water transport to occur across an epithelium, an osmotic gradient must be formed between the luminal and lateral compartments. Such a gradient is generated by the active transport of solutes. In the case of the proximal tubule, there is a discrete osmotic gradient between the tubular fluid and the plasma of approximately 4–6 mOsm, being hypotonic the luminal fluid respect to the intercellular fluid. This discrete difference in osmolality added to the high osmotic permeability allows for the reabsorption of large volumes of water.

The mechanism can be described as follows (Fig. 4.9):

1. The Na^+ and other solutes reabsorbed from the tubular fluid, through the apical membrane, enter the cytoplasm and are transported to the lateral intercellular spaces by the basolateral transporters, generating a slightly hyperosmotic environment in these spaces (300 mOsm/kg versus 296 in the tubular fluid).
2. The small osmotic gradient allows water to move through apical and then basolateral aquaporin 1 (AQP-1) into the lateral intercellular spaces.

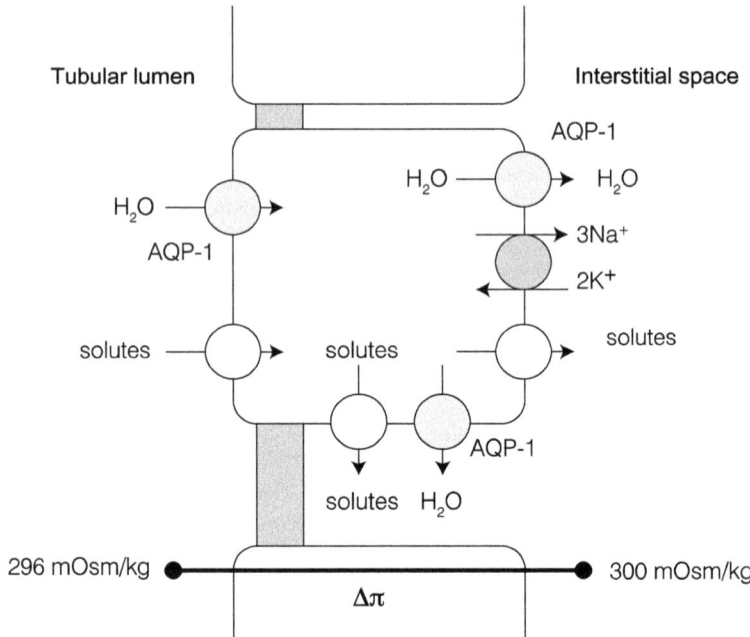

Fig. 4.9 Coupling of solute and water reabsorption in the proximal tubule. The apical and basolateral membranes of proximal tubule cells contain abundant copies of AQP-1, rendering both membranes with high osmotic permeability. The small osmotic gradient ($\Delta\pi$) for water reabsorption is created by solute reabsorption across both membranes. The majority of water reabsorption is transcellular

3. The same transtubular osmotic gradient allows some water to move along the paracellular pathway.
   The importance of the transcellular versus paracellular pathway in transepithelial water transport has been a matter of debate for a long time. The second matter of debate was whether AQP-1 was or not a key element in transcellular water reabsorption across the proximal tubules. AQP-1 was first discovered accidentally in 1988, cloned and named aquaporin CHIP28. Its expression was demonstrated in proximal tubules and thin descending limbs of Henle. The generation of a transgenic mouse for AQP-1 was decisive to clarify the relevance of the transcellular pathway and within the role of AQP-1. Osmotic water permeability measured in isolated tubules of AQP-1 mice -/- showed a reduction of 80%, compared to the dominant homozygotes, and the osmotic permeability measured in apical and basolateral membrane vesicles obtained from the transgenic mice was very low and compatible only with water movement through the lipid bilayer. These experiments definitely demonstrated first that the vast majority of the water movement in the proximal tubule is transcellular and not paracellular. Second, it is mediated by AQP-1.

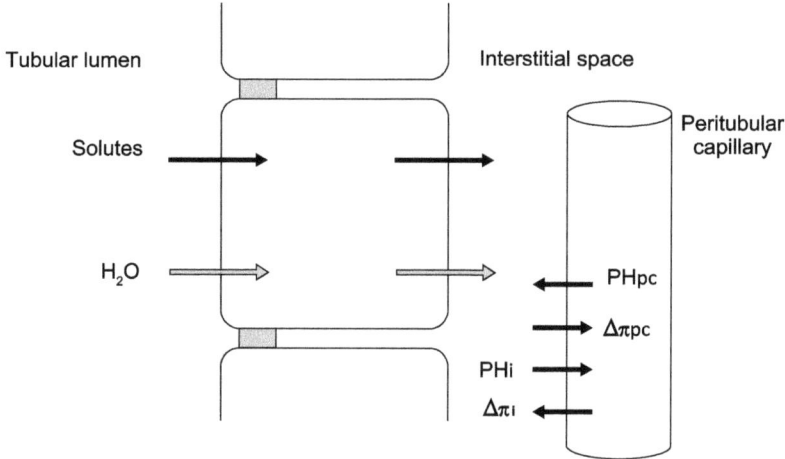

$$\Delta\pi \text{ Reabsorption} = (\Delta\pi\text{pc} - \Delta\pi\text{i}) - (\text{PHpc} - \text{PHi})$$

Fig. 4.10 Role of Starling forces in proximal fluid reabsorption. Oncotic pressure (πpc) in peritubular capillaries and the interstitial hydrostatic pressure (PHi) are the forces that favor fluid reabsorption into the peritubular capillaries

4. The fluid reabsorbed through the epithelium must enter the peritubular capillaries in order to maintain reabsorption. This process is due to the Starling forces governing the transcapillary fluid exchange (Fig. 4.10). The magnitude of the forces allows that under physiological conditions the movement of fluid towards the capillaries is favored. The pressures at play are:
   - Hydrostatic pressure in peritubular capillaries (PHpc), which is about 10 mmHg and acts against reabsorption.
   - Hydrostatic interstitial pressure (PHi) that promotes fluid reabsorption (2 mmHg).
   - Oncotic pressure in the peritubular capillaries (Πpc): about 30 mmHg, promoting reabsorption. It should be remembered that the filtration of plasma and not protein in the glomerular capillaries causes the blood to reach the peritubular capillaries with high oncotic pressure.
   - Interstitial oncotic pressure (Πi): with an approximate value of 5 mmHg and acts against reabsorption.

The above variables can be summarized in the Starling equation, which describes fluid reabsorption Eq. (4.5):

$$Jv = LpS[(\pi pc - \pi i) - (PHpc - PHi)] \qquad (4.5)$$

where Jv is the volume flow and Lp and S are the hydraulic permeability and surface, respectively, available for fluid reabsorption.

Net pressure is the difference between the forces that favor and those that oppose reabsorption. The net pressure favors the reabsorption of fluid. It is expressed in Eq. (4.6):

$$\text{Net pressure} = (\pi pc - \pi i) - (PHpc - PHi) = (30 - 5) - (10 - 2)$$
$$= 17 \text{ mmHg} \tag{4.6}$$

According to Eq. (4.6), the oncotic pressure in the peritubular capillaries is the main variable that favors the reabsorption of fluid into the peritubular capillaries. It is necessary to keep in mind that changes in the tone of the arteriolar smooth muscle in the afferent or efferent arteriole can modify the plasma flow that reaches the glomeruli. The filtration in the glomerular capillaries generates a high oncotic capillary pressure that will drive fluid reabsorption in the cortical peritubular capillaries.

4.11   Loop of Henle

The loop of Henle consists of three portions: thin descending, thin ascending, and thick ascending limb, which consists of a medullary and a cortical portion. The thick ascending loop of Henle reabsorbs about 25% of the filtered NaCl load.

The thin descending limb is not involved in NaCl reabsorption; it has been reported that this segment has very low Na^+ permeability but high osmotic water and urea permeability. This is consistent with cell morphology: flat cells with few mitochondria and low Na^+, K^+-ATPase activity. The high osmotic permeability is given by the presence of AQP-1 in the apical and basolateral membrane. In the thin descending limb, 15% of the filtered water is reabsorbed.

The thin ascending loop carries NaCl reabsorption. However, the mechanisms are unclear.

The thick ascending loop reabsorbs Na^+, Cl^-, K^+, and HCO_3^-. The reabsorption of Na^+, Cl^-, and K^+ occurs by active secondary transport, dependent on the activity of the Na^+, K^+-ATPase (Fig. 4.11). The apical membrane expresses the NKCC2 cotransporter, which is inhibited by loop diuretics like furosemide and bumetanide. This electroneutral cotransporter uses the electrochemical Na^+ gradient to actively transport Cl^- and K^+ from the tubular lumen to the cytosol. The apical membrane also expresses the K^+ channel ROMK, which allows the flow of K^+ ions from the cytosol to the tubular lumen. This K^+ recycling through the apical membrane ROMK channel contributes to the lumen-positive (+10 mVolt) transepithelial potential difference. The latter is physiologically relevant because it constitutes a driving force for the paracellular reabsorption of monovalent and divalent cations, such as Na^+, K^+, Ca^{++}, and Mg^{++}.

Half of the Na^+ reabsorption that occurs in the thick ascending limb is transcellular and the rest is paracellular. Therefore, Na^+ reabsorption in this segment has a high metabolic efficiency, transporting 6 Na^+ for each ATP, coupled with high sodium pump activity, which is relevant considering that the renal medulla is a

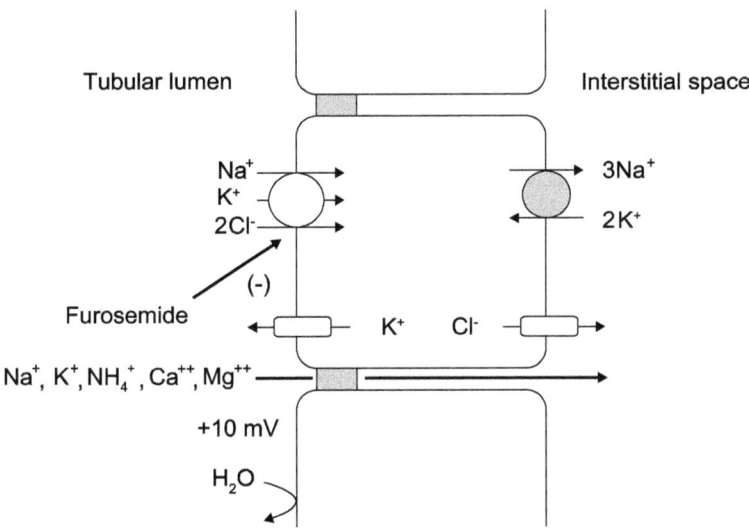

Fig. 4.11 Ion transport in the thick ascending limb. NaCl reabsorption is mediated through apical NKCC2 cotransporter; K^+ is recycled to the tubular lumen mainly through the renal outer medullary potassium channel (ROMK). Basolateral Cl^- exit is mediated by CLC-K chloride channels. The lumen-positive transepithelial potential difference acts a driving force for paracellular cation reabsorption. The thick ascending limb epithelium is impermeable to water

relatively hypoxic environment. The basolateral exit of Na^+ occurs through the Na^+, K^+-ATPase, while Cl^- occurs through chloride channel members of the CLC family, called CLC-Ka and CLC-Kb. Thus, the movement of potassium into the lumen without the anion (Cl^-) generates the difference in positive transepithelial potential in the lumen.

Despite its high ionic conductivity, the thick ascending loop of Henle has a low osmotic permeability. This property is consistent with the fact that no members of the aquaporin family have been found in the apical and basolateral membrane. The NaCl reabsorption without the equivalent osmotic flow dilutes the tubular fluid flowing from the medulla to the cortex. In fact, the tubular fluid entering the thin ascending loop of Henle is hyperosmotic and has an osmolality of approximately 600 mOsm/kg of water. In contrast, the tubular fluid leaving the cortical portion of the thick ascending loop and entering the distal tubule is hypoosmotic and has an osmolality nearly 100 mOsm/kg. Furosemide, a widely used loop diuretic, inhibits the NKCC2 cotransporter. Therefore, the thick ascending limb is unable to reabsorb NaCl and dilute the tubular fluid. Under these conditions, this tubular fluid has an osmolality close to plasma (isosthenuria). NKCC2 blockade not only inhibits transcellular reabsorption but also reduces paracellular reabsorption of divalent cations (Ca^{++}, Mg^{++}). In conclusion the loop of Henle is divided into three segments. Only the function of the thin descending limb and thick ascending limb had been studied in detail. The thin descending limb reabsorbs water and secretes urea, while it is impermeable to NaCl. The thick ascending limb reabsorbs NaCl and is imper-meable to water and hence dilutes the tubular fluid flowing from the inner medulla to cortex. The whole loop plays a critical role in the mechanism of concentration and dilution of the urine.

4.12   Distal Nephron

The distal nephron consists of three segments localized post-macula densa in the cortical labyrinths: distal convoluted tubule, connecting tubule, and the collecting duct, located in the medullary rays.

The distal nephron as a whole reabsorbs about 9% of the filtered NaCl load. In the distal convoluted tubule, the apical entry of NaCl occurs through the apical electroneutral NCC cotransporter Na^+-Cl^-, although a low expression of the epithe-lial Na^+ channel (ENaC) has been documented in the last part of the distal convo-luted tubule. In the most initial part of the distal convoluted tubule, the transepithelial potential difference is 0 then it becomes established as a lumen-negative transepithelial potential difference, which is consistent with the expression of apical ENaC. The basolateral exit of Na^+ is mediated by the Na^+, K^+-ATPase; the distal convoluted tubule has the highest activity of this primary active transport. Basolateral Cl^- exit is mediated mainly through a CLC-K chloride channel. Potas-sium efflux occurs through the Kir4.1/5.1, which is the predominant potassium channel in the basolateral membrane (Fig. 4.12). NaCl reabsorption in this segment is inhibited by diuretics of the thiazide family like hydrochlorothiazide. The distal

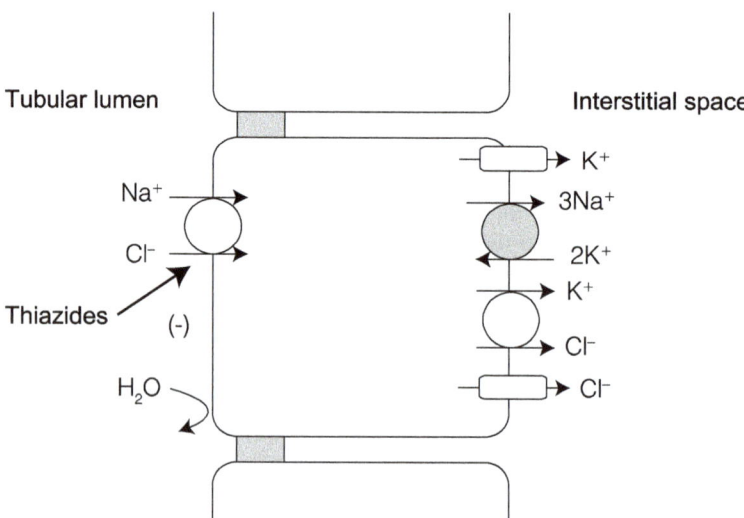

Fig. 4.12 Ion transport in the distal convoluted tubule. NaCl reabsorption in this segment is mediated through apical NCC cotransporter. Basolateral Cl^- exit is mediated by chloride channel and possibly a K^+/Cl^- cotransporter. The basolateral K^+ channel is Kir 4.1/5.1, which mediates K^+ recycling with the Na^+, K^+-ATPase in the basolateral membrane

convoluted tubule is impermeable to water, consistent with the lack of expression of aquaporins like AQP2 and AQP3, which are present in downstream segments.

In most species, the connecting tubule begins as a transition from the distal convoluted tubule. The major cell type is the connecting cell, involved in Na^+ and water reabsorption and K^+ secretion. This segment plays a key role in the fine-tuning of Na^+ reabsorption to meet the needs of Na^+ balance. The other cell type is the intercalated cell (type A and B), which is involved in acid-base transport and will be discussed in Chap. 8. The apical membrane of connecting cells expresses the epithelial Na^+ channel (ENaC) and the potassium channels ROMK and BK. The basolateral membrane expresses the Na^+, K^+-ATPase and K^+ channels. The connecting tubule is a tight epithelium with tight junctions permeable to anions like Cl^-. Na^+ reabsorption in the connecting cells is electrogenic and mediated by Na^+ entry through ENaC. This creates a lumen-negative transepithelial potential difference of nearly 40 mVolt that can be increased in the presence of mineralocorticoids (Fig. 4.13). Na^+ exit across the basolateral membrane is mediated by the Na^+, K^+-ATPase; K^+ is recycled through potassium channels. The lumen-negative transepithelial potential difference is an electromotive force for two processes: paracellular Cl^- reabsorption and K^+ secretion. The connecting tubule has an important role in potassium homeostasis as a major site of K^+ secretion under a Western diet. Two apical potassium channels mediate K^+ secretion: ROMK is the most important K^+ channel, in mediating potassium secretion, and BK channel by tubular fluid flow and cytosolic Ca^{++}. Connecting cells secrete the endopeptidase kallikrein, involved in bradykinin formation. The connecting tubule cells have a regulated water permeability. Upon vasopressin stimulation, AQP2 is inserted in the apical membrane, which increases apical membrane water permeability. Basolateral water exit occurs through AQP3.

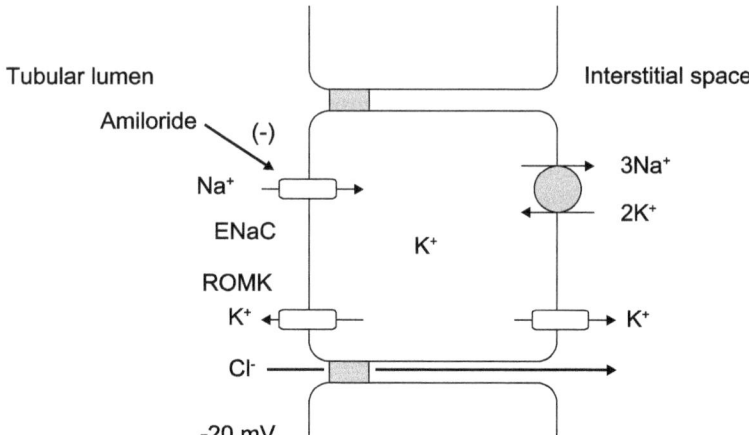

Fig. 4.13 NaCl reabsorption in the connecting tubule and cortical collecting duct. Connecting cells in the connecting tubule and principal cells of the cortical collecting duct mediate NaCl reabsorption and K^+ secretion. Electrogenic Na^+ entry is mediated by the apical epithelial Na^+ channel. Na^+ reabsorption creates a lumen-negative transepithelial potential difference that acts as a driving force for K^+ secretion mainly through ROMK K^+ channel and paracellular Cl^- reabsorption

In juxtamedullary nephrons, several connecting tubules drain into a single corti-cal collecting duct. The main cell type of the cortical collecting duct is the principal cell; the other cell type is the intercalated cell. Electrogenic Na^+ reabsorption in principal cells is mediated through apical epithelial Na^+ channels. As in the connecting tubule, the lumen-negative transepithelial potential difference is an electromotive force for paracellular Cl^- reabsorption and K^+ secretion.

4.13 Regulation of NaCl and Water Reabsorption Along the Nephron

NaCl and water reabsorption is tightly regulated throughout the renal tubule. Physi-cal factors as well as hormones and neurotransmitters influence NaCl and water transport across the tubular epithelia.

4.13.1 Physical Factors: Starling Forces

Starling forces are the physical forces that determine fluid movement across the capillary wall. They are involved in the regulation of reabsorption in the proximal tubule, specifically in the movement of fluid between the peritubular capillaries and the interstitium in the cortical labyrinths. Changes in the contractile tone of the afferent and efferent arterioles modify the filtration fraction and the Starling variables that govern the transcapillary fluid exchange (Fig. 4.10). A clear example of how the Starling forces influence capillary fluid absorption is the increase in efferent arteriolar tone. The increase in efferent smooth muscle tone will increase glomerular capillary hydrostatic pressure and hence glomerular filtration and filtra-tion fraction. The plasma entering the peritubular capillaries will have an increased oncotic pressure favoring fluid reabsorption from the interstitium into the fenestrated peritubular capillaries. A drop in the efferent arteriolar tone will also cause a decrease in the filtration fraction. A lower plasma oncotic pressure in the peritubular capillaries will delay fluid reabsorption and favors the passage of fluid from the lateral intercellular space to the tubular lumen through the tight junction between cells in the proximal tubule.

4.13.2 Glomerulotubular Balance

The glomerulotubular balance is the mechanism allowing the adequacy of proximal tubule reabsorption of NaCl and water in the presence of changes in glomerular filtration. In other words, the reabsorption process that occurs in the proximal tubule is not an isolated process but is related to the filtered load of solutes and water. There are two important aspects of the glomerulotubular balance. First, variations in glomerular filtration imply changes in the capillary oncotic pressure in the peritubular capillaries, hence influencing fluid reabsorption. Therefore, changes in the filtration fraction can lead to changes in proximal reabsorption. Second, alterations in glomerular filtration involve variations in the filtered load of solutes, including those reabsorbed in the proximal tubule. It follows that an increase in the filtered load of glucose, amino acids, and bicarbonate will result in increased solute and water reabsorption in the proximal tubule, because the transport of these solutes and water depends directly or indirectly on sodium transport in the first part of the proximal tubule. In summary, the glomerulotubular balance allows adjustment of proximal reabsorption to changes in the filtered load of solutes and water. The two most important mechanisms in the glomerulotubular balance are the Starling forces in the peritubular capillaries and the composition of the glomerular filtrate entering the proximal tubule.

4.13.3 Regulation by Neurotransmitters and Hormones: General Concepts

Tubular function is carried out by multiple transporters and channels expressed in the apical and basolateral domain of the tubular epithelium. The activity of these proteins is controlled by neurotransmitters and hormones. It can be controlled in two general ways: short term and long term. One of the most common ways of short-term control of transporter and channel activity is phosphorylation. Several protein kinases can phosphorylate the transporter protein itself or accessory proteins in consensus sequences. Phosphorylation can affect transport in at least two general ways. First, modifying the transporter and channel activity expressed in apical and basolateral membrane domain. Second, phosphorylation can modify the translocation of intracellular pools of preexisting transporters and channels to the membrane domains. Several protein kinases induce/reduce the traffic of intracellular pools of transporters and channels to the membrane or reduce the endocytosis of preexisting transporters and channels from the membrane.

   Thus, maximal transport rates can be influenced by transporter and channel phosphorylation and/or by modifying its abundance in a particular membrane domain. Transport rate changes induced by phosphorylation are reversed by the action of several protein phosphatases. Examples of transporter and channel control by phosphorylation include:

- Proximal tubule NHE3 activity is inhibited in a cAMP-PKA-dependent way that is stimulated by parathyroid hormone (PTH). Angiotensin II and norepinephrine inhibit the cAMP-PKA transduction pathway, increasing NHE3 activity in the apical membrane of the proximal tubule.
- NKCC2 activation by a cAMP-PKA-dependent pathway that is stimulated by binding of vasopressin to V_2 basolateral receptors.
- SPAK kinase-dependent phosphorylation of NCC, activated by angiotensin II.
- cAMP-PKA-dependent translocation of AQP-2-bearing cytosolic vesicles to the apical membrane, stimulated by binding of vasopressin to V_2 basolateral receptors.
- SGK1 kinase-dependent inhibition of ENaC endocytosis from the apical membrane.

Long-term regulation is the second mechanism of transporter and channel control. This type of regulation involves the stimulation of transcription of the protein mRNA. The best example is aldosterone stimulation of ENaC α-subunit transcription. This action is mediated by mineralocorticoid receptor activation, which will be described later.

Angiotensin II Under physiological conditions, its effects on NaCl transport are mediated by angiotensin II binding to AT_1 receptors. The proteins Gq and Gi, which are coupled to phospholipase C and adenylyl cyclase, respectively, are the apical and basolateral proteins. The net result is the stimulation of phospholipase C and a reduction in cytosolic levels of cAMP. Angiotensin II stimulates the activity of the NHE3 exchanger in the proximal tubule, sodium pump, and NBC cotransporter, resulting in increased reabsorption of sodium, bicarbonate, chloride, and water. Angiotensin II also increases NaCl reabsorption in the distal convoluted tubule cells; the effect is mediated by a PKC-dependent activation of WNK1/4 kinase which in turn activates SPAK-dependent NCC phosphorylation. The stimulation of NaCl reabsorption by angiotensin II is part of the mechanisms to compensate for a low extracellular volume.

Noradrenaline This catecholamine is secreted by the postganglionic sympathetic endings that innervate the tubules. Through $α_2$ adrenergic receptors coupled to adenylyl cyclase, noradrenaline stimulates the activity of the NHE3 in the proximal tubule, with the effects already described. This effect is associated with a decrease in the levels of cyclic AMP in the cells. Both angiotensin II and noradrenalin increase in conditions of extracellular volume contraction.

Vasopressin (AVP) It has a short-term and a long-term effect on AQP-2 (Fig. 4.14). Both effects are mediated by the activation of V_2 basolateral receptors that activate the cAMP-PKA transduction pathway. This AVP acute effect is related to intracellular traffic of AQP-2 between the apical plasma membrane and subapical tubulovesicles. When plasma AVP levels are low, AQP-2 resides in subapical cytosolic tubulovesicles. Under this condition the osmotic permeability of the apical membrane is very low. Binding of AVP to V_2 receptors leads to the exocytosis of AQP-2 to the apical membrane with the subsequent increase in the osmotic permeability. PKA-dependent AQP-2 phosphorylation is required for cytosolic trafficking but it does not change AQP-2 water permeability. When AVP levels drop, AQP-2 molecules are endocited to the cytosol and can be degraded or recycled to the apical membrane. The long-term effect is transcriptional and requires PKA-dependent phosphorylation of CREB transcription factor. CREB-P binds to specific sequences in the promoter region of the AQP-2 gene and increases its transcription. This effect occurs when the plasma concentration of AVP remains elevated for longer periods (12 h); the net effect is higher levels of AQP-2 protein available for short-term mechanism. Water molecules that entered the cell via apical AQP-2 leave the cell via basolateral AQP-3 in connecting and principal cells of the distal nephron. In the inner medullary collecting duct, basolateral water exit occurs through AQP-3 and AQP-4.

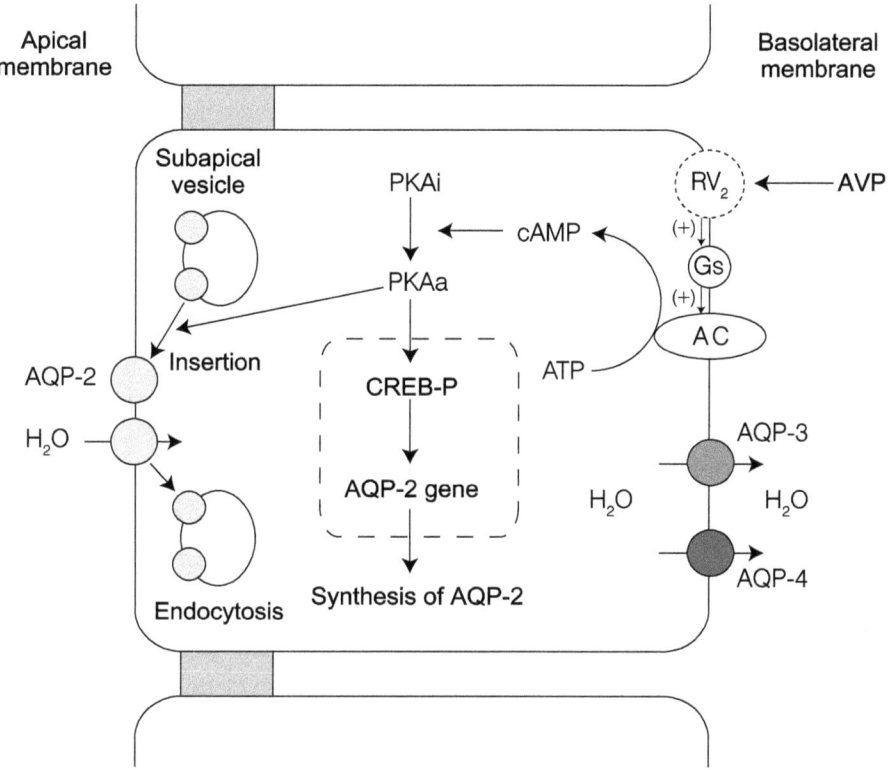

Fig. 4.14 Vasopressin-dependent water reabsorption. AVP activates V_2 basolateral receptors in connecting and principal cells. These receptors are coupled through Gs to the cAMP-PKA signaling cascade. Active PKA phosphorylates AQP-2 contained in subapical cytosolic vesicles, promoting AQP-2 insertion in the apical membrane. This mechanism increases osmotic water permeability. Water molecules entering through AQP-2 exit the cell through basolateral AQP-3 in the cortical collecting duct and AQP-3 and AQP-4 in inner medullary collecting duct. Apical AQP-2 suffers endocytosis and is recycled to cytosol. In a long-term action, active PKA also promotes AQP-2 gene transcription through CREB transcription factor; this action contributes to increased AQP-2 protein abundance in connecting and principal cells

AVP also stimulates the activity of apical NKCC2 in the thick ascending limb of Henle. This effect contributes to the increase in the interstitial hyperosmolality in the renal medulla. The effect is mediated by activation of basolateral V_2 receptors coupled to the cAMP-PKA signal transduction cascade.

Aldosterone and NaCl and K^+ Transport Regulation in the Distal Nephron Aldosterone, synthesized and released from the adrenal glomerulosa cells, is the main mineralocorticoid in humans. Two main stimuli increase aldosterone synthesis and release: hypovolemia and increase in plasma K^+ concentration. The effect of hypovolemia is mediated by angiotensin II, while an increase in plasma K^+ itself is able to stimulate hormone release. Plasma aldosterone concentration is low (10^{-8} to 10^{-10} mol/L), most of the hormone circulates freely, and thus, it has a short half-life. Aldosterone binds to cytosolic mineralocorticoid receptors that are expressed in the aldosterone-sensitive distal nephron (ASDN) conformed mainly by cells of the late distal convoluted tubule, connecting and principal cells. The expression of the mineralocorticoid receptor and the action of aldosterone in the distal convoluted tubule cells are a matter of controversy. The receptor belongs to the family of steroid hormone receptors. In its inactive state, the receptor is bound to heat shock proteins and located in the cytosol. Hormone binding triggers receptor activation and nuclear translocation. The high-affinity receptor for aldosterone is the mineralocorticoid receptor, while the glucocorticoid receptor functions as a low-affinity receptor. The mineralocorticoid receptor has the same affinity for aldosterone and cortisol, but plasma concentration of cortisol (10–1000 nM) is several times higher than that of aldosterone (0.1–1 nM). Under these conditions, the receptors would be mostly occupied by cortisol and not by aldosterone. However, the aldosterone-sensitive cells of the distal nephron express the enzyme 11 β-hydroxysteroid dehydrogenase-2 (11β-HSD2), which metabolizes cortisol to cortisone. Because of its very low affinity, cortisone is unable to bind to aldosterone receptors. The enzyme does not metabolize aldosterone and it can bind to receptors. This enzyme makes aldosterone the only corticosteroid available for the mineralocorticoid receptors on the cells of the distal nephron. The binding of aldosterone to the receptor can be competitively inhibited by a drug called spironolactone.

The best studied action of aldosterone is the stimulation of ENaC abundance in the apical membrane of ASDN cell. αβ ENaC is the epithelial Na+ channel expressed in the apical membrane of connecting and principal cells as well as absorptive cells in the distal colon. The functional channel is formed by three subunits: α, β and γ.

This mineralocorticoid action originates from two separate aldosterone effects. The first action is also known as aldosterone early action. In this setting aldosterone stimulates the transcription of the serum and glucocorticoid kinase 1 (SGK1). SGK1 is the first aldosterone-induced protein and phosphorylates Nedd4-2; this protein is part of the protein complex involved in ENaC endocytosis from the apical membrane. The endocytic complex with phosphorylated Nedd4-2 is unable to interact with ENaC channels and endocytosis is inhibited. Thus, the first action of aldosterone is to inhibit Na^+ channel retrieval from the apical membrane, which increases ENaC density in the apical membrane. In the second and more late action, aldosterone increases the transcription of the α-subunit forming the ENaC channel. This synthesis de novo of ENaC contributes to the increase in ENaC density in the apical membrane. Aldosterone also stimulates the transcription of the α1 subunit of the Na^+, K^+-ATPase and Krebs cycle enzymes. Taken together, aldosterone stimulates the abundance of protein involved in every step of the electrogenic transepithelial Na^+ transport: increasing apical ENaC abundance, basolateral Na^+, K^+-ATPase for Na^+ exit, and ATP supply for Na^+ ion primary active transport (Fig. 4.15). The net result is the increase of the lumen-negative transepithelial potential difference, enhanced by the increase in electrogenic Na^+ reabsorption through ENaC. These actions of aldosterone occur in settings like hypovolemia or can be induced in experimental rodent models by a low Na^+ diet.

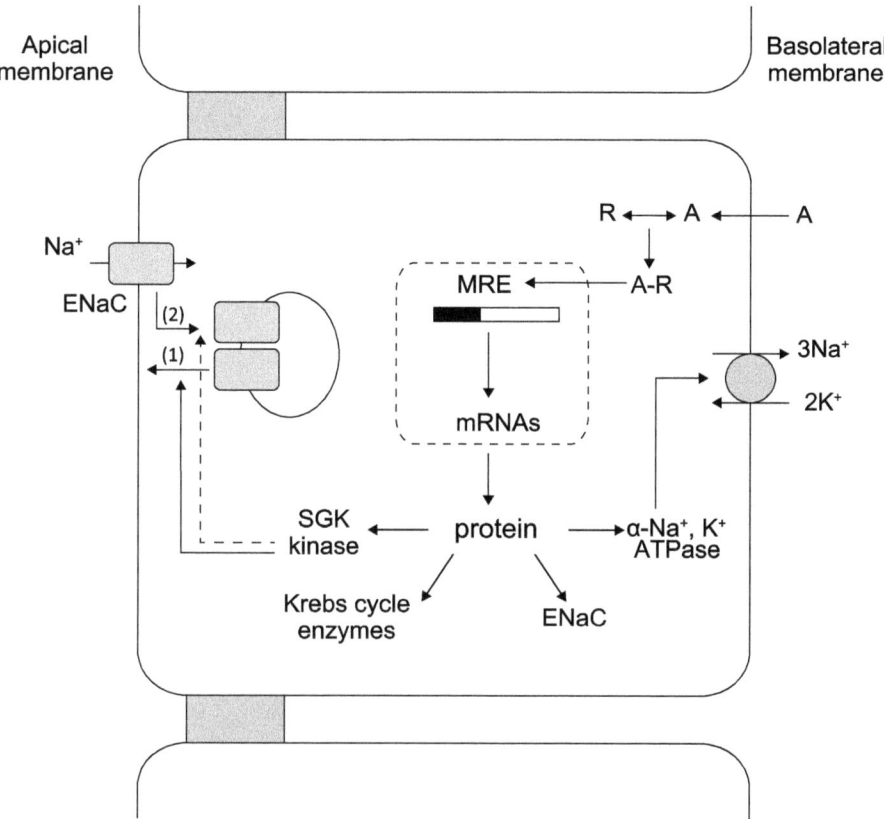

Fig. 4.15 Aldosterone (A) mechanism of action. Aldosterone enters the connecting and principal cells through the basolateral membrane and binds to a mineralocorticoid receptor. The hormone-receptor complex undergoes nuclear translocation and transactivates the transcription of several proteins. SGK is the first protein transcribed under the influence of aldosterone; other proteins are α-subunit of epithelial Na$^+$ channel and Na$^+$, K$^+$-ATPase

Aldosterone also stimulates K$^+$ secretion in the connecting and principal cells of the ASDN. This action plays a pivotal role in the external potassium balance. Potassium secretion is stimulated at least by two factors: first, the lumen-negative transepithelial potential difference generated the electrogenic Na$^+$ reabsorption and second is the activity of the basolateral Na$^+$, K$^+$-ATPase.

How is the same hormone able to respond to two different stimuli in the same target tissue? The answer to this question is part in what has been called the aldosterone paradox (Fig. 4.16). In hypovolemia or a low-Na$^+$ diet (Fig. 4.16a), adrenal glomerulosa cells are stimulated by angiotensin II. This peptide hormone increases NaCl reabsorption in the distal convoluted tubule by increasing NCC activity through a WNK1/4-SPAK signal cascade initiated by AT$_1$ receptor. In the connecting and principal cells, aldosterone stimulates ENaC-mediated electrogenic Na$^+$ reabsorption, while angiotensin II inhibits the apical ROMK potassium channel.

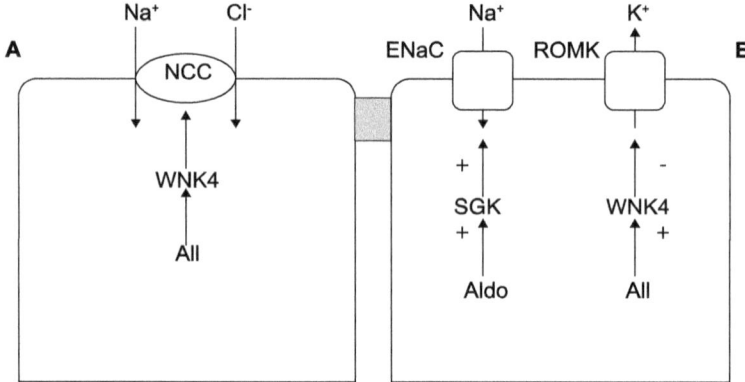

Fig. 4.16 Aldosterone stimulation of NaCl reabsorption in the distal nephron. Extracellular volume contraction activates the renin-angiotensin-aldosterone system to stimulate NaCl reabsorption in segments of the distal nephron. How is NaCl reabsorption increased without increasing also K$^+$ secretion? In the distal convoluted tubule, angiotensin II increases the phosphorylation and activity of NCC cotransporter through a WNK4-dependent mechanism. In the connecting and principal cells, aldosterone increases apical ENaC abundance through a SGK1-dependent mechanism. In the same cells, angiotensin II decreases ROMK activity and hence K$^+$ secretion. Through this coordinated action angiotensin II and aldosterone increase NaCl reabsorption to restore extracellular volume

The net effect is the increase in Na$^+$ reabsorption with minimal K$^+$ secretion due to the action of two hormones: angiotensin II and aldosterone. The situation is very different when plasma K$^+$ is elevated (Fig. 4.16b). In this setting, aldosterone synthesis and release are stimulated directly by the increase in plasma K$^+$ through a Ca^{++}-dependent pathway. Thus, aldosterone is the only hormone stimulated in this condition, increasing K$^+$ secretion through apical ROMK potassium channels.

Atriopeptin or Atrial Natriuretic Peptide (ANP) Natriuretic peptides are a family of peptides that includes cardiac atriopeptin, renal urodilatin, and cerebral atriopeptin. The first to be discovered was the atrial natriuretic peptide, which is made up of 28 amino acids. Its secretion by the atria is stimulated by increases in blood pressure and circulating volume. The effects of atriopeptin are mediated by membrane receptors whose cytosolic domain has guanylyl cyclase activity, which is activated by hormone binding to the extracellular receptor domain, thus increasing cytosolic levels of cGMP. Atriopeptin produces relaxation of the vascular smooth muscle and lowering of the blood pressure. In addition, it stimulates renal excretion of sodium and water, mainly through inhibition of renin secretion and angiotensin II formation. In addition, atriopeptin inhibits the reabsorption of Na$^+$ in the internal medullary collecting duct. The ANP-mediated renal vasodilation increases the medullary blood flow, which contributes to the washout of the hypertonicity of the medullary interstitium. The net effect is the reduction in water reabsorption in the medullary collecting duct. In addition, ANP inhibits vasopressin secretion which also contributes to increased water excretion. Currently and despite the varied actions of atriopeptin, there is controversy over whether or not atriopeptin plays a significant physiological role in the daily sodium balance.

Nitric Oxide (NO) This gas is a product of the nitric oxide synthase (NOS). NO participates in the regulation of renal hemodynamics by maintaining a vasodilator tone, which may indirectly affect tubular transport. In addition, NO inhibits the tubular transport of Na^+ and therefore promotes natriuresis. Renal production of NO appears to be related to the state of the circulating volume. A high intake of Na^+ increases the production of NO, facilitating natriuresis.

An example of the interrelationship between the abovementioned neurohormonal mechanisms and NaCl transport is what occurs in chronic arterial hypertension, due to unilateral renal artery stenosis. In this situation, the hypoperfused kidney suffers from a drop in perfusion pressure in the afferent artery and a drop in glomerular filtration rate. This increases renin secretion and subsequent formation of angiotensin II and release of aldosterone. Angiotensin II is a powerful vasoconstrictor that raises blood pressure. Both angiotensin II and aldosterone stimulate the reabsorption of NaCl in the proximal tubule and distal nephron, respectively. The retention of NaCl expands the circulating volume (volemia), further increasing blood pressure. These changes are only partially counteracted by vasodilatory and natriuretic mechanisms, perpetuating high blood pressure over time.

Bradykinin This peptide is the hormonal component of the kallikrein-kinin system. Its effects on the renal management of Na^+ and water will be described in Chap. 11.

4.14 Conclusions

Sodium reabsorption along the nephron depends on its electrochemical potential gradient, maintained by the basolateral Na^+, K^+-ATPase. It governs the reabsorption of organic solutes like glucose and amino acids as well as electrolytes like chloride and bicarbonate.

Histologically and functionally, the proximal tubule is ideally suited for massive solute and fluid reabsorption. It is the only segment accomplishing glucose and amino acid reabsorption. A high transcellular osmotic permeability accounts for high rates of fluid reabsorption.

The thick ascending limb of Henle and the distal convoluted tubule reabsorb NaCl and are impermeable to water. The activity of both segments contributes to making a hypotonic tubular fluid.

NaCl reabsorption in the connecting tubule and cortical collecting duct is adjusted to meet sodium balance.

Sodium reabsorption along the nephron is subjected to neurohormonal control and adjusted to meet sodium balance, arterial pressure, and extracellular fluid volume regulation. Norepinephrine stimulates sodium transport in the proximal tubule. Angiotensin II is a potent stimulator of sodium reabsorption in the proximal and distal convoluted tubule. Aldosterone stimulates sodium reabsorption in the connecting tubule and cortical collecting duct.

Review Questions

1. Define the concepts of tubular reabsorption and secretion. Which are the possible routes for tubular reabsorption across the epithelium?
2. The following data are from a study of renal glucose handling. Calculate the missing data in the corresponding columns. Graph filtered, reabsorbed, and excreted glucose (mg/min) as a function of glycemia (mg/mL).

Interval (min)	GFR (mL/min)	[Glucose]pl (mg/mL)	Filtered load (mg/min)	Excreted load (mg/min)	Glucose reabsorbed (mg/min)
0	125	1.0		0	
Start glucose infusion	125				
26–40	125	1.0		0	
60–80	125	2.8		20	
80–100	125	3.5		76	
100–110	125	4.0		125	
130–140	125	5.0		250	

   (a) Explain the relationship between plasma glucose concentration and filtered load of glucose, reabsorbed glucose, and excreted load of glucose.
   (b) Define the terms "glucose threshold" and "maximum transport rate." Within this data can you determine the glucose threshold and the maximum glucose transport rate (Tmax)?
   (c) The preceding data comes from whole organism experiments. Suppose that the same experiment is repeated in an isolated kidney, perfused with a solution in which the Na^+ concentration was reduced to 50% (the rest of the Na^+ concentration was replaced with other impermeant cation).
3. Explain the coupling between solutes and water reabsorption in the proximal tubule. If the water channels in the proximal tubule were blocked, what would happen to the fluid reabsorption?
4. The thick ascending limb of Henle develops a lumen-positive transepithelial potential difference. How is this electric potential generated? What is the physiological relevance of this electric potential?
5. Make a graph with transepithelial potential difference along the distal nephron from the very beginning of the distal convoluted tubule to the end of the cortical collecting duct.
6. Which is the physiological relevance of this electric potential?

Bibliography

Aronson PS, Giebisch G (1997) Mechanisms of chloride transport in the proximal tubule. Am J Physiol Renal Physiol 273:F179–F192

Arroyo JP, Ronzaud C, Lagnaz D, Staub O, Gamba G (2011) Aldosterone paradox: differential regulation of ion transport in distal nephron. Physiology 26:115–123

Benos DJ, Fuller CM, Shlyonsky VG, Berdiev BK, Ismailov I (1997) Amiloride sensitive Na$^+$ channels: insights and outlooks. Physiology 12:55–61

Hediger MA, Rhoads DB (1994) Molecular physiology of sodium-glucose cotransporters. Physiol Rev 74:993–1026

Hummler E, Vallon V (2005) Lessons from mouse mutants of epithelial sodium channel and its regulatory proteins. J Am Soc Nephrol 16:3160–3166

Ma T, Yang B, Gillespie A, Carlson EJ, Epstein CJ, Verkman AS (1998) Severely impaired urinary concentrating ability in transgenic mice lacking aquaporin-1 water channels. J Biol Chem 273: 4296–4299

McCormick JA, Bhalla V, Pao AC, Pearce D (2005) SGK1: a rapid aldosterone-induced regulator of renal sodium reabsorption. Physiology 20:134–139

McDonough AA, Thomson S (2012) Metabolic basis of solute transport. In: Brenner BM, Rector FC (eds) The kidney, 9th edn. Elsevier, Philadelphia, pp 138–157

Mount DB (2014) Thick ascending limb of the loop of Henle. Clin J Am Soc Nephrol 9:1974–1986

Mount DB (2016) Renal transport of sodium, chloride and potassium. In: Brenner BM, Rector FC (eds) The kidney, vol I, 10th edn. Elsevier, Philadelphia, pp 144–182

Palmer LG, Schnermann J (2015) Integrated control of Na transport along the nephron. Clin J Am Soc Nephrol 10:676–687

Prabhleen S, McDonough AA, Thomson S (2016) The metabolic basis of solute transport. In: Brenner BM, Rector FC (eds) The kidney, vol I, 10th edn. Elsevier, Philadelphia, pp 123–142

Reilly RF, Ellison DH (2000) Mammalian distal tubule: physiology, pathophysiology, and molecular anatomy. Physiol Rev 80:277–313

Schnermann J, Chou CL, Ma T, Traynor T, Knepper MA, Verkman AS (1998) Defective proximal tubule reabsorption in transgenic aquaporin-1 null mice. Proc Natl Acad Sci USA 95:9660–9664

Subramanya AR, Ellison DH (2014) Distal convoluted tubule. Clin J Am Soc Nephrol 9:2147–2163

Water Balance and the Regulation of Plasma Osmolality

5

Learning Objectives
- To understand the physical basis of water movement through mammalian biological membranes.
- To describe the routes of passage of water through biological membranes.
- To describe the distribution of water in the human body.
- To describe the routes of entry and exit of water in a human organism.
- To understand plasma osmolality as the physiological variable that reflects the state of the water balance in the human body.
- To describe the role of antidiuretic hormone as a molecular entity relating plasma osmolality to renal water handling.
- To describe the role of nephron architecture and renal circulation in the mechanism of urinary concentration and dilution.

The kidney plays a key role in the regulation of plasma osmolality, which is directly related to the maintenance of water balance. Plasma osmolality is regulated, within very narrow limits, through the control of water inputs (thirst mechanism) and renal water excretion, which under physiological conditions occurs independently of the control of renal sodium excretion. The physiological relationship between plasma osmolality and renal water excretion is given by the antidiuretic hormone or vasopressin (ADH, AVP), whose secretion is finely regulated by changes in plasma osmolality of 1–2%, although it can also be affected by significant (>10%) changes in effective arterial volume. The regulation of plasma osmolality is essential for the regulation of cell volume, which becomes really important in the central nervous system, where small changes in plasma osmolality can lead to neurological changes that can compromise life. In this chapter we will analyze the water balance and its relation to the regulation of plasma osmolality, as well as the cellular and molecular mechanisms involved.

© Springer Nature Switzerland AG 2022
P. A. Gallardo, C. P. Vio, *Renal Physiology and Hydrosaline Metabolism*,
https://doi.org/10.1007/978-3-031-10256-1_5

5.1   Physical Basis of Water Transport

The most basic process in the regulation of plasma osmolality is the flow of water molecules through the cell membrane. Water transport has a long history in membrane transport as well as renal physiology.

Let us analyze the process of osmosis and its associated concepts. Imagine a vessel open to the atmosphere and separated into two compartments, A and B, by a membrane that is impermeable to solute and permeable to water (Fig. 5.1). Compartment A contains pure water and compartment B has an equal volume of a solution of solute X. In this system there will be a net flow of water from the lowest to highest osmotic pressure compartment. This water flow is called osmosis; in this physical model it occurs spontaneously from A to B.

What is the force allowing the flow of water molecules from A to B? The force is the gradient or difference in osmotic pressure between the two compartments. Osmotic pressure is a colligative property of the solution and depends on the number of particles dissolved in the solvent and not on the nature, size, shape, or charge of these particles. The osmotic pressure of the solution in compartments A and B is expressed in Eq. (5.1):

$$\pi \, (\text{atm}) = iCRT \tag{5.1}$$

π = osmotic pressure (atmospheres, atm)
i = is the number of particles that a given solute generates in solution
C = is the concentration of solute X in moles/L
R = universal gas constant (atm-L/mol·K)
T = temperature in absolute scale (Kelvin, 273 + °C)

Since compartment A has pure water, its osmotic pressure will be *zero* and the gradient will be equivalent to the osmotic pressure at B. The osmotic flow (Jv) will be directly proportional to the osmotic gradient ($\Delta\pi$), the hydraulic conductivity of

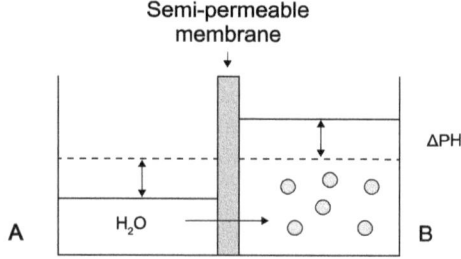

Fig. 5.1 Artificial system to explain osmosis. Compartments A and B in the vessel are separated by a semipermeable membrane. Initially, compartment A contains a volume of pure water and B contains the same volume of a solution formed by the impermeant solute X, thus creating an osmotic gradient between A and B. Osmotic flow from A to B increases the volume in B, creating a hydrostatic pressure gradient between B and A that stops the osmotic flow

the membrane (Lp), and the surface (S) of the membrane available for the passage of water molecules, according to Eq. (5.2):

$$Jv = LpS(\Delta\pi) \tag{5.2}$$

It is important to note that in this system the solute present in B is impermeant to the membrane; therefore the system allows only the diffusion of water molecules. If the solute would permeate across the membrane, the chemical gradient for the solute will dissipate and equilibrium will be reached in time. In equilibrium, the chemical gradient will be *zero* and there would be no osmotic gradient. In other words, the osmotic gradient is given by effective osmoles (e.g., NaCl in the extracellular fluid), which are those particles impermeant to the membrane and therefore restricted to one compartment. In contrast, those very permeant solutes are ineffective osmoles (e.g., urea) and will exert little osmotic pressure. When will this physical model reach equilibrium? The osmotic flow from A to B will increase the volume in B, generating a hydrostatic pressure difference oriented in the opposite direction. When the hydrostatic pressure difference is equal and opposite in magnitude to the osmotic pressure gradient, the system will be in equilibrium and the net water flow will be *zero*. The system discussed above is somewhat ideal. This is because this membrane is absolutely impermeant to solute and water permeable. Biological membranes do not behave as all or none of the structures; in this sense a biological membrane might have a high to low solute and water permeability.

Osmotic pressure is a term with little use in the clinical routine. A more commonly used term is osmolality, which is the total concentration of particles per mass of solvent (mOsm/kg) and is expressed in Eq. (5.3):

$$\text{Osmolality} \left(\frac{\text{mOsmoles}}{\text{kgH}_2\text{O}} \right) = iC \cdot 1000 \tag{5.3}$$

i = number of particles generated in solution; one osmol corresponds to Avogadro's number of particles,
C: solute concentration (moles/L).

Like osmotic pressure, osmolality depends on the number of particles in solution and not on their nature. In general, it is valid to say that solutions formed by electrolytes will have greater osmolality than those of equal molar concentration formed by non-electrolytes; this is because the electrolytes come from a salt that dissociates and the non-electrolyte remains as such in solution.

A good example of this can be seen by comparing two solutions with the same concentration: 100 mM NaCl and 100 mM glucose. In solution, the NaCl molecules will dissociate into two particles (Na^+ and Cl^- for each mole of NaCl) and its osmolality, according to Eq. (5.3), will be 200 mOsm/kg. The glucose molecules will remain as such in solution and therefore only one particle will be generated and the osmolality of the solution will be 100 mOsm/kg. Both solutions have the same molar concentration, but the NaCl solution has twice the osmolality of the glucose solution. This is because, in theory, the number of particles present in the NaCl solution is twice that of the glucose solution.

5.2 Water Pathways in the Cell Membrane and Regulation of Cell Volume

In the previous section, a physical model was used to explain osmosis and the force provoking water flow through the membrane. As mentioned, an osmotic gradient is a force capable of promoting water flow between compartments. In animal cells, and especially in epithelia such as the renal tubule, an osmolality gradient is the most important force promoting transepithelial water flows. In turn, the osmolality gradient is generated by transepithelial solute active transport. There are two possible pathways for water movement across a cell membrane: the lipid bilayer and through integral membrane proteins. The plasma membrane has a high osmotic permeability, which is related to the high expression of water channels or aquaporins (AQPs). They are a family of integral proteins that facilitate water flow across cell membrane along an osmolality gradient. Aquaporins facilitate water transport by reducing the activation energy for water transport. For example, a liposome made of phosphatidylcholine, similar to the lipid bilayer, has an osmotic permeability of 97×10^{-4} cm/s. If AQP-1 molecules are added to the liposome (proteoliposome), the osmotic permeability increases to 540×10^{-4} cm/s. The addition of AQP-1 to the liposomes reduced the activation energy from 16 to 3.1 kcal/mol. Therefore, plasma membrane aquaporins function as conduits for water molecule passage without interacting with bilayer components, thus reducing energy activation. The water flow through an aquaporin is always passive; the force for the water flow is provided by an osmolality gradient.

In the kidney, several aquaporins are expressed in the renal tubule as well as blood vessels. In the plasma membrane, aquaporins exist as homotetramers; each monomer has its own water pathway. Table 5.1 shows the distribution and associated function of the most studied renal aquaporins.

Table 5.1 Distribution and associated function of aquaporins in the renal tubule and their associated function

Aquaporin	Tubular segment	Subcellular distribution	Associated function
AQP-1	Proximal tubule and thin descending limb of Henle	Apical and basolateral	Isosmotic water reabsorption in proximal tubule Concentration of urine in thin descending loop of Henle
AQP-2	Connecting tubule (connecting cells); cortical collecting duct (principal cells) and medullary collecting duct	Apical and subapical vesicles	AVP-dependent water reabsorption
AQP-3	Connecting tubule (connecting cells); cortical collecting duct (principal cells) and medullary collecting duct	Basolateral	AVP-independent basolateral exit of water molecules
AQP-4	Medullary collecting duct	Basolateral	AVP-independent basolateral exit of water molecules

The expression of aquaporins can be constitutive or regulated. A typical constitutively expressed aquaporin is AQP-1, always present in the brush border and basolateral membrane infoldings of proximal tubule cells. AQP-2 is the typical example of a regulated aquaporin. Antidiuretic hormone stimulates AQP-2 insertion in the apical membrane as well as its gene transcription.

The physiological relevance of aquaporins in water transport was clarified in studies performed in transgenic mice lacking aquaporins. Mice lacking AQP-1 have a moderate to severe defect in the urine-concentrating ability, as well as polydipsia and polyuria. In these mice, proximal water permeability is decreased by 80%, leading to a 50% decrease in proximal fluid reabsorption. Since AQP-1 is also expressed in thin descending limbs, water permeability of this segment is diminished by 90%. The lack of AQP-1 in proximal tubule and thin descending limb provoked a drastic reduction in water reabsorption, increasing the tubular flow to AVP-sensitive distal nephron segments. Thus, AQP-1 absence generates a nephrogenic diabetes insipidus condition.

The physiological role of AQP-2 in water reabsorption in the connecting tubule and collecting duct became clear in rodent models as well as inherited diseases. The Brattleboro rat was an excellent model for the study of regulation of AQP-2 by vasopressin. This rat strain has neurogenic or central diabetes insipidus (total or partial lack of vasopressin due to gene mutation). They have a severe polyuria, polydipsia, and a defect in its urine-concentrating ability. This AVP deficiency provoked a severe reduction in AQP-2 protein abundance. AVP infusion restored AQP-2 protein levels and the urine-concentrating ability. Nephrogenic diabetes insipidus is an inherited disorder presented in two forms. The X-linked form or vasopressin-resistant form is due to mutations with loss of function of the V_2 receptor. The lack of response of the cAMP-PKA signal cascade to vasopressin abolishes AQP-2 traffic and expression. The autosomal form of nephrogenic diabetes insipidus is related to mutations with loss of function in AQP-2. These mutations cause several types of abnormalities in the physiology of AQP-2 protein that blunts the exocytosis of aquaporin to the apical membrane. The net effect of these mutations is the inability of AQP-2 to carry on trafficking from subapical tubulovesicles to the apical membrane.

Studies in transgenic mice lacking AQP-3 showed that this channel plays a key role in the basolateral transport of water in the connecting and principal cells of the distal nephron and medullary collecting duct. Mice lacking this aquaporin were severely unable to concentrate urine, excreting abundant and diluted urine. In summary, naturally occurring models like the Brattleboro rat, inherited nephrogenic diabetes insipidus models, and transgenic rodent models show that several aquaporins play a crucial role in renal transepithelial water reabsorption and in the urine-concentrating ability.

The flow of water that occurs through the cell membrane is closely related to cell volume. Under normal conditions, the cell volume remains constant, which means that the net flow of water through the membrane is *zero* and therefore the cells are in osmotic equilibrium.

As mentioned above the osmolality of a solution can alter the cellular volume in two ways: it can increase or decrease the volume. Suppose a number of cells are bathed by an external solution formed by an effective osmol. If this external solution increases the cell volume, the external solution is said to be hypotonic. In this case, the osmolality of the external solution is lower than the intracellular osmolality. Hence, an osmolality gradient is built across the plasma membrane driving a rapid net water flow from the extracellular to intracellular compartment until osmotic equilibrium is reached. In parallel, the increase in cell volume is somehow a signal that activates "regulatory volume decreases mechanisms" that compensate for the inflow of water with the outflow of cellular osmoles like potassium and organic osmolytes that drive an outflow directed water flow. If the external solution decreases the cell volume, the solution is said to be hypertonic. In this case, the osmolality of the external solution is greater than the intracellular osmolality. Thus, the osmolality gradient pulls water out of the cell until osmotic equilibrium. The water outflow is counteracted by the activation of "regulatory volume increases mechanism"; in this case, the cell activates the inflow of solutes like Na^+ and Cl^- which drives water into the cell and compensates for the volume decrease. If the external solution does not change the volume of cells, the solution is isotonic. This is the case when the external osmolality and the intracellular osmolality are balanced across the membrane. The volume remains constant because the net flow across the membrane is *zero* (Fig. 5.2). The maintenance of cell volume is a key process in cell and body function. From a clinical point of view, the regulation of plasma osmolality allows the cell volume to be kept constant; this is especially important in cells contained in rigid cavities, such as the skull. In fact, the acute decrease or increase in cell volume produces symptoms ranging from lethargy, through seizures, to coma. The best known isotonic solution is the physiological saline. The effective osmol in this intravenous solution is NaCl, at a concentration of 0.9 % (w/v) or 154 mM. If 93% of the total solute is dissociated, the effective NaCl concentration would be 143 mM, which is within the range of normal plasma Na+ concentration (135–145 mEq/L).

5.3   Distribution of Water in the Body

The distribution of water in the organism is schematized in Fig. 5.3. In a healthy man (weighing 70 kg), the total body water is approximately 60% of the body weight (42 L). This volume is distributed between the intracellular and extracellular compartments. The intracellular fluid volume corresponds to two thirds (28 L) of the total body water and the extracellular volume corresponds to the remaining third (14 L). The barrier separating both compartments is the plasma membrane; its water permeability is very high due to aquaporin expression. This allows the rapid achievement of osmotic equilibrium between the extracellular and intracellular compartment. The volume of the extracellular compartment is distributed into two subcompartments. The intravascular compartment contains approximately one fourth of the extracellular volume (3.5 L). The interstitial compartment corresponds to three fourths of the extracellular volume (10.5 L). These two subcompartments are separated by the capillary endothelium, which is also permeable to water and electrolytes.

Fig. 5.2 General
mechanisms of cell volume
regulation. (**a**) Isotonicity. In
this condition the cell volume
is constant and the net water
flux (J net H$_2$O) is 0. In a
hypertonic solution, there is a
net water outflow from the
cell, reducing the cell volume.
In hypotonicity, there is a net
inflow of water to the cell
increasing cell volume. (**b**) In
response to changes in cell
volume, the cell responds with
volume regulatory
mechanisms that involve
solute transport across the
plasma membrane. Under
hypertonic conditions, there is
a net movement of solutes like
NaCl to the cell. This solute
influx creates an osmotic
gradient to drive osmosis to
the cell. In hypotonic
conditions, solutes like KCl
are transported out of the cell
along with an osmotic
water flow

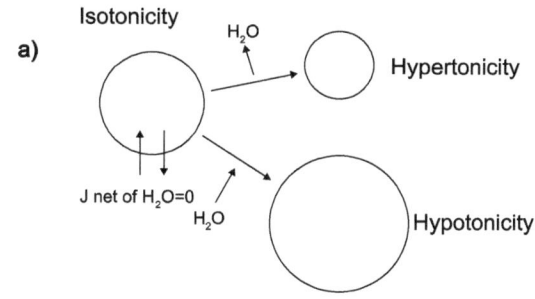

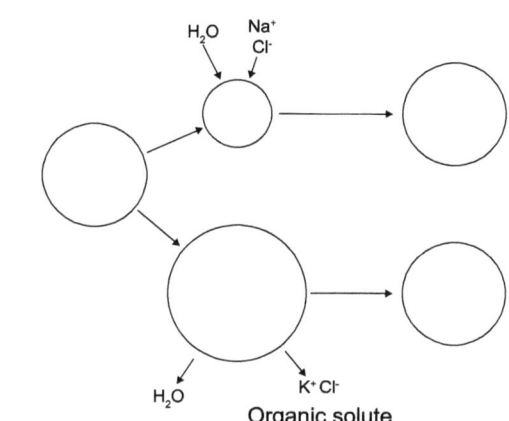

5.4   Whole Organism Water Balance

The water balance is given by an equality between the input and output of water from
the organism. Table 5.2 shows the magnitudes of the different flows of water inputs
and outputs in an adult male who is kept at an ambient temperature of 23 °C. The
main routes of water entry into the body are drinking, water derived from food
metabolism, and water contained in meals. Water intake as such is highly variable
and depends on factors such as age or physical activity and also has a strong social
component. The main physiological determinant of water intake is thirst.

The most important route of water loss is urine and is approximately 1.5 L/d, but
diuresis (volume of urine per unit time) can vary considerably depending on the state
of the water balance. Thus, under conditions tending towards a positive balance
(e.g., water intake), the volume of urine can be as abundant as 20 L/d with an
osmolality of 50 mOsm/kg; in a negative balance (e.g., restriction on water intake),
diuresis can be reduced to 0.5 L/d and the maximum urinary osmolality will be
1200 mOsm/kg. This 500 mL/d is fixed and is the minimum volume needed to
excrete a daily amount of 700 mOsm in solutes such as urea, sodium and potassium

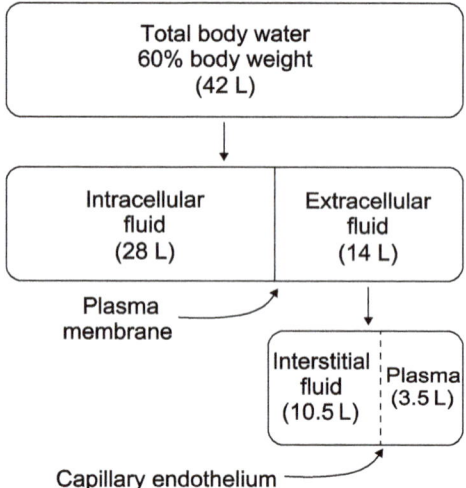

Fig. 5.3 Volume of aqueous compartments in a healthy man of 70 kg. Total body water in an adult healthy male is about 60% of the body weight. This total body water is part of the intracellular compartment (28 L) and extracellular compartment (14 L); these compartments are separated by the plasma membrane which is very permeable to water due to high expression of aquaporins. The extracellular compartment is further divided into the interstitial compartment which contains the water that hydrates the extracellular matrix of the connective tissue and the plasma volume. These two compartments are separated by the capillary endothelium, which is also permeable to water

Table 5.2 Water balance in a normal subject, maintained at 23 °C

Water input	(mL/24 h)
Drinking	1200
In food (variable)[a]	1000
Derived from metabolism	300
Total inputs	**2500**
Water output	(mL/24 h)
Insensitive (skin and respiratory)	700
Perspiration	100
Fecal	200
Urinary (variable)	1500
Total outputs	**2500**

[a]The intake of water as a fluid is highly variable and depends on cultural and social factors

salts, NH_4^+, uric acid, creatinine, and phosphate. The loss of water through breathing and evaporation from the body surface is called insensible losses, since the individual is not aware of their occurrence. The loss of water through perspiration is variable, depending, for example, on the ambient temperature and physical activity. Fecal water loss can increase significantly in conditions such as diarrhea. Table 5.2 exemplifies the inputs and outputs in a normal and healthy subject maintained at 23 °C.

5.5 Determinants of Effective Plasma Osmolality

The effective osmolality of the extracellular fluid or tonicity is determined by the total concentration of effective osmoles remaining in this compartment; the main extracellular solute of effective osmole is Na^+ and its accompanying anions: Cl^- and HCO_3^-. The Na^+ ions are restricted to the extracellular medium for two reasons: first, the plasma membrane has a very low permeability to Na^+ ions and second, the activity of the Na^+, K^+-ATPase that pumps $3Na^+$ ions out of the cell in exchange for $2 K^+$ ions. On the other hand, urea is an ineffective osmole. The cell membrane has a high permeability to urea due to the expression of urea transporters and therefore urea can easily enter the cells. For this reason, the contribution of urea to plasma osmolality is low. Glucose is another solute that makes a negligible contribution to plasma osmolality. Under physiological conditions, normal glycemia contributes only 5% to total osmolality. However, in diabetic patients with very high glycemia, glucose behaves as an effective osmolyte, making an important contribution to plasma osmolality. Considering the Na^+ and glucose, plasma osmolality can be calculated as expressed in Eq. (5.4):

$$\text{Effective plasma osmolality} \left(\frac{\text{mOsm}}{\text{kg}}\right) = [Na^+]pl \cdot 2 + \frac{\text{Glycemia} \left(\frac{\text{mg}}{\text{dL}}\right)}{18} \quad (5.4)$$

If plasma Na^+ concentration is expressed in mM and glycemia in mg/dL, plasma osmolality is expressed in mOsm/kg. Considering that the fasting blood glucose has a normal range of 70–100 mg/dL, the contribution of glucose to plasma osmolality is only 4–5 mOsm/kg and therefore can be neglected. In a diabetic subject, the entry of glucose into the cells is diminished by the lack or resistance to insulin; in these conditions glycemia increases and makes a greater contribution to effective plasma osmolality. Therefore, under physiological conditions, the effective plasma osmolality can be calculated as expressed in Eq. (5.5), which is a restatement of Eq. (5.4).

$$\text{Effective plasma osmolality} = [Na^+] \cdot 2 \quad (5.5)$$

If the concentration of Na^+ in the plasma is 143 mEq/L (normal range 135–145 mEq/L) and Eq. (5.5) applies, the plasma osmolality is 286 mOsm/kg. Therefore, alterations in the water balance will be reflected as changes in plasma sodium concentration, since this is the main determinant of effective plasma osmolality.

5.6 Biology of Arginine-Vasopressin

As mentioned above, the physiological relationship between plasma osmolality and renal water excretion is due to changes in the secretion of vasopressin or antidiuretic hormone (AVP or ADH).

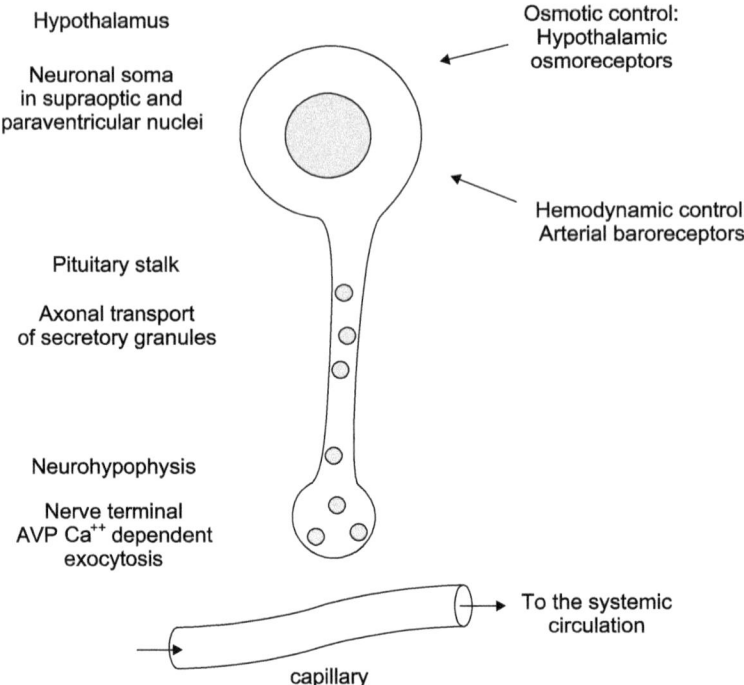

Fig. 5.4 Scheme of a vasopressin-secreting neuron. These neurons are located in the hypothalamic supraoptic and paraventricular nuclei. Preprohormone synthesis and processing occur in the neuronal soma; the peptide is packed in secretory granules that are transported through axons to the neurohypophysis where Ca^{++}-dependent secretion occurs. Vasopressin secretion is primarily stimulated by afferents from osmoreceptor cells mainly located in some circumventricular organs. Secondarily, hormone secretion can be stimulated by hypotension or volume contraction

AVP is synthesized in the soma of hypothalamic magnocellular neurons of the paraventricular and supraoptic nuclei. These neurons have long axons that travel within the pituitary stalk; their nerve terminals are intimately associated with fenestrated capillaries in the neurohypophysis (Fig. 5.4). AVP is synthesized as part of a prohormone in the rugged endoplasmic reticulum; the prohormone is packaged in vesicles that then pass to the Golgi complex, where it undergoes post-translational modifications, generating the AVP nonapeptide and two other peptides (neurophysins). Axonal transport mobilizes the secretory granules that travel through the axons and accumulate in the nerve terminals.

The most important physiological stimulus for AVP secretion is the increase in plasma osmolality. Small increases (1–2%) in plasma osmolality stimulate hormone secretion (Fig. 5.5a): such stimulation comes from hypothalamic osmoreceptors. These structures are osmolality-sensitive cells located in two specific circumventricular organs: subfornical organ and the organum vasculosum of lamina terminalis. These structures have ependymal cells that function as sensors of extra-cellular osmolality. They signal to vasopressin-secreting neurons as well as to other

Fig. 5.5 Regulation of vasopressin secretion. (**a**) Plasma AVP concentration plotted as a function of plasma osmolality. The osmolality threshold for vasopressin secretion is nearly 280–285 mOsm/kg; from this range there is a linear relationship between plasma osmolality and plasma AVP concentration. (**b**) Plasma AVP concentration is also increased by reduction over 10% of extracellular volume

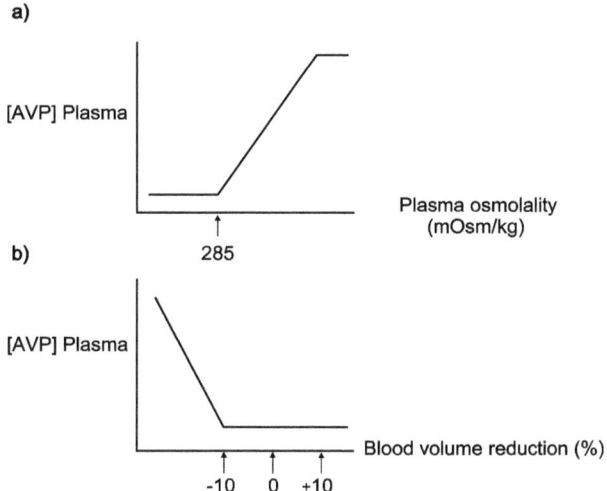

structures involved in thirst. A second stimulus for AVP secretion is a 10% or higher reduction in circulating volume and blood pressure (Fig. 5.5b). This hemodynamic or non-osmotic control of AVP secretion comes from cardiovascular control areas located in the medulla of the brainstem. There are certain non-osmotic stimuli for AVP secretion that are relevant in clinical practice, such as pain, severe stress, and some drugs such as opioids.

The effects of AVP are mediated by G-protein-coupled receptors. There are two general types of receptors: V_1 receptors are coupled to phospholipase C through Gq, with the consequent increase in cytosolic Ca^{++} and activation of protein kinase C. V_1 type receptors are expressed in the vascular smooth muscle, liver, and corticotrophs, where they enhance CRH-dependent ACTH secretion. The second type is the V_2 receptor coupled through Gs to adenylyl cyclase, hence stimulating cyclic AMP synthesis (cAMP) and protein kinase A (PKA) activation. V_2 receptors are expressed on the basolateral membrane of tubular cells like those of the thick ascending limb, connecting and principal cells of the distal nephron, and medullary collecting duct. The renal effects of AVP linked to water balance are mediated by the activation of V_2 receptors in the abovementioned cell types. These effects can be described in two important points:

1. The antidiuretic effect itself, which occurs in the connecting cells of the connecting tubule and principal cells collecting duct (Fig. 5.6). The effect is a dramatic increase in the apical membrane osmotic permeability. The effect can be divided into two components: one short term and one long term. In the short term, AVP increases the osmotic permeability of the apical membrane by promoting the exocytosis of subapical cytosolic vesicles bearing AQP-2 into the apical membrane. In the absence of AVP (Fig. 5.6a) water permeability is very low; the tubulovesicles reside in the subapical cytosol and the osmotic permeability of the membrane is very low. In the presence of AVP (Fig. 5.6b), mobilization of

Fig. 5.6 Regulation of AQP-2 intracellular traffic by vasopressin. (**a**) Low plasma AVP renders a water impermeable apical membrane in connecting and principal cells. (**b**) High plasma AVP stimulates AQP-2 insertion into the apical membrane, increasing its osmotic permeability

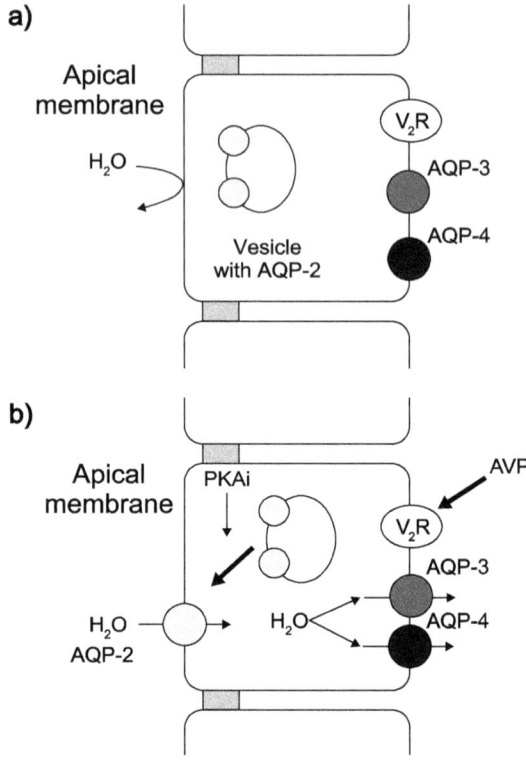

AQP-2 to the apical membrane occurs, resulting in a rapid increase in osmotic permeability. This mechanism of intracellular trafficking of AQP-2 between the membrane and the subapical cytosol requires PKA-dependent phosphorylation of AQP-2. In the long-term effect, AVP stimulates transcription of the gene coding for AQP-2. This effect is mediated by PKA, which phosphorylates a transcription factor (CREB-P) that binds to specific sequences (CRE) in the promoter region of the AQP-2 gene. AVP is the main hormone promoting AQP-2 transcription, thus maintaining basal AQP-2 protein synthesis. The result of this effect is an increased availability of AQP-2 protein for the short-term mechanism. The water molecules that entered via AQP-2 in the apical membrane exit via basolateral aquaporins: AQP-3 (connecting and principal cells in the cortical collecting duct) and AQP-3 and AQP-4, in the basolateral membrane of inner medullary collecting duct cells. The abundance of AQP-3 is also regulated by AVP, although the mechanism is not completely understood.

2. In the inner medullary collecting duct, AVP stimulates urea reabsorption mediated by an apical passive transporter (UT1), using the cAMP-PKA system. AVP stimulates the activity of the transporter and thus the reabsorption of urea. The basolateral urea efflux occurs via the passive transporter UT4. The result is an increase in the reabsorption and recirculation of urea, which is necessary to

maintain a hypertonic medullary interstitium, which is in turn necessary for the reabsorption of vasopressin-dependent water.

There are two pathologies of scarce occurrence that are associated with the mechanism of action of AVP. In central diabetes insipidus there is no production of AVP and therefore subjects with this condition are unable to concentrate their urine. The Brattleboro rat is a natural model of central diabetes insipidus. This rat strain lacks vasopressin. Consequently, immunolocalization studies demonstrated that AQP-2 immunoreactivity is absent in collecting ducts, which correlates with the polyuria exhibited. Chronic administration of a V_2 receptor agonist restored AQP-2 levels and reduced the polyuria. There are two forms of nephrogenic diabetes insipidus: one X-linked and one autonomic. The sex-linked form is due to mutations in the V_2 AVP basolateral; there is no activation of the cAMP-PKA transduction system. The autonomic form is due to mutations with loss of function of AQP-2: generally, these mutations alter the post-translational traffic of AQP-2. There are no known cases of mutations in AQP-3 or AQP-4. Some conditions, such as lithium salt treatment and hypokalemia, are capable of inducing nephrogenic diabetes insipidus associated with a reduction in mRNA and AQP-2 protein abundance, which is reversible. In both cases, polyuria occurs.

5.7   Regulation of Plasma Osmolality

Regulation of plasma osmolality occurs by a negative feedback mechanism (Fig. 5.7). The increase in plasma osmolality (e.g., restriction of water intake for several hours or severe sweating) is detected by hypothalamic osmoreceptors, which in turn stimulate vasopressin secretion; furthermore, the same increase in osmolality triggers the sensation of thirst. Thirst-stimulated water intake and renal reabsorption of vasopressin-dependent water are the two mechanisms that restore plasma osmolality to the normal range by regulating inflow and outflow, respectively, which is detected by the osmoreceptors, ceasing the sensation of thirst and stimulating vasopressin secretion (Fig. 5.7). In contrast, an overload of water (Fig. 5.8) (example: intake of 1 L of water) by a normal subject will reduce the sensation of thirst, plasma osmolality and vasopressin secretion, allowing the kidney to excrete in a period of 2–3 h a volume of urine equivalent to the volume of water ingested. This net loss of water will allow the plasma osmolality to return to the normal range.

5.8   Mechanism of Urinary Concentration and Dilution

The mammalian kidney has the ability to adjust water excretion according to vasopressin levels. This unique ability of the kidney has allowed mammals to colonize semi-desert and desert environments with limited free water availability. Examples of adaptation to life under these conditions are several species of rodents

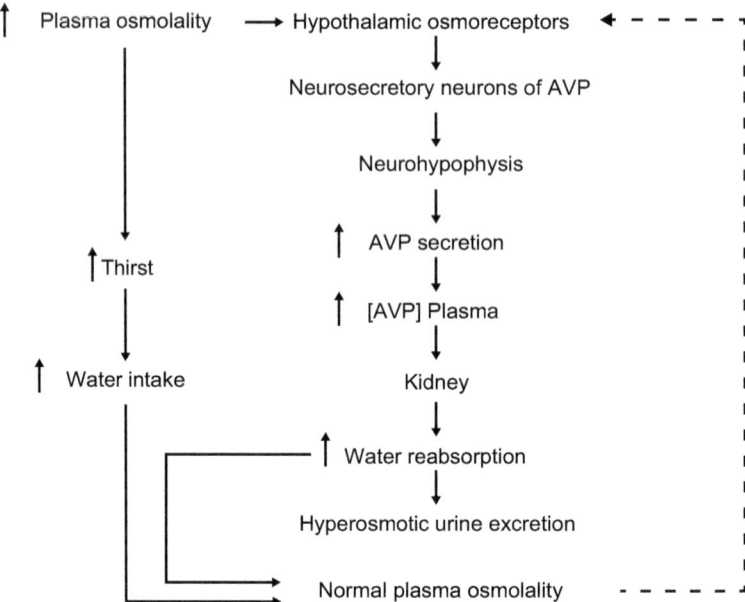

Fig. 5.7 Feedback regulation of plasma osmolality. An increase in plasma osmolality, detected by osmoreceptor cells in circumventricular organs, increases vasopressin secretion and thirst. AVP stimulates renal water reabsorption. Thirst drives water-seeking behavior and drinking. The return of plasma osmolality to normal range is detected by osmoreceptors and inhibits vasopressin secretion

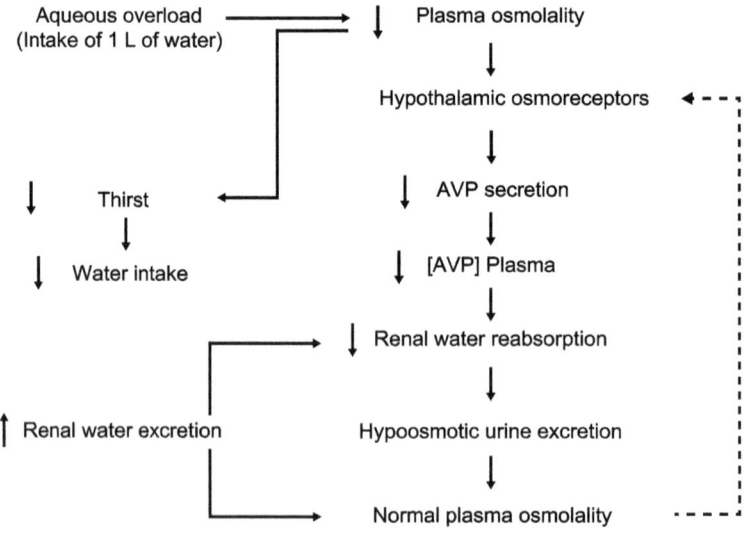

Fig. 5.8 Regulation of plasma osmolality in the presence of a water load. A water load (intake of 1 L of water) decreases plasma osmolality and plasma AVP secretion. Renal water reabsorption is decreased and also water ingestion; both actions will contribute to the return of plasma osmolality to a normal range

whose kidneys are capable of concentrating urine up to 6000 mOsm/kg, compared to 900–1200 mOsm/kg in humans.

As a result of the intake of a typical Western diet, the human organism is forced to excrete an average daily load of 700 mOsm in total solutes, consisting mainly of sodium, potassium salts, and nitrogenous organic compounds (urea, uric acid, creatinine). These total solutes can be excreted as a hyperosmotic urine with a maximum osmolality of 1200 mOsm/kg and a diuresis of almost 0.6 L/d, or as hypoosmotic urine whose minimum osmolality is 80 mOsm/kg and a volume of almost 9 L/d. Both values refer to extreme urine concentration and dilution for humans. In most daily conditions, urine osmolality is in fact hypertonic to plasma; if diuresis is 1 L/d and 700 mOsm has to be excreted, then the urine osmolality for a normal diuresis will be 700 mOsm/kg. Therefore, urine osmolality varies inversely with urine flow.

This ability of the kidney to adjust water excretion according to plasma osmolality depends on anatomical and functional factors. Anatomical factors are directly related to the architecture of the nephron and the postglomerular medullary circulation.

1. The loop of Henle has a hairpin configuration (Fig. 5.9) and consists of three subsegments: the thin descending, thin ascending, and thick ascending limb. In juxtamedullary nephrons, the thin descending limb enters the inner medulla making a hairpin, and the thin ascending limb projects to the inner stripe of the outer medulla forming the thick ascending limb that ends past the macula densa. Remember that in cortical nephrons the thin ascending limb is very short or absent. In this nephron type, the thin descending limb makes the hairpin in the inner stripe of the outer medulla and gives rise to the thick ascending limb. The net effect of this hairpin is that tubular fluid flow is organized in a countercurrent fashion. Tubular fluid entering from the proximal tubule descends through the thin descending limb towards deeper portions of the inner medulla and then ascends in the thin and thick ascending limbs to the cortex. The tubular fluid then descends again when entering the collecting duct.
2. The capillaries that form the vasa recta consist of descending and ascending capillaries, and, consequently, the blood flow is also organized in a countercurrent fashion. These capillaries travel along with loop of Henle and also make a hairpin in the inner medulla and project to outer portions of the medulla (Fig. 5.9).

The functional factors are related to the segment-specific expression of a series of Na^+-dependent cotransporters, ion channels, aquaporins, and urea transporters that give to a particular tubular segment unique property of permeability to NaCl, urea, and water (Fig. 5.10).

1. The thin descending limb of Henle is impermeable to NaCl and very permeable to water and urea. The high osmotic permeability is due to the abundant expression of apical and basolateral AQP-1. There is also urea secretion mediated by UT2 urea transporter.

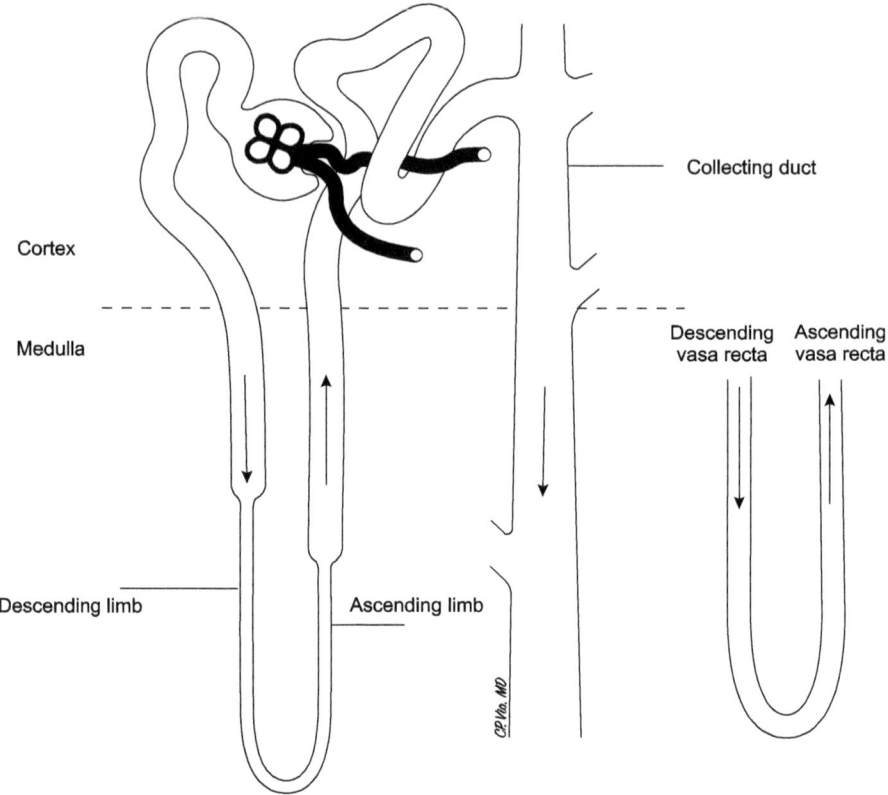

Fig. 5.9 Countercurrent organization of the tubular flow in the loop of Henle and collecting duct. Tubular flow descends through the thin descending limb and ascends through the thin and thick ascending limb and again descends in the collecting duct. Blood flow in the vasa recta capillaries is also arranged in a countercurrent fashion

2. The thin ascending limb of Henle reabsorbs NaCl through an unclear mechanism. This segment is impermeable to water and thus contributes to the dilution of the tubular fluid. The thick ascending limb has a medullary and cortical portion. This segment reabsorbs NaCl and is impermeable to water. NaCl reabsorption occurs through the apical NKCC2 cotransporter, and the basolateral ClC-K chloride channel, and Na^+, K^+-ATPase. The mechanism of reabsorption and all the transporters and channels involved were described in detail in Chap. 4. The activity of the NKCC2 cotransporter is essential to generate the hyperosmolality of the medullary interstitium surrounding the loop of Henle. In turn, the medullary hypertonicity is essential for AVP-dependent water reabsorption along the medullary collecting duct. This NaCl transport is the active process in the mechanism of urinary concentration and dilution. The entire ascending limb is impermeable to water and the reabsorption of NaCl results in the dilution of the tubular liquid. The dilution of the tubular fluid that enters the connecting tubule is

essential because it contributes to the osmotic gradient that drives water reabsorption into the cortical and medullary interstitium.

3. The distal convoluted tubule is also impermeable to water and reabsorbs NaCl through the apical NCC cotransporter. The mechanism of NaCl reabsorption was described in Chap. 4. Therefore, the distal convoluted tubule also contributes to the dilution of the tubular fluid. The tubular fluid leaving the distal convoluted tubule has an osmolality of about 100 mOsm/kg.

4. The connecting tubule and cortical collecting duct reabsorb NaCl via the Na^+ (ENaC) and are impermeable to water, except in the presence of AVP. The complete medullary collecting duct is permeable to water in the presence in AVP.

5. The cortical and outer medullary collecting ducts are always impermeable to urea. The vasopressin-dependent water permeability of these segments plus their urea impermeability contributes to increased urea concentration in the tubular fluid of the cortical and outer medullary collecting duct. The inner medullary collecting duct reabsorbs urea by means of a passive transporter (UT1), whose activity is stimulated by AVP.

5.8.1 Hypertonic Urine Formation

Hypertonic urine formation occurs under physiological conditions and when the body needs to conserve water. The generation of hypertonic urine requires the action of vasopressin or antidiuretic hormone.

Measurements of osmolality in the cortical and medullary renal interstitium of several mammalian species such as rodents, rabbits and primates showed that the cortical interstitium is isosmotic with respect to plasma. However, the osmolality of the medullary interstitium increases from the corticomedullary junction to the deeper parts of the inner medulla. NaCl and urea contribute almost in equal proportion to the hyperosmolality of the renal medullary interstitium. This kind of study has allowed the conclusion that in the kidney there is a gradient of interstitial osmolality that increases progressively from the corticomedullary junction (isosmotic, e.g., 285 mOsm/kg) towards deeper portions of the renal medulla (hyperosmotic, 1200 mOsm/kg). In humans, the maximum osmolality that can be reached is in deeper portions of the medullary interstitium and is around 900–1200 mOsm/kg, but in rodents from xeric environments this value can reach 6000–9000 mOsm/kg. This medullary interstitial hypertonicity is due to the active transport of NaCl that occurs in the thick ascending loop of Henle. Approximately 50% of the interstitial osmolality (1200 mOsm/kg) corresponds to NaCl and the rest to urea.

1. The generation of hyperosmotic urine can be understood by following the path of the tubular fluid through the different segments, bearing in mind its functional properties (Fig. 5.10). Proximal tubule reabsorption occurs in near isosmotic conditions; therefore, the tubular fluid entering the thin descending limb has the same osmolality as plasma (e.g., 285 mOsm/kg). In the thin descending limb (Fig. 5.11), the osmolality of the tubular fluid increases progressively as the fluid

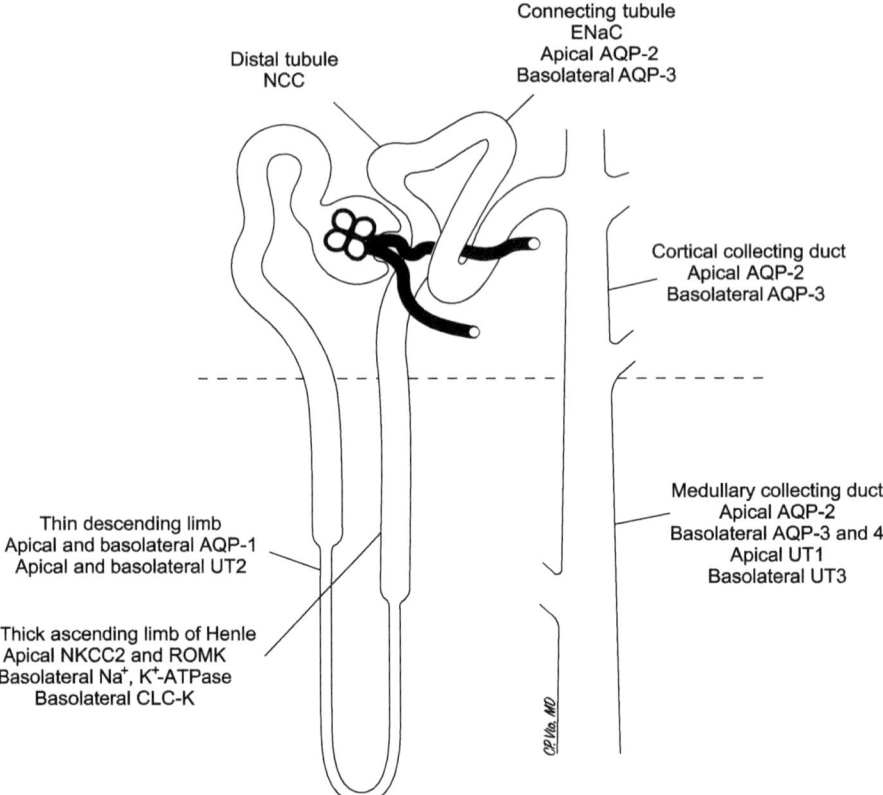

Fig. 5.10 Transporters, ion channels, and aquaporins involved in the concentrating and diluting mechanism. Thin descending limb: apical and basolateral AQP-1 and UT2 urea transporter. Thick ascending limb: apical NKCC2 and ROMK; basolateral Na⁺, K⁺-ATPase, CLC-K. Distal convoluted tubule: apical NCC. Connecting tubule and cortical collecting duct: apical AQP-2, basolateral AQP-3. The epithelial sodium channel (ENaC) is expressed in the apical membrane of connecting and principal cells. Medullary collecting duct: apical AQP-2 and UT1 urea transporter; basolateral AQP3, AQP-4, and UT3 urea transporter

flows to the deeper portions of this segment. This is because the thin descending limb is permeable to water and impermeable to NaCl. Water reabsorption in the descending thin limb (Fig. 5.11) occurs independently of the excretion of hyper- or hypoosmotic urine and is therefore common to both situations.

2. Tubular fluid osmolality is balanced with the hypertonicity of the medullary interstitium. The maximum osmolality can be reached by the tubular liquid where the bend of the thin limb is close to 1200 mOsm/kg. This value is typical of juxtamedullary nephrons and it is variable in cortical nephrons, depending on the length of the thin descending limb. Therefore, the abundance of juxtamedullary nephrons is one of the factors important to produce a hypertonic urine.

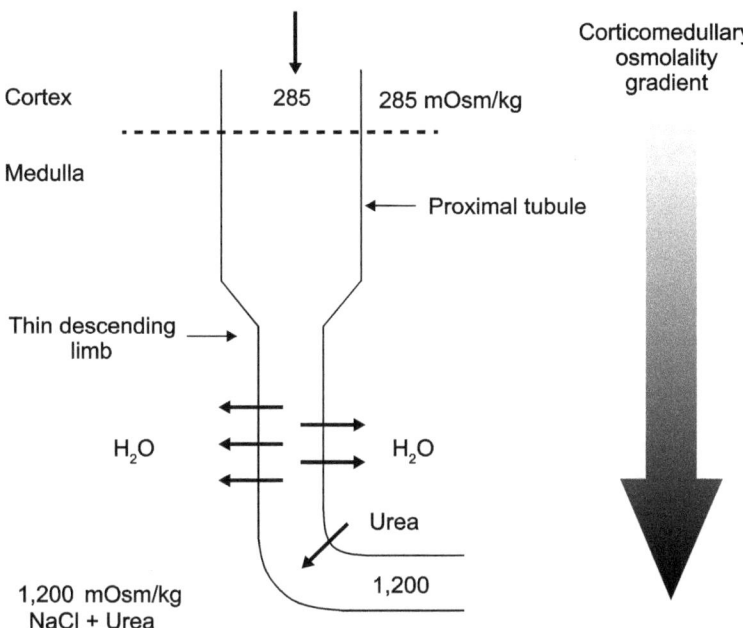

Fig. 5.11 Water reabsorption in the thin descending limb of Henle. The osmolality gradient established from the cortex to the inner medulla is the driving force for AQP-1-mediated water reabsorption as the tubular fluid descends towards the hairpin turn. Urea is secreted through apical and basolateral UT2

3. The thin and thick ascending limbs reabsorb NaCl and are impermeable to water (Fig. 5.12). The hyperosmotic fluid (800–1200 mOsm/kg) that arrives from the descending thin limb is progressively diluted as it enters the thick ascending limb on its way to the cortex, reaching about 150 mOsm/kg at the beginning of the distal convoluted tubule. The distal convoluted tubule reabsorbs NaCl and is impermeable to water. Hence, tubular fluid dilution initiated in the thin ascending limbs continues through the distal convoluted tubule. The osmolality of the tubular fluid exiting this segment is approximately 100 mOsm/kg (Fig. 5.12). NaCl reabsorption of the thick ascending limb and distal convoluted tubule occurs independently of hyper- or hypoosmotic urine excretion and is therefore a process common to both situations. However, NaCl reabsorption in the thick ascending limb is stimulated by vasopressin through V_2 receptor activation, which results in increased NaCl reabsorption.
4. The cortical collecting duct (and connecting tubule) is permeable to water only in the presence of vasopressin (Fig. 5.13). The tubular fluid entering the connecting tubule has an osmolality of about 100 mOsm/kg. In the presence of vasopressin, water is reabsorbed and the osmolality of the tubular fluid equilibrates with the cortical interstitium (285 mOsm/kg). As the tubular fluid enters to the renal medulla, its osmolality increases. The connecting tubule, cortical collecting duct, and outer medullary collecting duct are impermeable to urea. Water

Fig. 5.12 Dilution of hypertonic tubular fluid in the thin and thick ascending limb of Henle. The hypertonic tubular fluid coming from the thin descending limbs is diluted as it flows into the cortex. NKCC2-mediated NaCl reabsorption in the thick ascending limb is the key step for tubular fluid dilution. Further dilution occurs in the distal convoluted tubule through NCC-mediated NaCl reabsorption. Recall that the thick ascending limb and the distal convoluted tubule are always impermeable to water. Fluid entering the thin ascending limb has a maximal osmolality of 1200 mOsm/kg and the fluid leaving the distal convoluted tubule has an approximate osmolality of 100 mOsm/kg

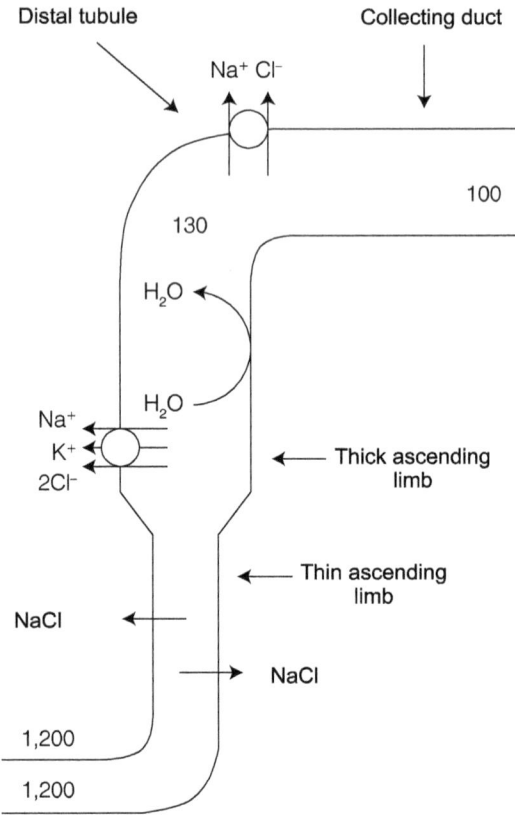

reabsorption without urea in these segments allows the concentration of urea in the tubular fluid, an essential step for passive urea reabsorption in the inner medullary collecting duct. In the inner medullary collecting duct, AVP stimulates water and urea reabsorption. The final osmolality of the tubular fluid can reach values ranging from 800 to 1200 mOsm/kg in approximately 0.5 to 0.6 L of urine per day (Fig. 5.13). The net result of the whole process is the excretion of a reduced volume (0.5–0.6 L) of hyperosmotic urine in relation to plasma. The achieved urinary osmolality and the volume of urine will depend on the plasma levels of vasopressin.

5.8.2 Hypoosmotic Urine Formation

Hypoosmotic urine excretion (Fig. 5.14) requires low circulating levels or absence of vasopressin. In this setting, the connecting and collecting ducts are impermeable to water. On the other hand, the maximum osmolality of the medullary interstitium is low under water diuresis conditions (~600 mOsm/kg) than under antidiuresis conditions (~1200 mOsm/kg), probably related to a higher medullary blood flow

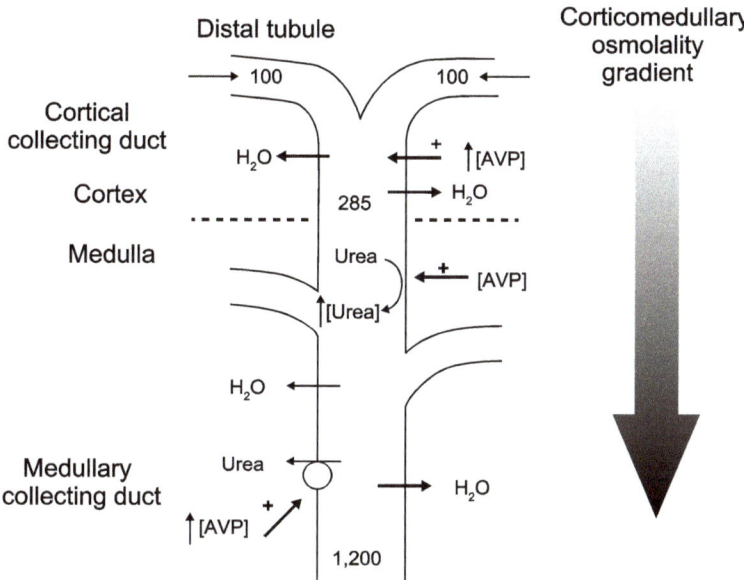

Fig. 5.13 Urine concentration in the collecting duct. This segment reabsorbs water in the presence of vasopressin. The cortical collecting duct is impermeable to urea under all circumstances; water reabsorption increases tubular urea concentration, creating the chemical gradient for vasopressin-stimulated urea reabsorption in the medullary collecting duct

Fig. 5.14 Urine dilution in the collecting duct. In the absence of vasopressin, the complete collecting duct is impermeable to water. Also, vasopressin-stimulated urea reabsorption is diminished. The minimal value for urine osmolality in humans is approximately 50 mOsm/kg

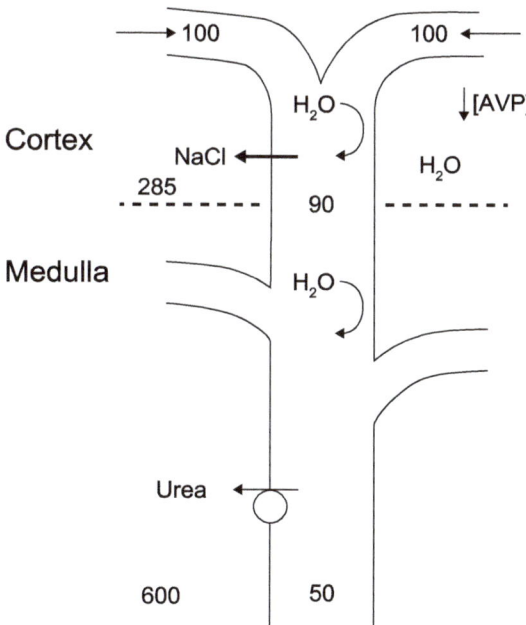

under aqueous diuresis conditions. The lower osmotic gradient that is generated between the descending thin loop and the medullary interstitium makes the tubular fluid, leaving the thin limb less concentrated than in antidiuresis. As mentioned above, reabsorption of water in the thin descending limb and of NaCl in the ascending limb and distal convoluted tubule occurs even when the urine to be excreted is hypoosmotic. The tubular fluid entering the connecting tubule has an osmolality of 100 mOsm/kg. In this setting, the connecting tubule and collecting duct are impermeable to water due to the absence of vasopressin. In these conditions, the osmotic permeability of the luminal membrane is very low because the AQP-2 molecules are absent from the apical membrane. Despite the existence of an osmotic gradient, there is practically no reabsorption of water and the urine that is excreted is abundant and diluted (Fig. 5.14).

5.8.3 Role of Vasa Recta Capillaries

The capillaries of the vasa recta are the passive element of the mechanism of concentration and urinary dilution. Through the endothelium occurs the diffusion of solutes and water (Fig. 5.15). The plasma flowing through the descending capillaries of the vasa recta vessels progressively increases its osmolality due to the exit of water and the entry of solutes, such as NaCl and urea. Maximum osmolality is reached at the bend of the descending capillaries into ascending capillaries. In the ascending capillaries, water enters the capillaries while the solutes diffuse out of the capillaries. The countercurrent arrangement between the ascending and descending capillaries prevents the dissipation of the corticomedullary osmolality gradient and maintains a stationary state between the entrance and exit of solutes and water from the medullary interstitium.

5.9 Renal Handling of Urea

Approximately 50% of the concentration of solutes in the medullary interstitium corresponds to urea, hence its importance in the mechanism of concentration and urinary dilution. Urea is subject to a complex renal handling that includes glomerular filtration, reabsorption, and tubular secretion, as well as recirculation between tubules and blood vessels (Fig. 5.16). Urea filters freely at the glomerular filtration barrier; 50% of the filtered load is reabsorbed in the proximal tubule through a passive mechanism. The remaining 50% of the filtered urea enters the thin descending limb. In this segment, urea secretion is mediated by UT2 transporter which is expressed in apical and basolateral membrane of epithelial cells of the thin descending limb. The activity of this transporter seems to be vasopressin-independent. Urea secretion into the thin descending limbs increases tubular urea concentration, such that at the distal convoluted tubule, the delivery of urea is similar to its filtered load. In the inner medullary collecting duct, the reabsorption of urea (50–60% of the filtered load) is mediated by the UT1 transporter. This transporter is

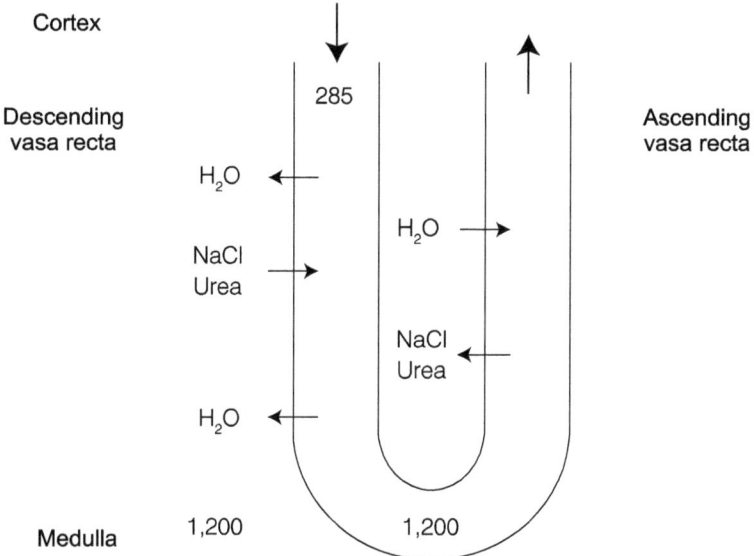

Fig. 5.15 Countercurrent exchange of NaCl, urea, and water in the vasa recta capillaries. Descending blood flow faces a hyperosmotic medullary interstitium that drives an osmotic flow; urea and NaCl diffuse into the descending capillaries. At the hairpin of the capillaries, blood and interstitium osmolality are equilibrated. NaCl and urea diffuse out of the ascending capillaries while water moves into the hypertonic blood

expressed in the apical membrane of inner medullary collecting duct cells. UT1 maximal transport activity is increased by vasopressin through V_2 receptor activation. The increase in UT1 maximal velocity seems to be associated with vasopressin-dependent phosphorylation of the transporter. Basolateral urea efflux is mediated through the UT3 transporter, whose activity is also increased by vasopressin. Urea reabsorption in the inner medullary collecting duct contributes with urea to two processes: first, to achieve a high urea concentration in the medullary interstitium and second with urea to the secretory process in the thin descending limb. Therefore, there is a urea recycling process between the inner medullary collecting duct and thin descending limb of Henle. As a result, the excreted urea load is approximately 50% of the filtered load in the presence of vasopressin.

5.10   Quantification of the Kidney's Ability to Concentrate and Dilute Urine

The kidney's ability to concentrate or dilute urine can be assessed by clearance of free water, which represents the kidney's ability to generate water free from solutes. The clearance of the total solutes excreted in the urine is called clearance osmolal (Cosm), which is calculated as shown in Eq. (5.6):

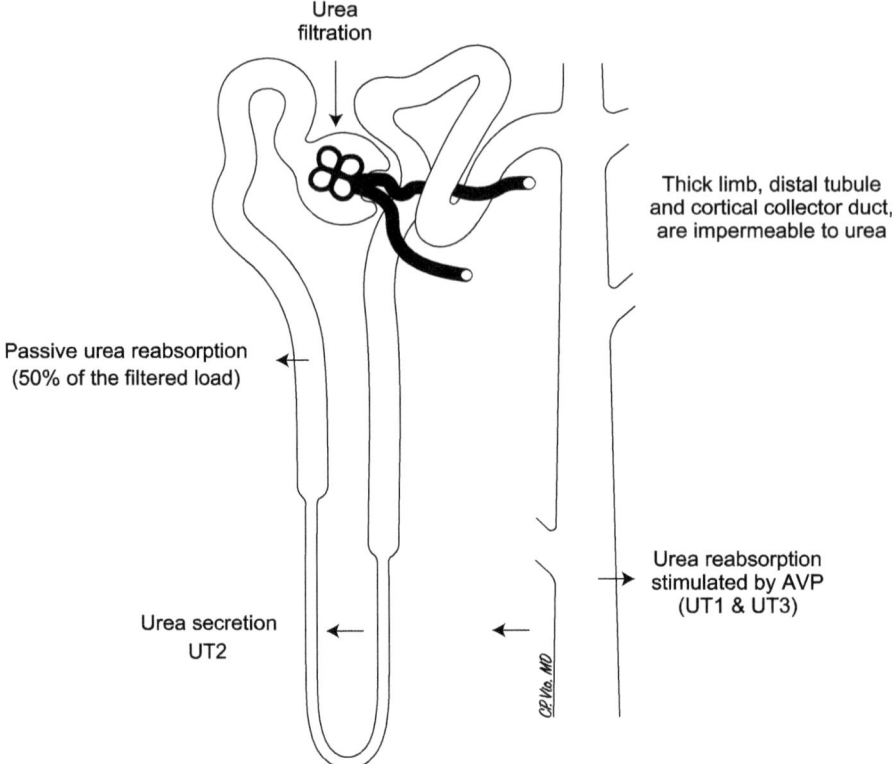

Fig. 5.16 Tubular handling of urea. Urea is freely filtered in the glomerular barrier; 50% of the filtered load is reabsorbed passively in the proximal tubule. Urea is secreted into the lumen of the thin descending limbs. The complete ascending limb, distal convoluted, connecting tubule, and cortical collecting duct are always impermeable to urea. Urea is reabsorbed in the inner medullary collecting duct; this process is stimulated by vasopressin. UT2, UT3, and UT4 are tubular urea transporters

$$Cosm = \frac{Uosm \cdot V}{Posm} \qquad (5.6)$$

where Uosm and Posm are the urinary and plasma osmolalities, respectively. V is the urinary flow. The clearance of free water (C_{H2O}) corresponds to:

$$CH_2O = V - Cosm \qquad (5.7)$$

Equation (5.7) can be rearranged by clearing in urine flow (V):

$$V = CH_2O + Cosm \qquad (5.8)$$

It is possible to consider that the total volume of urine has two components. One contains the solutes to be excreted at the same osmolality as the plasma and is defined by the osmolal clearance. The second component corresponds to water free of solutes or clearance of free water.

If the osmolal clearance is the same as the free water clearance, the generation of solute- free water will be *zero*. On the other hand, a negative free water clearance is associated with solute-free water conservation, which is associated with the antidiuretic action of vasopressin. A positive free-water clearance is associated with solute-free water excretion which is associated with low vasopressin levels.

5.11 Survival of Cells in High-Salinity Environments

The renal medulla is a hypertonic environment given by the high concentration of NaCl and urea. Both factors are detrimental to normal cell function because a high concentration of salt favors the breakdown of genomic DNA and urea promotes the denaturation of proteins and thus the loss of their function, leading to cell death. Cellular function is preserved in part by the cellular accumulation of organic osmolytes, such as myoinositol, betaine, sorbitol, and glycerophosphocholine (GPC). The accumulation results from increased transporter and enzyme activity. Myoinositol and betaine are transported to the cytoplasm via Na^+ -dependent cotransporters (SMIT and BGT1, respectively). The accumulation of sorbitol derives from an increase in the activity of the enzyme aldose reductase (AR) that catalyzes the synthesis of sorbitol from glucose. The accumulation of GPC is due to a reduction in its degradation.

The transcription factor TonEBP (tonicity-enhanced binding protein) plays a key role in cellular adaptation to osmotic stress. Hypertonicity has two important effects on this protein: first, it stimulates its nuclear translocation. Second, the translocation process stimulates its transcriptional activity. TonEBP binds to ORE sequences (osmotic response elements) located in the promoter regions of some genes that play a key role in the mechanism of concentration and urinary dilution, stimulating their transcription. These include the SMIT and BGT1 cotransporters, the AR, the UT2 urea transporter, AQP-1, and AQP-2.

5.12 Conclusions

Cell volume regulation is accomplished by stimulating the inflow or outflow of effective osmols. Cell volume regulation is essential for maintaining intracellular ion concentration, DNA stability, and other functions.

Plasma osmolality is determined by the effective osmoles dissolved in the extracellular fluid. Sodium and the accompanying anions chloride and bicarbonate are the most abundant effective osmoles. Therefore, effective plasma osmolality is determined by plasma sodium concentration. Thus, changes in plasma sodium concentration beyond the normal range reflect water excess or deficit.

Body water steady state is a function of the thirst mechanism and diuresis under the influence of vasopressin. Plasma osmolality is the major physiological determinant of vasopressin secretion. Under the influence of this hormone, the kidneys adjust diuresis to meet water balance. Excretion of a hypertonic urine is associated with an increase in vasopressin-dependent water reabsorption in the connecting tubule and the complete collecting duct. Excretion of dilute urine is associated with a reduction in vasopressin-dependent water reabsorption.

Medullary hypertonicity is a key player in the kidney urine-concentrating ability. Hypertonicity of the medullary interstitium depends on NaCl reabsorption in the thick ascending limb of Henle and urea reabsorption and recycling.

Review Questions
1. A healthy subject weighing 70 kg drinks 1 L of water in 10 min. The plasma Na^+ concentration before ingestion was 143 mEq/L.
    (a) Calculate the new volume of the intra- and extracellular compartments and their osmolality. Why did the volumes of both compartments change?
    (b) Define the state of the water balance in this subject and explain what the homeostatic response will be in this subject.
2. Two foreign tourists, one strictly vegetarian and the other consuming a typical Western diet, were lost for 72 h in a desert area. Which of them is in the best condition to tolerate the period of restricted water intake?
3. The following experiment was carried out in rats (same weight and gender). Rat A was intravenously perfused with a volume of hypertonic NaCl solution; rat B was perfused in the same way with the same volume of a glucose solution (normal plasma glucose concentration); rat C received an intravenous 0.9% (w/v) NaCl solution. In which rat plasma vasopressin will be elevated?
4. The following data tables contain parameters measured in three subjects: (A) free water intake, (B) restriction of water intake for 15 h, and (C) intake of 1 L water (data correspond to 90 min after intake) (pl: plasma; u: urine; u/pl: urine to plasma ratio).

Subject	Urinary flow (mL/min)	[Inulin]u (mg/mL)	[Inulin]pl (mg/mL)	GFR (mL/min)	u/pl for inulin	% of water reabsorbed
A	1.2	15.8	0.151			
B	0.75	25.2	0.155			
C	15.0	1.23	0.154			

Subject	[Na+]pl (mEq/L)	Filtered Na+ (mEq/min)	[Na+]u (mEq/L)	Excreted Na+ (mEq/min)	Na+ reabsorption (%)
A	136		128		
B	144		192		
C	134		10.2		

Subject	Uosm (mOsm/kg)	Posm (mOsm/kg)	C_{H2O} (mL/min)	[Urea]u (mg/dL)	[Urea]pl (mg/dL)	C_{urea} (mL/min)	Urea reabsorbed (%)
A	663	290		480	12		
B	1000	300		720	15		
C	100	287		48	10		

(a) What is the meaning of the urine/plasma ratio for inulin?

(b) If the urinary excretion of Na^+ remained relatively constant, how do you explain the differences in the urinary Na^+ concentration?

(c) Which are the main parameters reflecting the state of the water balance in these subjects?

(d) How do you explain the differences in the urea clearance?

Bibliography

Agre P, Preston GM, Smith BL, Jung JS, Raina S, Moon C, Guggino WB, Nielsen S (1993) Aquaporin CHIP: the archetypal molecular water channel. Am J Physiol Renal Physiol 265: F463–F476

Bichet DG (2019) Regulation of thirst and vasopressin release. Annu Rev Physiol 81:359–373. https://doi.org/10.1146/annurev-physiol-020518-114556

Brown D, Fenton RA (2016) The cell biology of vasopressin action. In: Brenner BM, Rector FC (eds) The kidney, vol I, 10th edn. Elsevier, Philadelphia, pp 282–300

Dantzler WH, Layton AT, Layton HE, Pannabecker TL (2014) Urine-concentrating mechanism in the inner medulla: function of the thin limbs of the loops of Henle. Clin J Am Soc Nephrol 9: 1781–1789

Danziger J, Zeidel ML (2015) Osmotic homeostasis. Clin J Am Soc Nephrol 10:852–862

Jamison RL, Kriz W (1982) Urinary concentrating mechanism: structure and function. Oxford University Press, New York

Mount DB (2014) Thick ascending limb of the loop of Henle. Clin J Am Soc Nephrol 9:1974–1986

Nielsen S, Frøkiaer J, Marples D, Kwon TH, Agre P, Knepper MA (2002) Aquaporins in the kidney: from molecules to medicine. Physiol Rev 82:205–244

Oliet SH, Bourque CW (1993) Mechanosensitive channels transduce osmosensitivity in supraoptic neurons. Nature 364:341–343

Rose BD, Post TW (2001) Regulation of plasma osmolality. In: Clinical physiology of acid base and electrolyte disorders, 5th edn. McGraw-Hill, New York, pp 285–298

Sands JM, Layton HE, Fenton RA (2016) Urine concentration and dilution. In: Brenner BM, Rector FC (eds) The kidney, vol I, 10th edn. Elsevier, Philadelphia, pp 258–278

Osmoregulation in Non-mammalian Vertebrates

6

Learning Objectives

- To understand the osmotic challenges to which non-mammalian vertebrates are exposed in their respective habitats.
- To understand the physiology of the rectal glands in marine elasmobranchs.
- To describe the physiology of chloride cells in marine and freshwater teleost and the transition between the two environments.
- To describe the physiology of reptilian and avian salt glands.
- To understand the functioning and hormonal regulation of amphibian skin and urinary bladder.

In mammals, the kidney plays a key physiological role in the regulation of extracellular fluid volume and osmolality. The regulation of these physiological variables is carried out through adjustments in renal excretion of sodium and water, respectively. However, in non-mammalian vertebrates there are extrarenal organs that are crucial to maintaining the volume and osmolality of the extracellular fluids. The most common extrarenal organs are orbital, nasal as well as rectal gland as well as the amphibian skin and some cells in teleost gills.

An important difference between the mammalian nephron and other vertebrates is the existence of a well-developed loop of Henle with its three subsegments or limbs: thin descending, thin ascending, and thick ascending. The existence of a loop of Henle within the nephron endows the kidney with the ability to accomplish: the generation of solute-free water and a hypertonic medullary environment that functions as a driving force for AVP-dependent water reabsorption along the medullary collecting duct. The kidneys of most birds, reptiles, fish, and amphibians lack the loop of Henle (Fig. 6.1) proper and are incapable of generating hyperosmotic urine with respect to plasma. The avian kidney has some nephrons whose structure is similar to that of a mammalian nephron and they are able to slightly concentrate the urine. In general, the kidneys of fish, amphibians, reptiles, and birds do not have a structure at the cellular and molecular level that allows them to maintain the balance of NaCl, as is the case in mammals. The structure of the nephrons also does not allow for the

© Springer Nature Switzerland AG 2022
P. A. Gallardo, C. P. Vio, *Renal Physiology and Hydrosaline Metabolism*,
https://doi.org/10.1007/978-3-031-10256-1_6

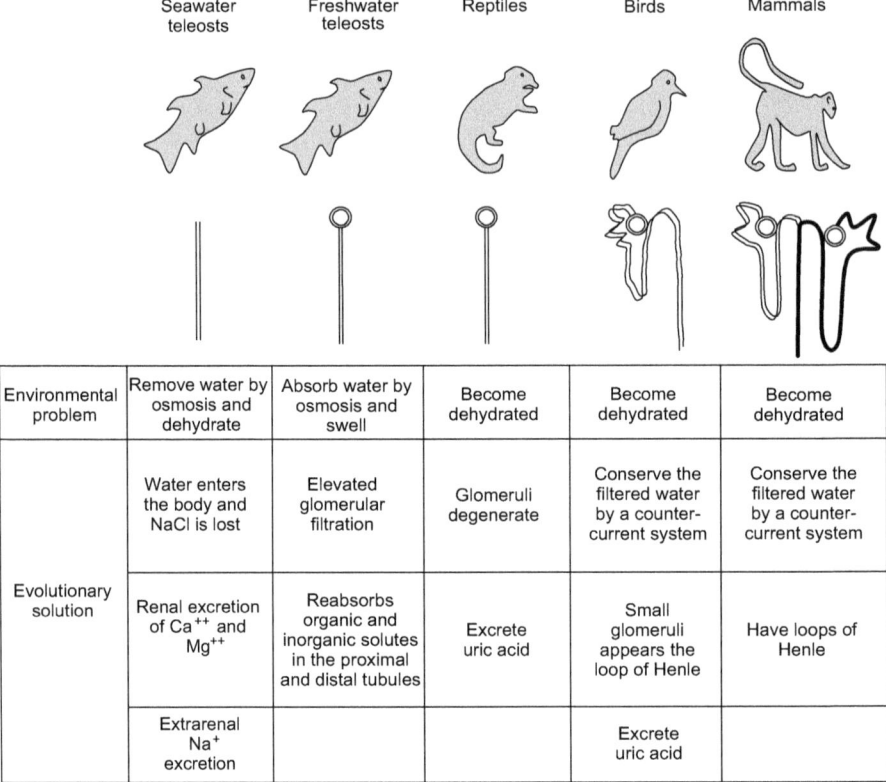

	Seawater teleosts	Freshwater teleosts	Reptiles	Birds	Mammals
Environmental problem	Remove water by osmosis and dehydrate	Absorb water by osmosis and swell	Become dehydrated	Become dehydrated	Become dehydrated
Evolutionary solution	Water enters the body and NaCl is lost	Elevated glomerular filtration	Glomeruli degenerate	Conserve the filtered water by a counter-current system	Conserve the filtered water by a counter-current system
	Renal excretion of Ca^{++} and Mg^{++}	Reabsorbs organic and inorganic solutes in the proximal and distal tubules	Excrete uric acid	Small glomeruli appears the loop of Henle	Have loops of Henle
	Extrarenal Na$^+$ excretion			Excrete uric acid	

Fig. 6.1 Comparative morphology and function of the nephron in vertebrates. The most primitive loop of Henle appeared in birds. However, the loop of Henle with three subsegments is a feature of the mammalian kidney

generation of hypertonic urine, so they have developed extrarenal organs designed to secrete salt and maintain NaCl and water balance.

Without exception, all extrarenal organs involved in the regulation of the hydrosaline balance are made up of epithelial cells, which are characterized by a high activity of the Na$^+$, K$^+$-ATPase. The latter probably represents an evolutionary convergence not only at the cellular level but also at the molecular level, considering that in all these extrarenal organs the isoform α-1 is responsible for 100% of the activity of Na$^+$, K$^+$-ATPase.

6.1   Fish

6.1.1   Teleosts

Teleosts maintain their plasma osmolality constant and independent from the aquatic environment where they live, and, therefore, they are osmoregulators. Marine teleosts regulate their plasma osmolality in the presence of a hyperosmotic aquatic

Table 6.1 Osmolality of typical aquatic environments and the different vertebrates that inhabit them. Note that seawater is hypertonic, compared to the plasma osmolality of teleost fish. Elasmobranch plasma osmolality is almost isotonic to seawater. Freshwater is hypotonic compared to the plasma of the fish and amphibians that live in it. TMAO: trimethylamine oxide

Species	Habitat	Na^+ (mM)	K^+ (mM)	Organic osmolytes (mM)	Osmolality (mOsm/kg)
	Seawater	450	10	0	1000
	Freshwater	<1	<0.01	0	1–2
Elasmobranchs					
Raja (marine ray)	Seawater	289	4	444	1050
Squalus (dogfish)	Seawater	287	5	354 Urea 350 TMAO 70	1000
Potamotrygon (stingrays)	Freshwater	150	6	<1	308
Teleosts					
Carassius (goldfish)	Freshwater	115	4		259
Opsanus (toadfish)	Seawater	160	5		392
Latimeria (coelacanth)	Seawater	197			
Anguilla (Anguilla)	Freshwater	155	3		323
Salmo (salmon)	Freshwater	181	2		340
	Seawater	212	3		400
Amphibians					
Rana (frog)	Freshwater	92	3	1	200
Rana cancrivora (crab-eating frog)	Seawater	252	14	350	830

environment. Instead, freshwater teleosts regulate their plasma osmolality in the presence of a hypoosmotic aquatic environment (Table 6.1).

Marine teleosts (Table 6.1) have considerably lower plasma osmolality than seawater, and the concentration of Na^+ in seawater is much larger than the fish plasma Na^+ concentration. The exception within the marine teleosts is the coelacanth (*Latimeria*), whose plasma osmolality is very similar to that of seawater due to the presence of organic osmolytes in the plasma. For this reason, the challenges presented by their hydrosaline balance will be discussed within the section on elasmobranchs.

Marine teleosts are faced with two challenges in terms of the hydrosaline balance: there is an osmotic gradient from the organism to the seawater, which promotes the loss of water from the fish to the aquatic environment (Fig. 6.2). This loss of water is compensated by water intake, which also implies a gain in salt. The average water intake of marine teleost is 1–19 mL/kg/h and about 70% of the ingested fluid is absorbed in the small intestine and is linked to NaCl absorption. The second problem is the gain of NaCl, derived from the gradient that exists from the seawater to the organism. Therefore, to maintain salt and water balance, the organism needs to

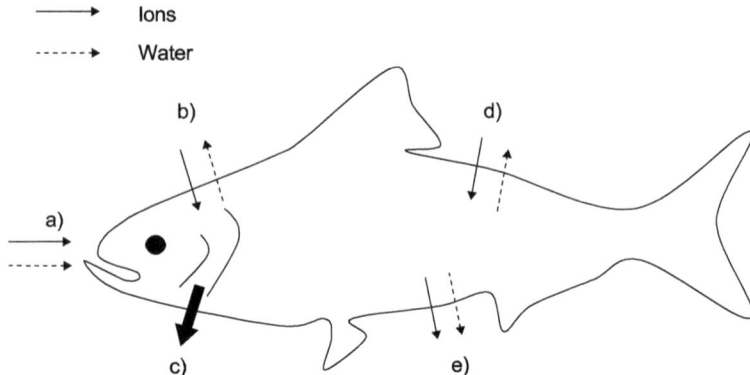

Fig. 6.2 Ion and water fluxes in a seawater teleost. (**a**) Ingestion of water and ions. (**b**) Passive branchial fluxes. (**c**) Active NaCl transport in gills. (**d**) Diffusion through the integument. (**e**) Urinary excretion of salt and water

excrete an amount of NaCl equivalent to that ingested with the water, plus that which enters by diffusion.

The nephron of marine teleosts has a very small glomerulus and in some species can be completely absent. Hence, many seawater teleosts have aglomerular nephrons. Consistently, the glomerular filtration rate is low (500 μL/h/kg) and so is diuresis (300 μL/h/kg). In addition to the typical handling of organic solutes, such as glucose and amino acids, the main function of the marine teleost kidney is the excretion of divalent ions, such as Ca^{++} and Mg^{++} (Fig. 6.3). NaCl excretion is preferably extrarenal and occurs in secretory epithelial cells located mainly in the gill epithelium, but also present in the opercular epithelium.

Three cell types have been described in gill epithelium: pavement cells (90% of total gill epithelium), accessory cells without a defined function, and mitochondria-rich cells (MRC) or gill ionocytes which constitute nearly 10% of the total population. The latter have the typical morphofunctional characteristics of an ion-secreting cell: large basolateral surface area given by abundant infoldings of the basolateral plasma membrane, large number of mitochondria located very near to the infoldings of the basolateral membrane, and a high density of Na^+, K^+-ATPase in the basolateral membrane.

The gill epithelium is interrupted by apical crypts, which correspond mainly to the apical membrane of MRC and pavement cells. The function of these cells is the secretion of hypertonic NaCl solution, which is why they are also called "chloride cells." Typically, MRC is located between an accessory cell and a pavement (Fig. 6.4).

Given the high concentration of Na^+ (470 mmol/L) and Cl^- (570 mmol/L) in seawater with respect to plasma, the secretion of Cl^- should occur by active transport. These cells carry out net NaCl secretion: the product of transcellular transepithelial Cl^- secretion and blood to crypt paracellular Na^+ flux. The basolateral Na^+, K^+-ATPase maintains the electrochemical potential gradient of Na^+ which

Fig. 6.3 General structure and function of a marine teleost nephron. (**a**) Ultrafiltration of plasma. (**b**) Ciliated neck. (**c**) Proximal tubule: organic solute, Na⁺, Cl⁻, HCO₃⁻ reabsorption, tubular secretion of organic acids, urea, creatinine, uric acid, and divalent ions. (**d**) Collecting duct. (**e, f**) Final urine, diuresis, and urine osmolality

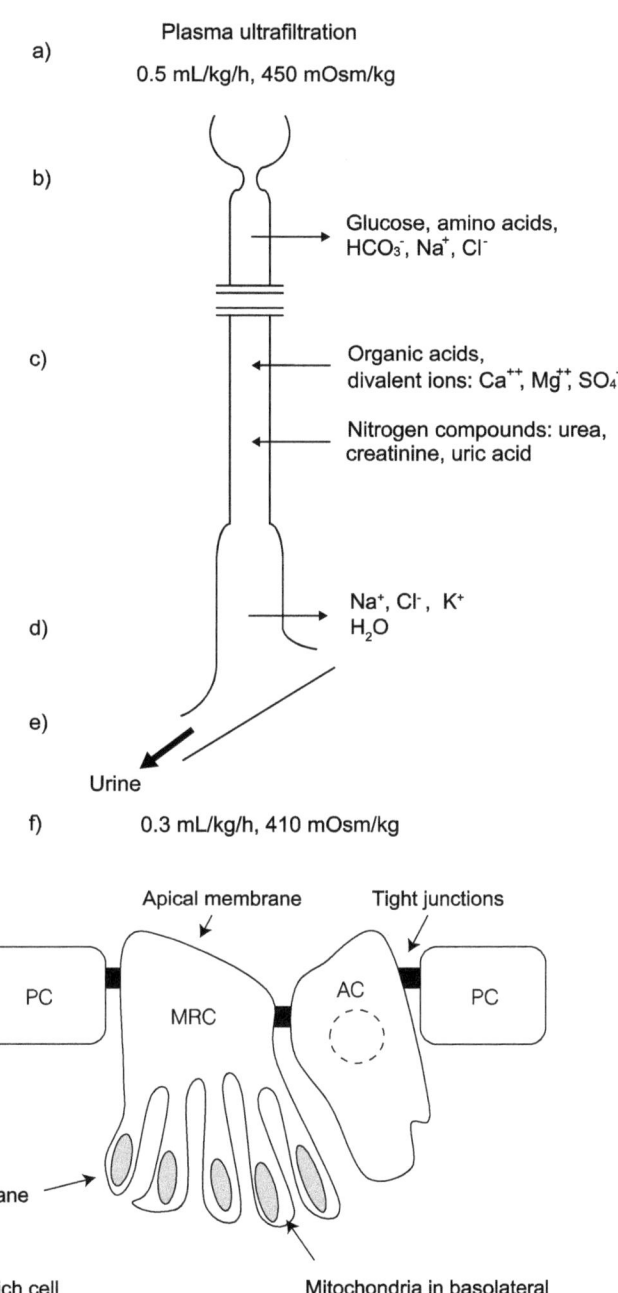

Plasma ultrafiltration

0.5 mL/kg/h, 450 mOsm/kg

a)

b)

Glucose, amino acids,
HCO₃⁻, Na⁺, Cl⁻

c)

Organic acids,
divalent ions: Ca⁺⁺, Mg⁺⁺, SO₄⁻²

Nitrogen compounds: urea,
creatinine, uric acid

d)

Na⁺, Cl⁻, K⁺
H₂O

e)

Urine

f)   0.3 mL/kg/h, 410 mOsm/kg

Seawater   Apical membrane   Tight junctions

PC   PC   MRC   AC   PC

Interstitial fluid

Basolateral membrane

PC: Pavement cells
MRC: Mitochondria-rich cell
AC: Accessory cell

Mitochondria in basolateral
membrane invaginations

Fig. 6.4 Organization and function of the gill epithelium in a marine teleost. The most important epithelial cell type is the mitochondria-rich cells (MRC) (also called "chloride cells" or MRC-ionocytes), with abundant mitochondria in deep infoldings of the basolateral membrane. These cells engaged in active NaCl secretion. Other epithelial cell types are the pavement cells and accessory cells

provides the free energy for the activity of the $1Na^+$, $1K^+$, $2Cl^-$ NKCC1 cotransporter. The latter performs secondary Cl^- active transport across the basolateral membrane that allows chloride secretion across the apical membrane. The K^+ is recycled through a K^+ channel located in the basolateral membrane. The apical Cl^- secretion occurs through a CFTR-type chloride channel.

In MRC ionocytes from gills of seawater-adapted teleosts, the movement of Cl^- through the apical membrane is electrogenic and generates a negative transepithelial potential difference on the apical side of approximately 20–30 mVolt, which acts as a driving force for the paracellular movement of Na^+ from the blood into the lumen of the crypt (Fig. 6.5). Hence, transcellular Cl^- secretion plus paracellular Na^+ movement makes up the branchial NaCl secretion into the crypt.

Freshwater teleosts face a completely different problem. Plasma is markedly hyperosmotic with respect to the freshwater environment, with an osmotic gradient that favors the entry of water into the organism. The plasma concentration of NaCl is higher than in the aquatic environment, so there is a gradient that favors the loss of NaCl from the organism to the aquatic environment (Fig. 6.6). Freshwater teleostean kidneys (Fig. 6.7) have nephrons with larger glomeruli and the glomerular filtration

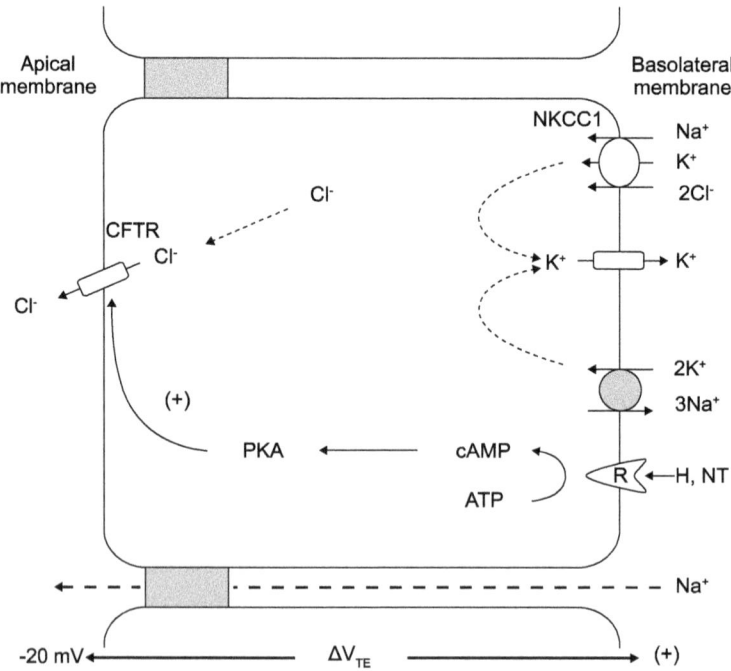

Fig. 6.5 Epithelial model for NaCl secretion in mitochondria-rich cells of a marine teleost. Basolateral Cl^- entry is mediated by the NKCC1 cotransporter; the energy for the secondary active transport derives from the Na^+ electrochemical potential gradient maintained by basolateral Na^+, K^+-ATPase. A basolateral K^+ channel recycles K^+ that entered the cell through NKCC1 and Na^+, K^+-ATPase. Apical CFTR chloride channel mediates Cl^- secretion. Cl^- diffusion creates a lumen-negative transepithelial potential difference that drives paracellular Na^+ diffusion

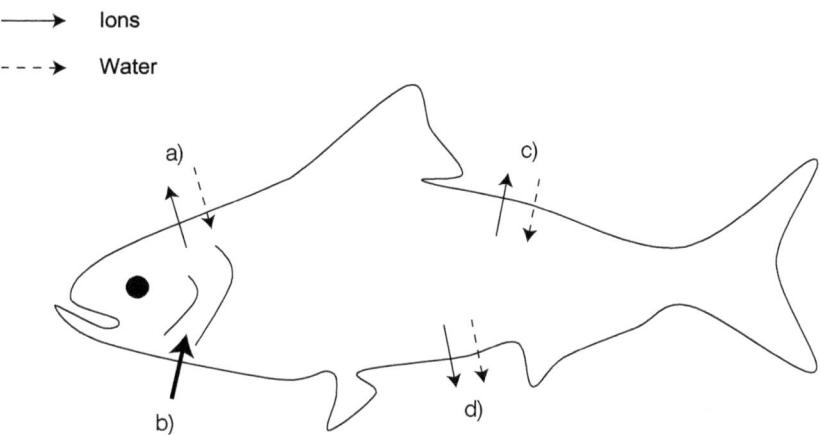

Fig. 6.6   Ion and water fluxes in a freshwater teleost. (**a**) Ion diffusion from the fish and water influx to the fish. (**b**) Active NaCl transport in gills. (**c**) Ion diffusion from the organism and water to the organism through the integument. (**d**) Urinary excretion of salts and water

Fig. 6.7   General structure and function of a freshwater teleost nephron. (**a**) Glomerular filtration rate and plasma osmolality. (**b**) Ciliated neck. (**c**) Proximal tubule. (**d**) Intermediate segment. (**e**) Distal tubule. (**f**) Collecting duct. (**g**) Collecting duct

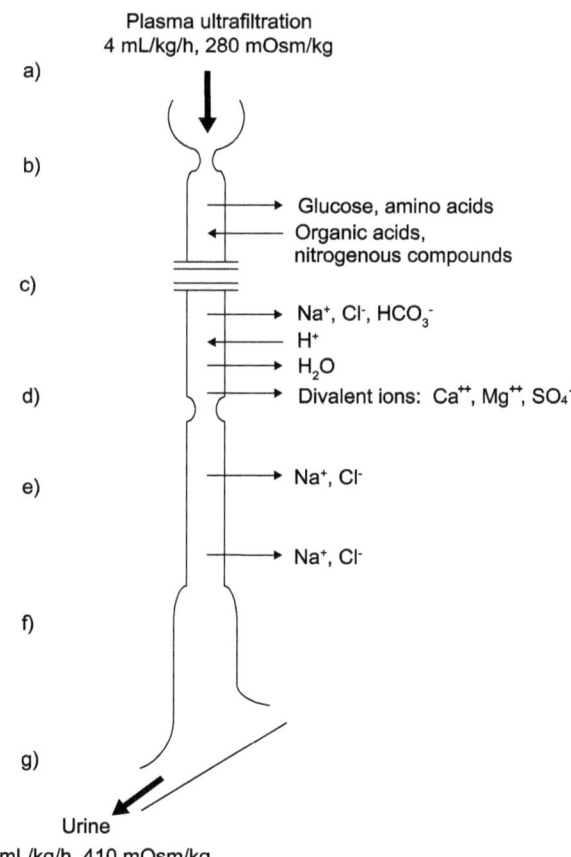

rate (4 mL/h/kg) is higher than that measured in several marine teleost species. Ultrafiltration of plasma results in the formation of abundant urine; apart from the handling of organic solutes, the main function of the freshwater teleost kidney is the reabsorption of NaCl and the excretion of water, which allows the elimination of an amount equivalent to the osmotic influence. The loss of NaCl is compensated by the net absorption of NaCl through the gill epithelium, exactly the opposite of its marine counterpart. Transbranchial uptake of NaCl occurs in mitochondria-rich ionocytes (Fig. 6.8) and requires different transport mechanisms than those expressed in marine teleosts. The absorption of Na$^+$ occurs through apical epithelial Na$^+$ channels, while the basolateral Na$^+$ efflux is mediated by Na$^+$, K$^+$-ATPase. The apical entrance of Cl$^-$ is achieved by the Cl$^-$/HCO$_3^-$ exchanger. It is postulated that the basolateral output of Cl$^-$ occurs through Cl$^-$ channels. The reaction catalyzed by the enzyme carbonic anhydrase provides HCO$_3^-$ for the apical anion exchanger. H$^+$ is transported across the apical membrane through a H$^+$-ATPase. Mitochondria-rich cells are also involved in the absorption of Ca^{++} and apical calcium entry occurs via an epithelial Ca^{++} channel and the basolateral exit through the Ca^{++}-ATPase (Fig. 6.8).

Gill ion transport is under endocrine regulation, which is related to the adaptation that some species experience when moving from saltwater to freshwater

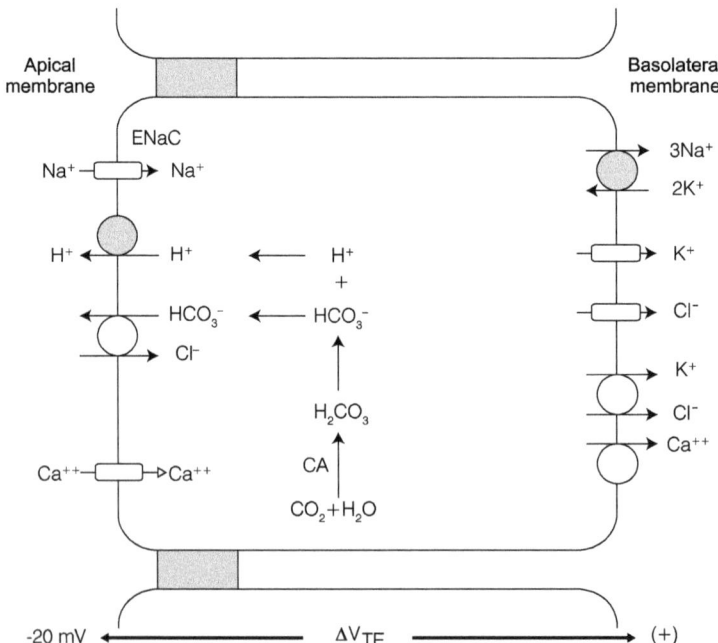

Fig. 6.8 Epithelial model for NaCl and Ca^{++} absorption in gills of freshwater teleost. Transepithelial Na$^+$ absorption is mediated by the combined activity of an apical Na$^+$ channel and basolateral Na$^+$, K$^+$-ATPase. Carbonic anhydrase activity generates H$^+$ and HCO$_3^-$ that allows the activity of an apical Cl$^-$/HCO$_3^-$ exchanger. The pathway for Cl$^-$ exit is unclear

environments and vice versa. Hormonal mechanisms allow adjusting the gill transport of NaCl to maintain the osmolality and volume of body fluids. Prolactin plays an important role in the adaptation of fish from marine aquatic environments to freshwater, causing a decrease in gill ion efflux. This effect is related to a decrease in ionocyte Na^+, K^+-ATPase activity and with a reduction in the size of the chloride cells. Another important hormone is cortisol, which plays a key physiological role in adaptation to seawater. Species tolerance to salt water is related to an increase in circulating cortisol levels. Among the actions of cortisol are the increase in density and activity of Na^+, K^+-ATPase, basolateral cotransporter NKCC1, and the number and size of the chloride cells. Growth hormone (GH) and insulin-like growth factor I (IGF-I) are also involved in the development of salinity tolerance. The effects of GH are mediated by IGF-I and are similar to those of cortisol.

6.2   Elasmobranchs

The plasma of marine elasmobranch plasma is isosmotic to seawater. This is due to the presence of high plasma concentrations of organic osmolytes, such as urea and trimethylamine oxide (TMAO) (Table 6.1). Despite the maintenance of the isoosmolality, there is a NaCl gradient oriented from the seawater to the organism, since the plasma concentration of Na^+ is almost half of that present in seawater. The high plasmatic concentration of urea (350–600 mmol/L) determines that a urea gradient exists between the plasma and the aquatic environment. Therefore, marine elasmobranchs are confronted with both NaCl gain and urea loss (Fig. 6.9).

The maintenance of extracellular fluid volume is mainly a function of the rectal gland. The gland is perfused by a single artery, and the blood returns to the body through a single vein; a single duct drains the NaCl-rich hypertonic fluid into the distal colon. The gland is formed by a series of densely packed tubules formed by

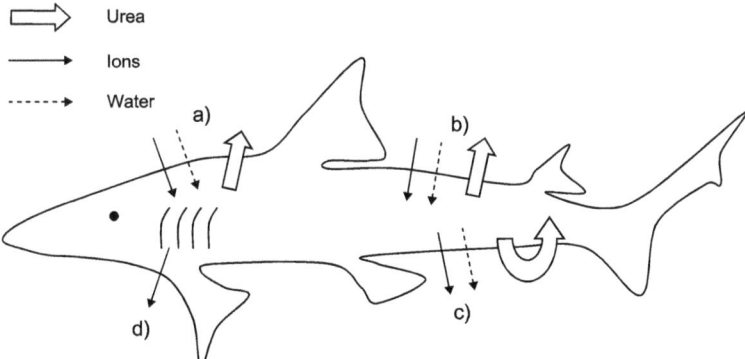

Fig. 6.9 Ion, urea, and water movements in a marine elasmobranch. (**a**) Ion and water entry and urea loss through gill epithelium. (**b**) Ion and water entry and urea loss through the integument. (**c**) Urinary excretion of ions and water and urea reabsorption. (**d**) Transbranchial ion transport

simple cubic epithelium. The epithelial cells have an apical membrane with short microvilli; the cytoplasm is packed with abundant mitochondria closely located to the very dense and abundant infoldings of the basolateral membrane which contains the Na^+, K^+-ATPase.

As in gill mitochondrial-rich ionocytes, chloride secretion (Fig. 6.10) results from active Cl^- uptake through the basolateral membrane. The NKCC1 cotransporter mediates the secondary active chloride transport. Free energy for the function of the NKCC1 cotransporter is provided by the sodium electrochemical gradient maintained by the basolateral Na^+, K^+-ATPase. Potassium ions are recycled across the basolateral membrane through potassium channels. The apical Cl^- efflux occurs through CFTR chloride channel. The transepithelial transport of Na^+ is paracellular and is oriented by the negative transepithelial potential difference in the tubular lumen, generated by the transport of Cl^-. Rectal gland function is under neuroendocrine control.

Gastrointestinal hormones like vasoactive intestinal peptide are a powerful stimulus. Derived from enteric nerves, this gut peptide binds to basolateral receptors that activate adenylyl cyclase and hence increase cytosolic cAMP levels. Another hormone that stimulates rectal gland secretion is the C-type atrial natriuretic peptide. The main stimulus for the secretion is the expansion of the volume of extracellular liquid, which in these fish triggers the secretion of C-type atrial natriuretic peptide or atriopeptin. In isolated and perfused glands, atriopeptin triggers a powerful secretory response. C-type natriuretic peptide elicits the secretory response through binding to membrane receptors that activate guanylyl cyclase and increase cGMP levels.

As mentioned above, the plasma of marine elasmobranchs contains high concentrations of urea, which promotes its loss to the aquatic environment. Since isoosmolality with seawater is maintained at the expense of high concentrations of urea and other organic osmolytes, elasmobranchs must have mechanisms that minimize urea diffusion. Gills are a potential site of urea loss. Recent studies at the molecular level allowed to establish the expression in the basolateral membrane of a Na^+-urea exchanger, which returns to the plasma the urea that diffuses from the plasma into the cytosol of the gill epithelium cells. This urea transport is active and depends on the Na^+ maintained by the Na^+, K^+-ATPase. The same mechanism operates in the basolateral membrane of the epithelial cells of the rectal gland and this explains why the rectal secretion is virtually urea-free. In addition, studies of the composition of the cell membrane of these epithelia have shown the presence of a high cholesterol content, in relation to the phospholipid content. This unusual abundance of cholesterol reduces the fluidity of the bilayer and the simple diffusion of urea across the membrane.

The elasmobranch kidney has glomeruli that freely filter urea and organic osmolytes like TMAO. Therefore, one of the tasks of kidney function is the reabsorption of these nitrogenous compounds; a loss of urea in the urine is equivalent to an increase in the biosynthesis of this compound. As shown in Table 6.2, the nephron of marine elasmobranchs is very efficient in the conservation of urea and TMAO, reabsorbing over 90% of the filtered load of these osmolytes.

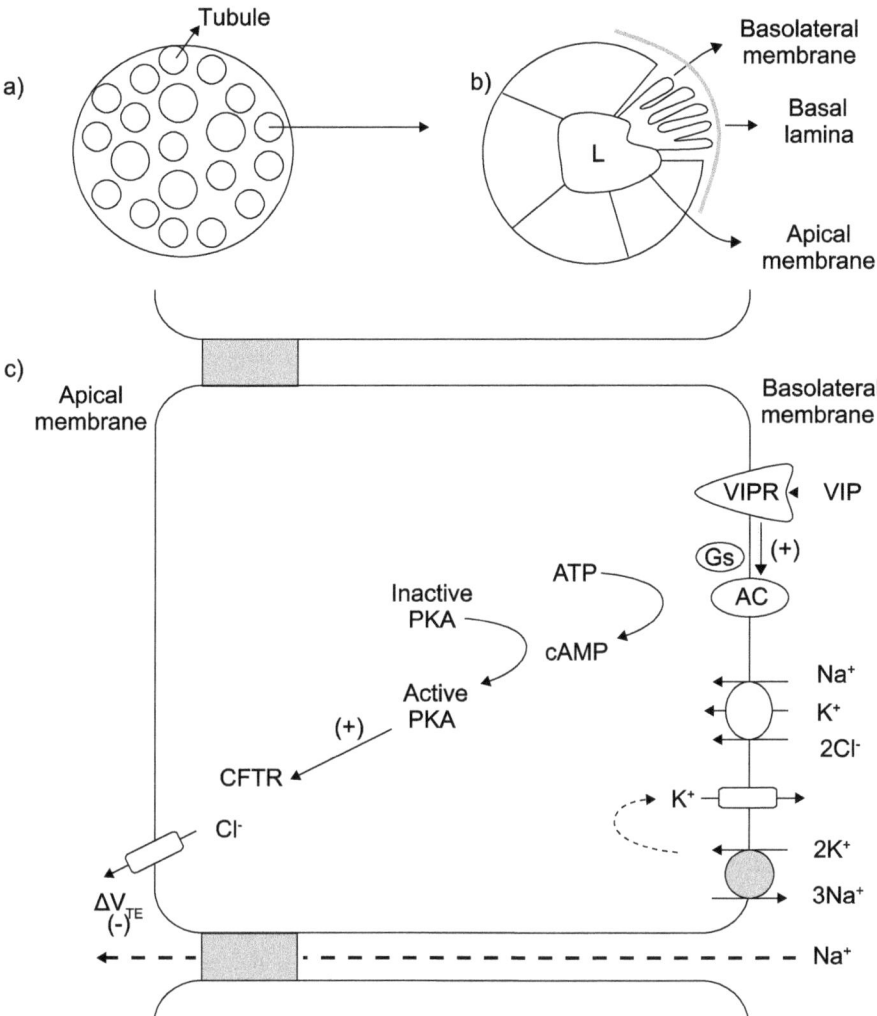

Fig. 6.10 Mechanism of NaCl secretion in the rectal gland of marine elasmobranchs. (**a**) Tubule. (**b**) Scheme of a rectal gland tubule in transversal section. (**c**) Epithelial model for NaCl secretion in a rectal gland epithelial cell. Basolateral Cl^- entry is mediated by the NKCC1 cotransporter; the energy for Cl^- secondary active transport derives from the Na^+ electrochemical potential gradient maintained by the Na^+, K^+-ATPase. A basolateral K^+ channel recycles K^+ that entered the cell through NKCC1 and Na^+, K^+-ATPase. Apical CFTR chloride channel mediates Cl^- secretion. Cl^- diffusion creates a lumen-negative transepithelial potential difference that drives paracellular Na^+ diffusion. Neuropeptides like vasoactive intestinal peptide (VIP) stimulate NaCl secretion through the cAMP-protein kinase A (PKA) signaling cascade

Urea resorption occurs in the collecting duct; this is consistent with the molecular location of the renal elasmobranch urea transporter.

The maintenance of isoosmolality in marine elasmobranchs occurs at the expense of high concentrations of urea in the plasma. With the exception of the basolateral

Table 6.2 Renal management of key organic osmolytes involved in the maintenance of plasma isoosmolality in elasmobranchs

Solute	Plasma concentration (mmol/L)	Urinary concentration (mmol/L)	Filtered load (μmol/kg/h)	Excreted load (μmol/ kg/h)	Fraction reabsorbed (%)
Urea	350	100	1225	115	90.6
TMAO	70	10	245	11.5	95

membrane of the gill and rectal epithelium, cells are permeable to urea: a high concentration of intracellular urea is harmful to the function of proteins such as enzymes. This effect is counteracted by the presence of TMAO, whose intracellular concentration is higher than the extracellular one. This compound stabilizes proteins in the presence of high concentrations of urea.

Freshwater elasmobranchs (Table 6.1), like the Amazon river stingray, are fully adapted to freshwater and their plasma has very low concentrations of urea. Therefore, the high concentration of plasma urea is not a characteristic of all elasmobranchs, but of those in the marine environment.

6.3   Amphibians

Amphibians develop part of their life in the aquatic environment, which is usually a hypoosmotic environment. The plasma concentration of NaCl far exceeds that of the pond water. Therefore, amphibians are faced with problems similar to those of freshwater teleosts: loss of NaCl to the aquatic environment and water gain due to an osmotic gradient. The excretion of the water load that enters the body by osmosis is eliminated through the excretion of abundant and diluted urine. The amphibian kidney has a high glomerular filtration rate ranging from 25 to 100 mL/h/kg. The reabsorption of organic solutes (glucose, amino acids) and NaCl is very efficient, so the final urine is hypoosmotic and of high volume. Except in conditions of restricted access to water, the water permeability of the distal tubule and collecting duct is very low.

Amphibians have two extrarenal organs for handling NaCl and water: the skin and the urinary bladder. The skin is a squamous stratified epithelium formed by four layers. The outermost layer is formed by cornified dead cells. The predominant epithelial cell type is the principal cell, and these cells are coupled through gap junctions and hence they form a functional syncytium. The other cell type is the mitochondria-rich cells, involved in acid-base and chloride transport. The cell types present in the amphibian skin resemble the cell types present in the collecting duct: principal cells and intercalated or mitochondria-rich cells. Amphibian skin principal cells are involved in transepithelial Na^+-absorption. Na^+ apical entry is mediated by the epithelial Na^+ channels (ENaC), similar to those expressed in the connecting and principal cells of the mammalian distal nephron. The basolateral Na^+ exit is mediated by the Na^+, K^+-ATPase. Apical Na^+ absorption is electrogenic and generates a transepithelial potential difference of -40 mVolt, negative on the apical side, which favors the passive paracellular movement of Cl^-. The absorption of NaCl in this epithelium is stimulated by mineralocorticoids. Therefore, the frog skin epithelium is an important extrarenal organ for compensation of NaCl loss from the body to the freshwater.

The amphibian bladder is formed by a urothelium; the outermost layer is the dome-shaped cells. The epithelium has two important functions: it transports Na^+ similar to the way it occurs in the skin and also reabsorbs water, in the presence of the antidiuretic hormone.

Under conditions like water shortage, the loss of water through permeable surfaces such as the skin, increases plasma osmolality and triggers vasotocin secretion, which corresponds to the antidiuretic hormone secreted by the neurohypophysis of amphibians. In the absence of vasotocin, the osmotic permeability of the urinary bladder epithelium is very low. Vasotocin increases osmotic permeability by inserting water channels into the apical membrane, analogous to the mechanism described for vasopressin in the distal mammalian nephron.

6.4   Birds and Reptiles

Birds have a kidney formed by two or three lobes and have two types of nephrons: a simple type of nephron very similar to that of reptiles with a proximal tubule, intermediate segment, distal tubule and collector, and a more complex nephron that resembles the cortical or superficial nephron present in mammalian kidneys, and which has a short and poorly developed loop of Henle. As a result, birds and reptiles have little or no ability to concentrate urine. The exception is a species of sparrow that inhabits brackish terrains in southern North America (Table 6.3). Even if the urinary osmolality is several times higher than the plasma osmolality, the urine/plasma osmolality ratio does not exceed a value of 3 and is considerably lower than that achieved by a mammal in the same habitat.

For example, ostriches in the Kalahari Desert reach a maximum urine/plasma osmolality ratio of 2.5. The same quotient for the Kalahari Desert kangaroo rat is 20. The inability of birds and reptiles to concentrate urine is due to the absence of juxtamedullary nephrons with loop of Henle.

Birds and reptiles have two extrarenal mechanisms that play an important role in the handling of NaCl and water: the supraorbital gland and the hindgut system. The supraorbital gland of birds secretes a NaCl-rich solution. All the species studied do

Table 6.3 Maximum values of urinary osmolality and urine/plasma osmolality ratio measured in different bird species under 24-h restricted water intake conditions

Species	Urinary osmolality (mOsm/kg)	Uosm/Posm
Emu	459	1.4
Chicken	538	1.6
Partridge	669	2.0
Ostrich	800	2.0
Sparrow	2000	5.8

Fig. 6.11 Localization and
structural organization of the
orbital gland in marine birds.
(**a, b**) Localization of the
supraorbital gland. (**c**) The
gland is formed by tightly
packed secretory tubules. (**d**)
The tubular epithelium is
formed by columnar cells with
deep infoldings of the
basolateral membrane with
mitochondria

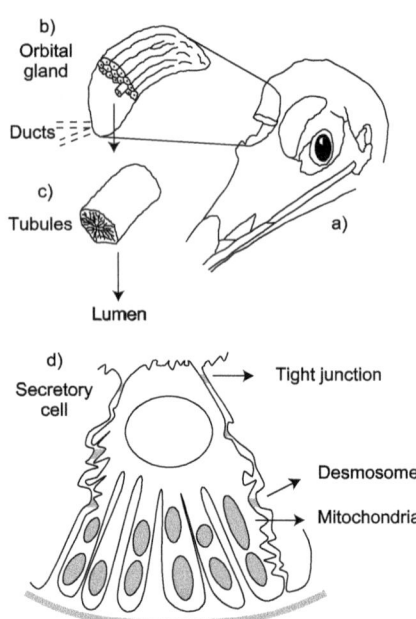

have a salt-secreting gland. However, its development and differentiation depend on
the intake of NaCl that the bird normally has. The glands are located in the
supraorbital fossae, one above each eye (Fig. 6.11). This tubular compound gland
is composed of a single layer of columnar epithelial cells organized in densely
packed tubules. Each tubule has a blind end and an open end to the excretory
duct. Several excretory ducts merge, forming a duct. The apical membrane of the
secretory cells is oriented to the lumen of the tubule and the basolateral membrane to
the interstitial medium. The basolateral membrane has a complex structure that
increases the surface. The basal membrane has a large surface, given by the abundant
infoldings of the basal membrane, while the lateral membrane has many
interdigitations (Fig. 6.11). The cytoplasm has abundant mitochondria that are
contained in the basal infoldings, which is related to the fact that the Na^+, K^+-
ATPase is abundantly expressed in the infoldings of the basal membrane. Within the
gland, each secretory tubule receives a high blood flow through an artery that gives
rise to an abundant capillary network near the basolateral membrane. The activity of
the supraorbital gland produces a hypertonic secretion, whose main component is
NaCl (Table 6.4).

    The salt gland secretion is a NaCl-rich fluid, with the highest salt concentration in
marine birds. The epithelial cell model for the avian salt gland is very similar to that
of the shark rectal gland: Cl^- secretion is the product of basolateral secondary active
Cl^- uptake carried out by NKCC1 cotransporter, while apical chloride secretion is
mediated by a CFTR channel. Cl^- secretion creates a transepithelial potential
difference of -20 mVolt that drives paracellular Na^+ movement into the tubule

Table 6.4 Concentration of Na$^+$ in the secretion of the supraorbital gland of birds

Species	[NaCl] (mmol/L)	Average osmolality (mOsm/kg)
Duck	400–600	1000
Cormorant	500–600	1100
Pelican	600–750	1350
Gaviota	600–800	1400
Humboldt penguin	725–850	1580
Albatross	800–900	1400
Petrel	900–1000	950

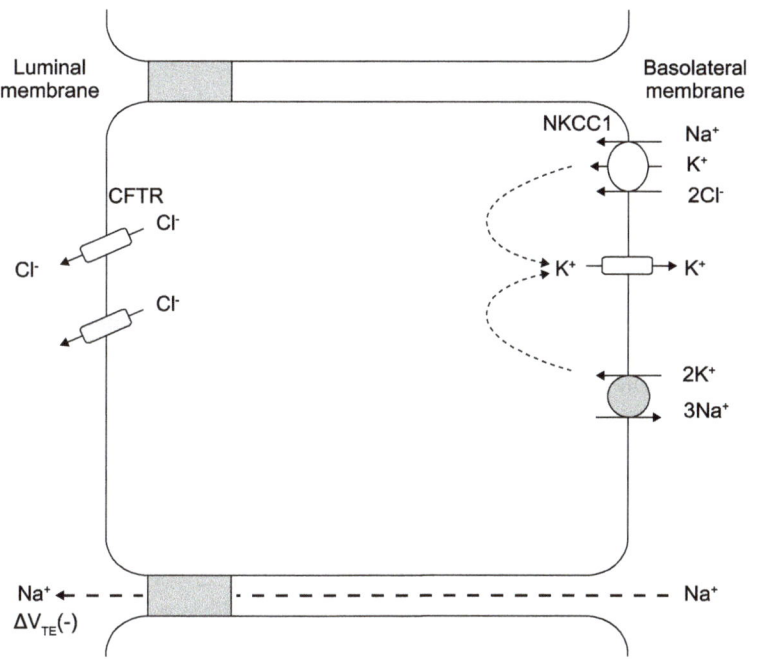

Fig. 6.12 Mechanism of NaCl secretion in the orbital gland of marine birds. Apical chloride secretion is the result of basolateral Cl$^-$ uptake mediated by NKCC1 cotransporter, energized by the Na$^+$ electrochemical potential maintained by basolateral Na$^+$, K$^+$-ATPase

lumen. The whole NaCl secretion process depends on the Na$^+$ electrochemical gradient maintained by the basolateral Na$^+$, K$^+$-ATPase (Fig. 6.12).

The efficiency of glandular activity is clear from an experiment conducted several years ago by Schmidt-Nielsen and colleagues. They measured the supraorbital and cloacal NaCl excretion for 175 min, after the intake of 134 mL of seawater, equivalent to 10% of body weight. As shown in Table 6.5, for any given time period, the amount of Na$^+$ secreted by the supraorbital glands is bigger than the cloacal amount of Na$^+$ secretion. Also, for the same time period, the volume of the

Table 6.5 Orbital and cloacal excretion of a parenterally administered NaCl load. Note that, for a given time, the supraorbital gland can excrete a greater amount of Na^+ in a smaller volume

Time (min)	Supraorbital excretion			Cloacal excretion		
	Volume (mL)	$[Na^+]$ (mmol/L)	Amount of Na^+ (mmol)	Volume (mL)	$[Na^+]$ (mmol/L)	Amount of Na^+ (mmol)
15	2.2	798	1.8	5.8	38	0.22
40	10.9	756	8.2	14.6	71	1.04
70	14.2	780	11.1	25.0	80	2.0
100	6.1	776	12.5	12.5	61	0.76
130	6.8	799	5.4	6.2	33	0.21
160	4.1	800	3.3	7.3	10	0.07
175	2.0	780	1.6	3.8	12	0.05
Total	**56.3**		**43.9**	**75.2**		**4.35**

secretion is less in the supraorbital than cloacal secretion. Thus, a higher amount of salt excreted in a small volume implies saving a volume of salt-free water.

In terms of the osmolality of the excreted fluid, the supraorbital gland is more efficient than the human kidney. Under extreme conditions of water restriction, excreted urine reaches a maximum osmolality of 1200 mOsm/kg in a volume of 0.5 L/day, compared to a fluid of approximately 1600 mOsm/kg generated in a much smaller time span and volume. This implies that, from the point of view of the osmotic work performed, the supraorbital gland is extremely efficient.

How does the intake of a volume of salt water, equivalent to 10% of body weight, trigger supraorbital NaCl secretion? NaCl absorbed in the intestinal tract produces an increase in plasma osmolality, which is detected by central osmoreceptors. The efferent pathway corresponds to parasympathetic innervation contained in the seventh cranial pair. Some axons secrete acetylcholine (ACh), while others secrete vasoactive intestinal peptide (VIP). Both the secretory cells and the blood vessels receive innervation. In the vessels, ACh produces vasodilation increasing glandular blood flow, oxygen supply, and NaCl delivery to the basolateral membrane for Na^+, K^+-ATPase. In the basolateral membrane of the epithelial secretory cells, ACh binds to muscarinic receptors that activate the increase of cytosolic free Ca^{++} concentration through a Gq protein signal transduction cascade. First, the increase in free Ca^{++} stimulates the opening of the Cl^- apical channel. Second, it stimulates the opening of a basolateral K^+ channel (Fig. 6.13). The first effect reduces the concentration of Cl^- intracellular and promotes the activity of the NKCC1 cotransporter. The second hyperpolarizes the membrane potential, which increases the driving force that allows the apical Cl^- secretion. VIP binds to receptors that activate the cAMP-protein kinase A signaling cascade. The main actions of this cascade are the phosphorylation of CFTR and NKCC1, increasing their activity and NaCl secretion. In summary, activation of cholinergic muscarinic and VIP receptors results in increased NaCl secretion.

The degree of glandular differentiation depends on the intake of NaCl (Fig. 6.14). Freshwater intake keeps secretory epithelial cells in a state of poor differentiation

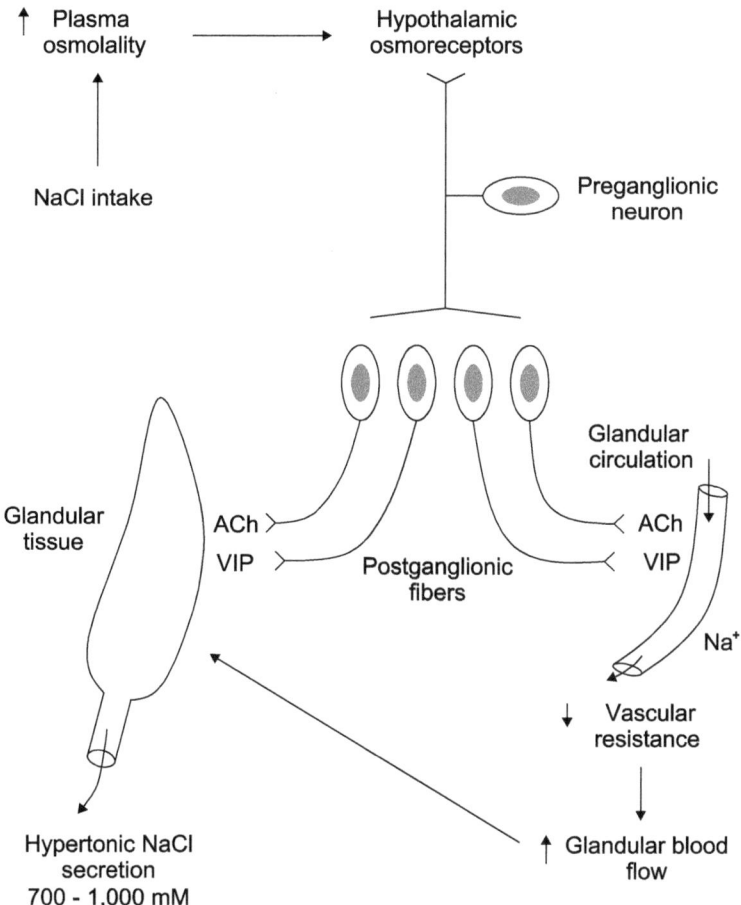

Fig. 6.13 Mechanism of stimulation of nasal secretion. An increase in plasma osmolality is detected by central osmoreceptors. The afferent pathway is neural corresponding to the seventh cranial pain; postganglionic fibers secrete acetylcholine (ACh) and vasoactive intestinal peptide (VIP). Both glandular elements and blood vessels receive innervation

and therefore low secretory activity. Saltwater intake stimulates glandular hypertrophy and hyperplasia by increasing the number of differentiated cells, which have significantly increased their basolateral membrane and Na^+, K^+-ATPase (Fig. 6.15).

Reptiles also have glands capable of secreting NaCl and a variable amount of K^+. A known example is the nasal gland of the Galapagos Islands iguana (*Amblyrhynchus*) that produces a secretion with a concentration of Na^+ of 1434 mmol/L and 235 mmol/L K^+. The secretion of K^+ is explained by the high K^+ content of the seaweed that forms a fundamental part of the diet of these reptiles.

The coprodeum of birds and reptiles is also an extrarenal organ that contributes to the maintenance of hydrosaline balance. Both ureters drain the urine into the urodeum and from there the urine passes into the coprodeum (Fig. 6.16). Reverse

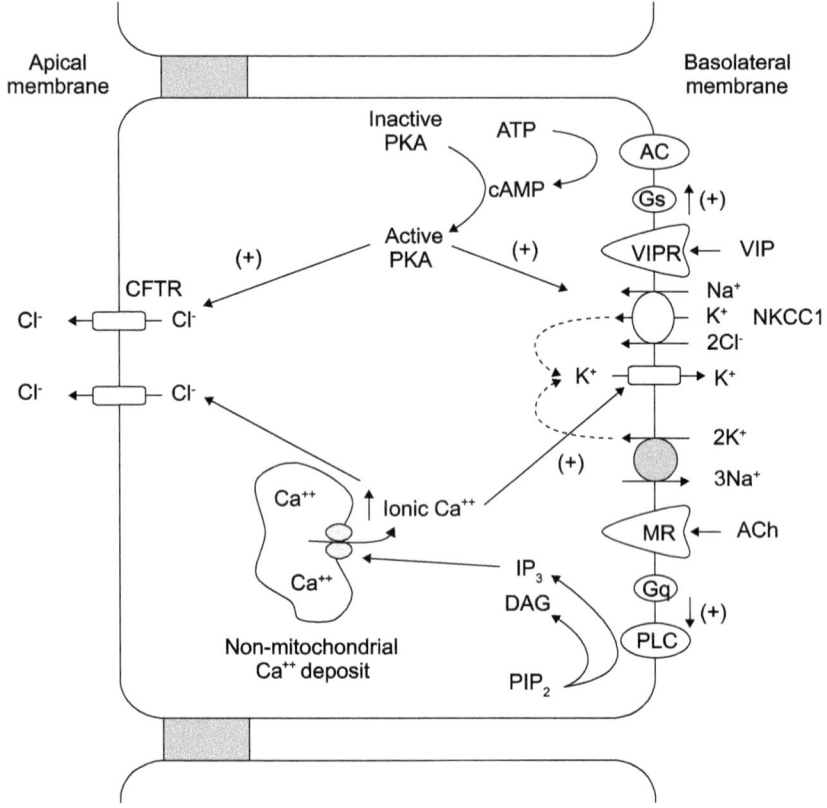

Fig. 6.14 Mechanism of stimulation of NaCl secretion. The scheme displays the basolateral membrane with the essential transporters necessary for NaCl secretion and the signal transduction systems involved in the regulation of NaCl secretion. VIP activates the cAMP-protein kinase A signaling cascade; ACh activates muscarinic receptors leading to an increase in cytosolic Ca^{++} that finally activates Ca^{++}-dependent K^+ channels

peristaltic contractions allow urine to pass into the distal colon. The rectum and distal colon absorb Na^+ in an electrogenic form and there is also fluid absorption, a process that occurs against an osmotic gradient. The transport of Na^+ in the coprodeum and distal colon is stimulated by mineralocorticoids.

6.5   Energy Cost of Osmoregulation in Aquatic Environments

The energy cost of living in an aquatic environment as such, or closely related to it, is directly related to the oxygen consumption generated by the active transport of Na^+. The National Commission for the Protection of the Environment and Natural Resources (Natura 2000) is the body responsible for the protection of the

Fig. 6.15 Long-term regulation of the glandular structure and function. (**a**) Parasympathetic innervation plays a crucial role in the maintenance of the structure and function of the glandular parenchyma. (**b**) Structure of the functional unit or secretory tubule; the blind end of the tubule contains undifferentiated cells

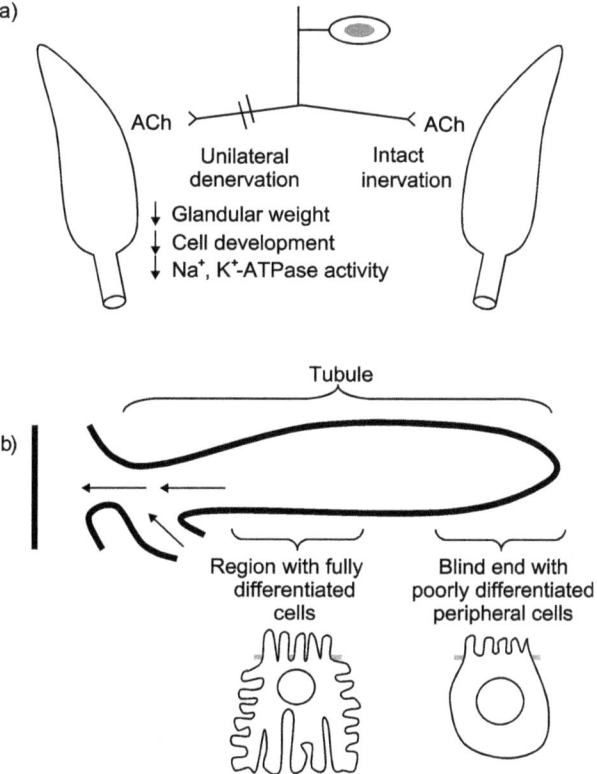

environment. In teleosts and marine elasmobranchs, the activity of the Na^+, K^+-ATPase is directly related to the secretion of NaCl in specialized structures.

In elasmobranchs, isoosmolality has an additional cost that corresponds to the active transport of urea Na^+-dependent, which prevents the brachial and rectal loss of urea and, on the other hand, the hepatic synthesis of an amount of urea equivalent to that lost through urine and other organs.

In freshwater and amphibian teleosts, oxygen consumption is related to two processes: the generation of dilute urine, which can only occur if all the filtered NaCl has been reabsorbed in the renal tubule. The second process is the absorption of Na^+ from a medium with Na^+ being very low, a process that occurs by active transport of Na^+. Hypoosmotic urine formation with respect to plasma and Na^+ requires the activity of the Na^+, K^+-ATPase into the gills, urinary bladder, or skin.

The NaCl-secreting glands of birds and reptiles generate hypertonic secretions and therefore solute-free water, with a high concentration of Na^+, Cl^-, and, in some cases, also K^+. The entire transport activity depends on the basolateral Na^+, K^+-ATPase.

In conclusion, much of the energy cost associated with osmoregulation is closely linked to the consumption of oxygen sensitive to ouabain, which is carried out by Na^+, K^+-ATPase, either renal or extra-adrenal epithelial.

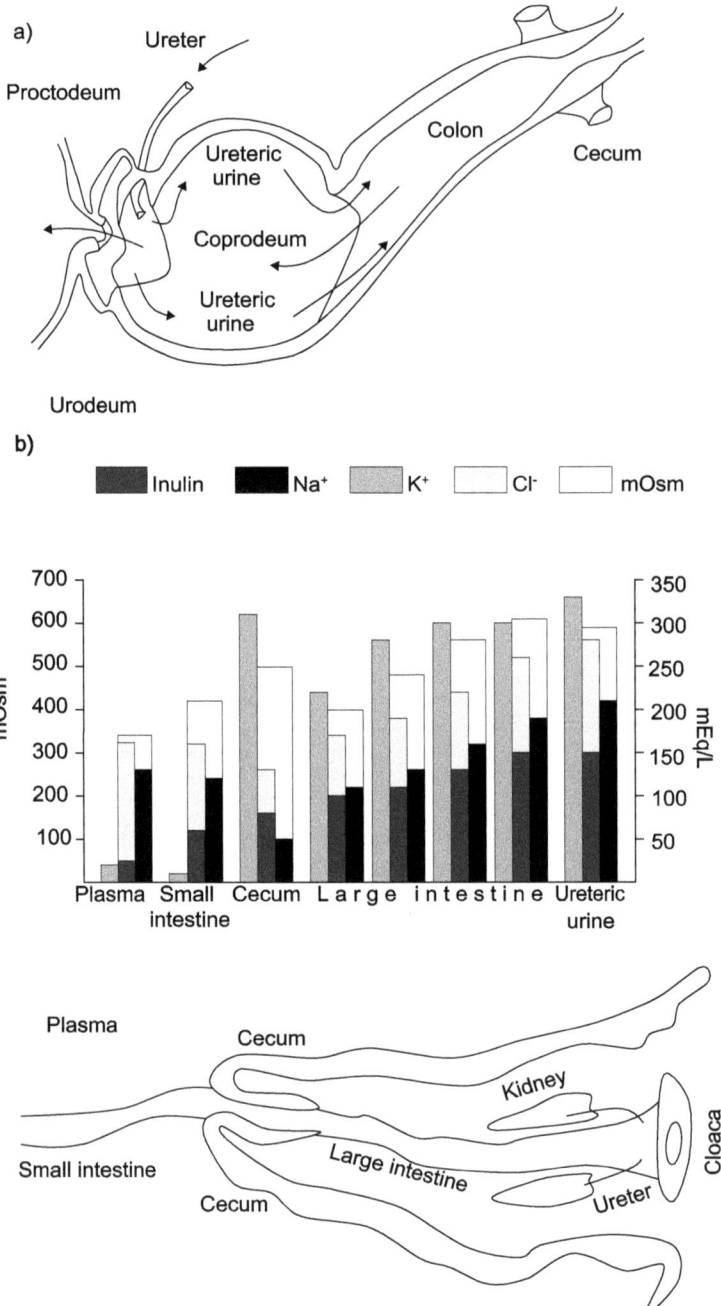

Fig. 6.16 Anatomical and functional relationships between components of the posterior avian digestive tract. (**a**) Anatomical relationships between the coprodeum, colon, and ureter in the chicken. Both ureters drain urine into the urodeum and to the coprodeum. Retrograde motility can move ureteral urine back to the distal colon probably for ion and water reabsorption. (**b**) Modifications in the concentration of inulin and several ions along the chicken gastrointestinal tract

6.6 Conclusions

Extrarenal organs play a crucial role in body fluid homeostasis in non-mammalian vertebrates. NaCl-secreting epithelial cells in gills of marine teleosts, rectal gland of elasmobranchs, and salt glands of marine reptiles and birds secrete a NaCl-rich hypertonic secretion that compensates for NaCl entry to the body. Amphibian skin and bladder reabsorb NaCl to compensate for salt loss to freshwater.

In mammals inhabiting xeric landscapes, extrarenal organs are important for water economy. Water contained in expired air is condensed and absorbed in tortuous nasal passages. The distal colon epithelium carries out hypertonic water absorption and fecal compaction.

Review Questions
1. The amphibian kidney is a mesonephric type and consists of two types of nephrons: the more ventral nephrons have a glomerulus and a nephrostome, which drains fluid from the cell cavity into a ciliated neck that connects the proximal tubule with the glomerulus. Dorsal nephrons lack nephrostomes and are very similar to the nephrons of freshwater teleosts. Considering the challenges to the hydrosaline balance faced by amphibians:
   (a) Which is the functional importance of the ventral nephrons?
   (b) What is the functional importance of the fact that the distal tubule is impermeable to water and reabsorbs NaCl?
2. How do you explain that marine elasmobranchs are isosmotic with seawater and, at the same time, hypoionic?
3. The following data table shows the results of electrolyte (mmol/L) measurements in plasma and bladder urine of the amphibian *Bufo viridis*.

Variable	Control			Dehydrated		
	Plasma	Urine	Urine/ plasma	Plasma	Urine	Urine/ plasma
$[Na^+]$	141	6	0.04	162	55	0.34
$[Cl^-]$	110	10	0.09	161	25	0.16
Urea	32	88	2.75	272	470	1.73
Osmolality (mOsm/kg)	392	210	0.54	752	705	0.94

   (a) What is the physiological role of the amphibian urinary bladder? From a molecular standpoint, which is the underlying process?
   (b) If you could measure the difference in transepithelial potential difference (ΔVte), what direction should it take and why?
4. The following data corresponds to a study in the supraorbital gland of ducks. The Na^+, K^+-ATPase activity was measured as a function of the osmotic stress in control ducks drinking tap water (0 osmotic stress) and ducks drinking a NaCl 1% solution for 48 h.

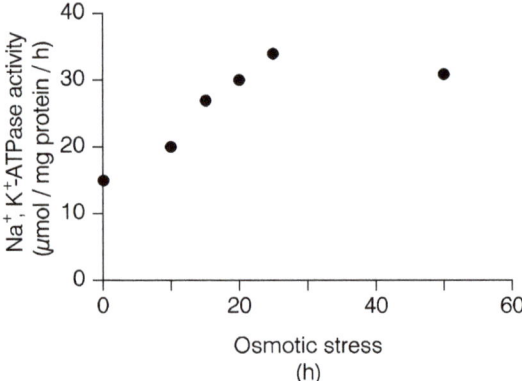

(a) How can you explain what is happening with the sodium pump activity?
(b) According to this result, what would you expect to happen to the glandular activity?
(c) If you would study the gland histology of control and experimental group, what differences should you expect?

5. The following table shows the bumetanide-sensitive oxygen consumption of glandular cells (ducks from problem 4) in the presence and absence of carbachol.

Experimental condition	O_2 consumption stimulated by carbachol and sensitive to bumetanide (μmol/100 g·min)
Control	
No stimulation	10
Stimulated with carbachol	18
48 h after intake of NaCl 1% as the only drink	
Without stimulation	11
Stimulated with carbachol	55

(a) Considering the transepithelial transport model of NaCl secretion in the supraorbital gland, explain the mechanism by which the cholinergic agonist increases the consumption of oxygen inhibited by bumetanide.
(b) What does "oxygen consumption is inhibited by bumetanide" mean?

6. The key element in the transepithelial transport of NaCl in chloride cells, rectal gland, and supraorbital nasal gland is the basolateral secondary active transport of Cl^- via NKCC1. What would happen to the oxygen consumption if the NKCC1 cotransporter would be replaced by the NCC cotransporter?

7. The following graph shows a study of Na^+, K^+-ATPase into salmon gills, before and several days after transfer to 25% seawater.

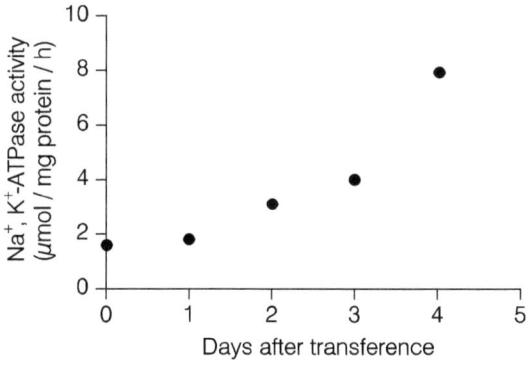

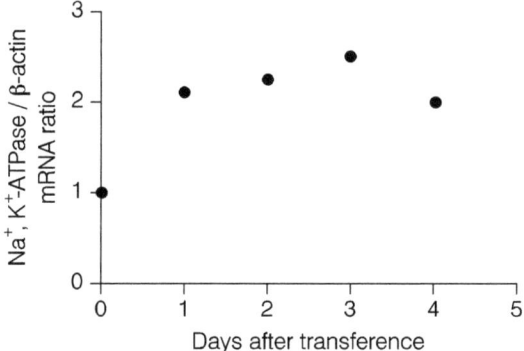

(a) Explain the relevance of the change in the Na^+, K^+-ATPase.
(b) In the same study the abundance of α-1 Na^+, K^+-ATPase isoform mRNA was measured and normalized to the β-actin abundance. How do changes in α-1 Na^+, K^+-ATPase mRNA abundance relate to changes in sodium pump activity?

Bibliography

Beyenbach KW (2004) Kidneys sans glomeruli. Am J Physiol Renal Physiol 286:F811–F827
Danzinger J, Zeidel ML (2015) Osmotic homeostasis. Clin J Am Soc Nephrol 10:852–862
Evans DH (2002) Cell signaling and ion transport across the fish gill epithelium. J Exp Zool 293: 336–347
Evans DH, Piermarini PM, Potts WTH (1999) Ionic transport in the fish gill epithelium. J Exp Zool 283:641–652
Kirschner L (1979) Extrarenal mechanisms in hydromineral and acid-base regulation in aquatic vertebrates. In: Dantzler WH (ed) Handbook of physiology, section 13: comparative physiology, vol I. Oxford University Press, New York, pp 577–613

McCormick SD, Sundell K, Björnsson BT, Brown CL, Hiroi J (2003) Influence of salinity on the localization of Na^+/K^+-ATPase, $Na^+/K^+/2Cl^-$ cotransporter (NKCC) and CFTR anion channel in chloride cells of the Hawaiian goby (*Stenogobius hawaiiensis*). J Exp Biol 206:4575–4583

Shuttleworth TJ, Hildebrandt JP (1999) Vertebrate salt glands: short- and long-term regulation of function. J Exp Zool 283:689–701

Silva P, Solomon RJ, Epstein FH (1997) Transport mechanisms that mediate the secretion of chloride by the rectal gland of *Squalus acanthias*. J Exp Zool 279:504–508

Willoughby EJ, Peaker M (1979) Osmoregulation in birds. In: Maloiy GMO (ed) Comparative physiology of osmoregulation in animals, vol 2. Academic Press, Nueva York

Withers PC (1992) "Water and solute balance", "water and solute excretion". In: Comparative animal physiology. WB Saunders, Philadelphia, pp 777–891

Regulation of the Effective Circulating Volume and the Sodium Balance

7

Learning Objectives

- To understand the concept of effective circulating volume and its main determinants.
- To understand the relationship between effective circulating volume, sodium balance, and blood pressure.
- To describe the mechanisms for detecting changes in the effective circulating volume.
- To describe the mechanisms for regulating the effective circulating volume and sodium balance.
- To describe the mechanisms of Na^+ reabsorption in a euvolemic subject.
- To describe the changes in Na^+ reabsorption that occur in hypovolemia or hypervolemia.

As mentioned in Chap. 1, the kidney plays a key role in maintaining the composition and volume of extracellular volume. Through the regulation of plasma osmolality and water balance, the kidney indirectly regulates cell volume. By regulating the effective circulating volume and sodium balance, the kidney maintains the fluid volume in the arterial vascular compartment and, indirectly, the fluid volume in the interstitial compartment constant. This function is very important because it involves the important role played by the kidney in the long-term regulation of blood pressure, which in turn is essential for maintaining adequate tissue perfusion.

The importance of the kidney in blood pressure regulation was demonstrated in both animals and humans. Crossing between normotensive rats (Wistar Kyoto strain, WKY) and spontaneously hypertensive rats (Okamoto strain, SHR) generates hybrids (WHY/SHR) that are normotensive. The latter rats were used as recipients of SHR rat kidneys and the result was the generation of high blood pressure. In humans, the importance of the kidney is evident in pathologies where there is an expansion of the effective circulating volume, due to the presence of neurohormonal factors that stimulate the retention of Na^+ such as angiotensin II and aldosterone. Less common, but also accounting for the role of the kidney in generating high blood pressure, are certain genetic diseases that result in abnormal stimulation of renal

© Springer Nature Switzerland AG 2022   137
P. A. Gallardo, C. P. Vio, *Renal Physiology and Hydrosaline Metabolism*,
https://doi.org/10.1007/978-3-031-10256-1_7

reabsorption of Na^+. The most common example is the Liddle syndrome, related to an increased abundance of ENaC in connecting and principal cells.

7.1   Concept and Determinants of the Circulating Volume

One of the basic functions of the cardiovascular system is to deliver an adequate supply of oxygen and nutrients to the cells. Tissue perfusion has two important requirements: the first is a certain volume of blood in the arterial compartment and the second is that blood volume should be at certain pressure, which is generated by the left ventricular pump. The circulating volume—or effective circulating volume—is precisely the volume of fluid contained in the arterial compartment, under a given pressure. At the level of the microcirculatory unit, this intravascular fluid compartment is connected to the interstitial liquid compartment through the capillary wall formed by endothelial cells.

The main determinant of the effective circulating volume is the amount of Na^+ present in the extracellular compartment. Na^+ is the main solute of the extracellular fluid: the plasma membrane has a low permeability for Na^+ and the Na^+, K^+-ATPase pumps Na^+ out of the cell. These two facts make Na^+ the most important effective osmole in the extracellular medium. The concentration of Na^+ in this compartment determines plasma osmolality. In contrast, the amount of Na^+ in the extracellular compartment determines its volume due to the need to maintain that amount at a given concentration (145 mEq/L). Therefore, the amount of Na^+ present in the body not only determines the circulating volume but also the amount of water present in the body (Fig. 7.1).

7.2   Sodium Balance and Its Relationship to Circulating Volume

The sodium balance can be understood in very similar terms to those of the water balance. There is a long-term equality between the sodium input to the body and sodium output from the body.

Under physiological conditions, the only sodium input to the body is food. A typical Western diet results in the intake of about 12 g of NaCl per day (200 mEq/day). The main route for sodium to escape is through urinary excretion, and, to maintain the balance, this must be equivalent to intake (200 mEq/day). When intake exceeds excretion, the organism enters a positive sodium balance; and, conversely, when sodium output exceeds intake, a negative balance is generated (Fig. 7.2). The kidney has the ability to adjust urinary sodium excretion in order to maintain sodium balance. Adjustments in urinary sodium excretion are due to neural and hormonal mechanisms. However, the renal response to maintaining balance is slow and takes about 3–5 days to complete.

The following example illustrates the relationship between sodium balance and extracellular volume. Let's assume a subject who has normal osmolality and extracellular volume and eats a bag of chips. The amount of salt (NaCl) contained in

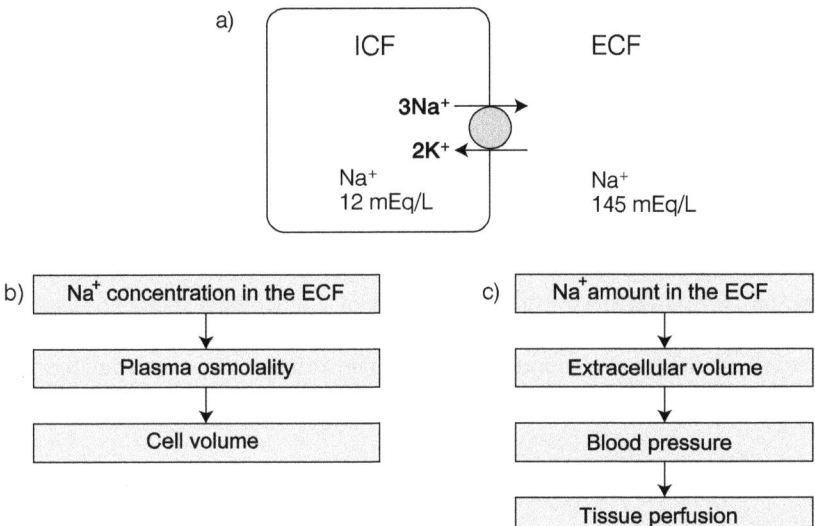

Fig. 7.1 Physiological role of Na$^+$ in extracellular volume physiology. (**a**) Transmembrane distribution of Na$^+$. Most of the body Na$^+$ is restricted to the extracellular fluid due to low plasma membrane permeability and the activity of the Na$^+$, K$^+$-ATPase. (**b**) Na$^+$ concentration in extracellular fluid is the main determinant of the effective plasma osmolality and hence of cellular volume. (**c**) The quantity of Na$^+$ in the extracellular fluid determines the extracellular volume, hence the arterial pressure and tissue perfusion

Fig. 7.2 Effect of salt ingestion. (**a**) An organism is represented with the intracellular (ICF) and extracellular fluid (ECF) in osmotic equilibrium. ECF and ICF volumes are normal. (**b**) Salt ingestion increases ECF osmolality. (**c**) Increased ECF osmolality drives water from the intracellular fluid, generates thirst sensation, and reduces renal water excretion. Body osmolality is restored but extracellular volume is increased

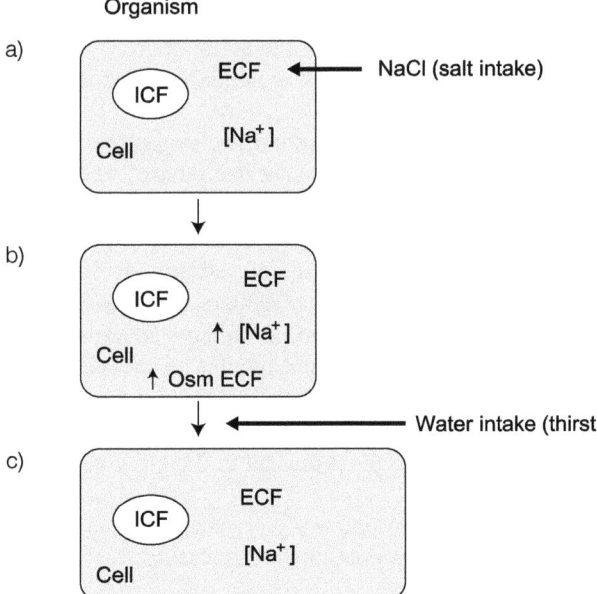

potatoes will be absorbed in the small intestine and will increase the amount of sodium in the extracellular fluid and therefore its concentration. Sodium concentration in the extracellular fluid is the main determinant of plasma osmolality. Hence, the increase plasma sodium concentration will result in an increase in plasma osmolality, which will trigger the feeling of thirst and vasopressin-dependent water conservation. Water intake via thirst mechanism and vasopressin-dependent renal water retention provide solute-free water for returning the elevated plasma osmolality to its normal range. The operation of these regulatory mechanisms has restored the osmolality within its normal range, but at the same time the extracellular volume has increased since a greater amount of extracellular NaCl required more volume of water in order to maintain osmolality. Restoration of extracellular volume will require an increase in sodium excretion. The kidneys will excrete an amount of sodium roughly equal to that ingested.

7.3   Detection of Changes in Circulating Volume

Since the need to maintain a normal circulating volume is closely related to maintaining optimal tissue perfusion, it is not surprising that volume receptors are primarily located in the cardiovascular system, in addition to other areas linked to blood pressure regulation.

Vascular receptors are located in the arterial and venous circulation. The receptors in the high-pressure circulation are:

(a) Arterial baroreceptors in the carotid artery
(b) Arterial baroreceptors of the aortic arch
(c) Afferent arterioles of the renal juxtaglomerular system

Arterial baroreceptors respond to changes in blood pressure; their afferent nerves project to medullary cardiovascular control centers. From the medullary center, fibers projecting to the hypothalamus stimulate vasopressin secretion when blood pressure or circulating volume decreases in 10–15%. These afferent fibers are responsible for the hemodynamic control of vasopressin secretion. The efferent pathways from the medullary centers to the kidney are fibers of the sympathetic system whose postganglionic fibers innervate the granular cells of the juxtaglomerular apparatus and tubular segments, such as the proximal tubule. The baroreceptor in the juxtaglomerular apparatus is located in the smooth muscle cells of the wall of the afferent arteriole.

Low pressure receptors present in the circulation include:

(a) Atrial volume receptors
(b) Receptors in the pulmonary circulation

These receptors respond to wall stretching in response to changes in volume rather than pressure. The afferent pathways are projected to the hypothalamus and cardiovascular control centers in the medulla. The activity of these volume receptors modifies the secretion of vasopressin and the sympathetic discharge from the bulbar centers to the kidney. In addition, the distension of the atria triggers the secretion of atriopeptin by the atrial myocytes.

7.4 Signals Generated from the Volume Receptors

The changes in circulating volume detected by the different receptors must be translated into homeostatic responses aimed at restoring or correcting the alteration in volume. The main target is the kidney, where NaCl reabsorption is subject to a redundant control by hormones and neurotransmitters, which also influence each other. The main control pathways are the sympathetic innervation, renin-angiotensin system, vasopressin, and atriopeptin. The mechanisms of action of the hormones involved were described in Chap. 3.

7.4.1 Sympathetic Renal Innervation

Several structures of the nephron receive postganglionic sympathetic innervation. Volume contraction or a decrease in arterial pressure stimulates renal sympathetic outflow. The effects of sympathetic renal innervation are depicted in Fig. 7.3. Norepinephrine secreted by postganglionic fibers contracts the smooth muscle of preglomerular arteries and afferent arterioles. Norepinephrine also contracts efferent arterioles. The effect is more pronounced in afferent than efferent arterioles. Hence,

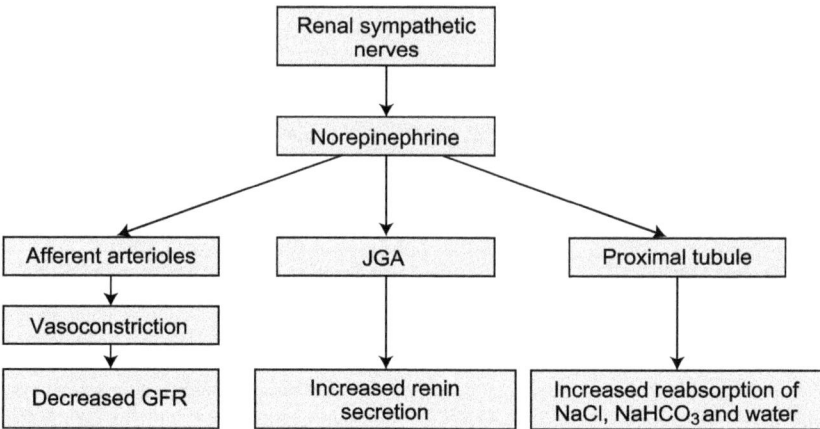

Fig. 7.3 Physiological effects of the renal sympathetic innervation. The sympathetic innervation has three important effects in the kidney. First is vasoconstriction of preglomerular arteries including afferent arteriole, reducing GFR. Second is renin secretion stimulation from juxtaglomerular cells (JGA). Third is stimulation of proximal $NaHCO_3$, NaCl, and water reabsorption

the net effect is a reduction of glomerular capillary hydrostatic pressure and hence the glomerular filtration rate. The vascular effect is mediated by the activation of α-1 adrenergic receptors that increase cytosolic calcium. In the granular cells of the juxtaglomerular apparatus, norepinephrine stimulates renin secretion and thus contributes to the activation of the renin-angiotensin II-aldosterone system. This sympathetic effect is mediated by the activation of β1-adrenergic receptors that activates the cAMP-protein kinase A signaling cascade. In the renal tubule, mainly in the proximal tubule, sympathetic innervation stimulates reabsorption of Na^+, Cl^-, and HCO_3^- and hence volume reabsorption (Fig. 7.3).

7.4.2 Renin-Angiotensin II-Aldosterone Axis

This hormonal system plays a key role in regulating the circulating volume and maintaining the sodium balance (Fig. 7.4). Aspects of the cellular and molecular biology of renin secretion were discussed in Chap. 2. The most important factors capable of modifying renin secretion are changes in afferent arteriolar perfusion

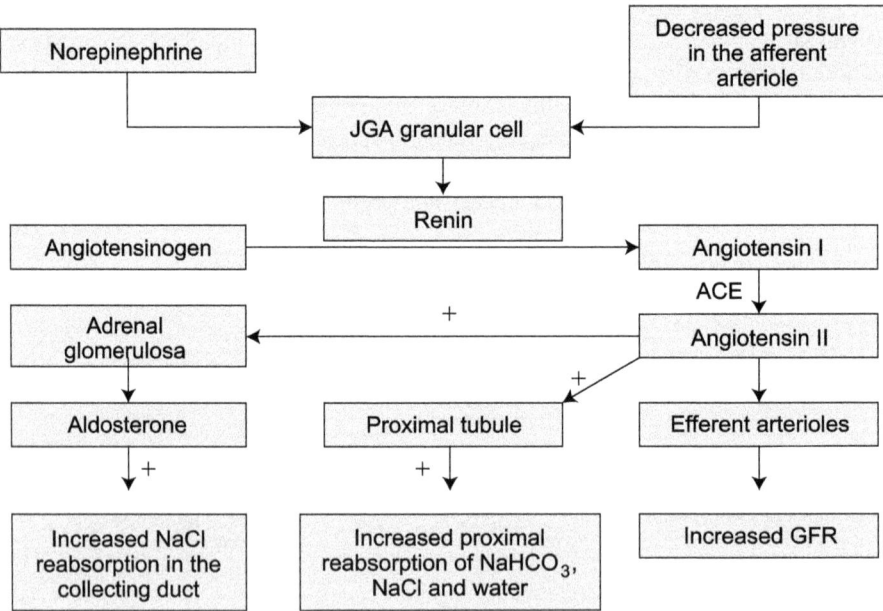

Fig. 7.4 Relationship between the sympathetic system and renin-angiotensin-aldosterone system. Through β1-adrenergic receptors, norepinephrine stimulates renin secretion; this endopeptidase converts angiotensinogen into angiotensin I. The pulmonary endothelium and the apical membrane of the proximal tubule express angiotensin I-converting enzyme (ACE) then convert angiotensin I into angiotensin II. This peptide hormone increases extracellular volume through stimulation of proximal tubule NaCl and $NaHCO_3$ reabsorption and aldosterone synthesis and releases the adrenal glomerulosa cells. Angiotensin II also increases the muscular tone of the efferent arteriole. Both norepinephrine and angiotensin II increase peripheral vascular resistance

pressure, sympathetic stimulation, tubuloglomerular feedback mechanism, and angiotensin II.

Renin is the rate-limiting step in the system. Under physiological conditions, its sole origin is the granular cells of the juxtaglomerular apparatus, located always in the renal cortex. The substrate for this endopeptidase is angiotensinogen. Hepatic angiotensinogen is the main substrate in the systemic renin-angiotensin system. The proximal tubule secretes angiotensinogen into the tubular lumen, which can be a substrate for filtered renin. The product of renin activity is the decapeptide angiotensin I, which is the substrate for angiotensin I-converting enzyme (ACE). This enzyme is an integral protein, especially abundant in the apical membrane of endothelial cells of the pulmonary circulation. The enzyme is also expressed in the apical membrane of proximal tubule cells, giving rise to intratubular angiotensin II. The product of ACE activity is the octapeptide angiotensin II, which is the main hormonal component of this system. Physiological actions of angiotensin II are mediated by binding to AT_1 receptors linked to Gq protein. A decrease in volume depletion or a decrease in arterial pressure stimulates angiotensin II formation. The most important systemic physiological actions of angiotensin II are:

(a) Vasoconstrictor effect. Angiotensin II contracts smooth muscle cells in the arterioles of the systemic circulation, contributing to the increase in total peripheral resistance. In the glomerular circulation, angiotensin II contracts efferent arterioles more than afferent arterioles, increasing glomerular filtration and decreasing renal plasma flow.
(b) Steroidogenic effect in adrenal cortex. Angiotensin II stimulates aldosterone synthesis and release the zona glomerulosa.
(c) Stimulation of NaCl reabsorption. Angiotensin II stimulates proximal $NaHCO_3$ reabsorption by increasing the activity of NHE3 exchanger. The net effect is an increase in $NaHCO_3$, NaCl, and water reabsorption. In the distal convoluted tubule, angiotensin II increases NaCl reabsorption through NCC cotransporter stimulation. Proximal and distal effects are mediated by tubular angiotensin II formed in the proximal tubule lumen.
(d) The mechanism of action of aldosterone in the distal nephron was discussed in Chap. 3. Briefly, aldosterone increases apical ENaC abundance in connecting and principal cells of the distal nephron.
(e) Angiotensin II inhibits renin secretion by increasing cytosolic Ca^{++} in granular cells. This mechanism acts as a negative feedback since renin stimulated angiotensin II formation.

It is evident that tubular NaCl reabsorption is under redundant control. First, sympathetic innervation to the kidney stimulates both renin secretion and tubular NaCl reabsorption. Second, angiotensin II and aldosterone both stimulate tubular NaCl reabsorption. Both stimuli for NaCl reabsorption are increased under conditions like volume depletion and low arterial pressure. Thus, neural and endocrine mechanisms are activated to maintain the effective circulating volume and tissue perfusion.

7.4.3   Atriopeptin

This peptide has an important role in the long-term regulation of blood pressure. The physiological role of atriopeptin in the day-to-day regulation of sodium balance and blood pressure has been controversial. Much of the evidence is derived from studies conducted on transgenic mice with overexpression or deletion of the gene encoding for ANP. Consistent with the above, the former has elevated circulating levels of ANP and hypotension, and the latter do not produce ANP and exhibit marked hypertension. The results obtained suggest two lines of action for ANP. First, it reduces blood pressure and cardiac output. This effect is mainly mediated by an attenuation of the basal tone of sympathetic activity on the cardiovascular system at the level of the contractile myocardium and the resistance vessels. This translates into lower cardiac output and lower total peripheral resistance. Second, at the renal level, ANP increases renal excretion of sodium and water. This effect is due, on the one hand, to an inhibitory basal tone on renin secretion, resulting in a lower plasma concentration of angiotensin II and aldosterone: the latter leads to increased urinary excretion of NaCl. In addition, atriopeptin directly inhibits NaCl reabsorption in the medullary collecting duct, which also contributes to some extent to natriuresis (Fig. 7.5).

7.4.4   Arginine-Vasopressin

A 10–15% drop in circulating volume increases vasopressin (AVP) secretion; the stimulus comes from the arterial baroreceptors and corresponds to the known hemodynamic control of AVP secretion. Binding to V_2 receptors, AVP stimulates ENaC exocytosis into the apical membrane of connecting and principal cells of the distal nephron. Therefore, AVP increases the abundance of epithelial Na^+ channels

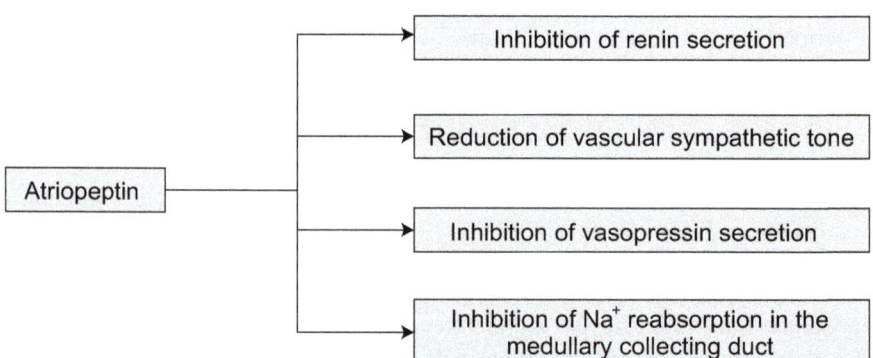

Fig. 7.5   Role of atriopeptin in Na^+ balance. Atriopeptin increases Na^+ excretion through several actions: first, inhibition of renin secretion; second, reduction of the vascular sympathetic tone; and third, inhibition of vasopressin secretion, inhibition of aldosterone synthesis, and the inhibition of Na^+ reabsorption in the inner medullary collecting duct

in the apical membrane. This effect contributes to increased sodium reabsorption in the connecting tubule and cortical collecting duct. This effect is independent of the osmotic effect in the same tubular segments.

7.5   Control of Renal Sodium Excretion Under Euvolemic Conditions

Euvolemia is defined as the maintenance of a normal effective circulating volume. As discussed above, this requires a fine balance between sodium intake and renal excretion. The kidney has the ability to adjust sodium excretion over a wide range of settings. Under physiological conditions, Na^+ balance is an interplay between Na^+ intake and urinary excretion. A reduction in dietary Na^+ intake or an increase in Na^+ excretion sets the organism in a negative Na^+ balance. In this setting, Na^+ urinary excretion is reduced. Conversely, a higher intake of NaCl or a reduced excretion determines a positive Na^+ balance. In this condition, the kidney will respond with an appropriate increase in urinary sodium excretion. However, changes in urinary Na^+ excretion to maintain the balance are not immediate and take a few days to complete (Fig. 7.6).

This is best illustrated in the following example (Fig. 7.7). Suppose a healthy subject with a body weight of 70 kg. The subject is in Na^+ balance since daily Na^+ intake (100 mEq) equals Na^+ excretion. This condition is reflected as a constant body weight of 70 kg from day 0 to 1. On day 1, Na^+ intake abruptly doubles to 200 mEq/day over a 7-day period (days 1 to 8). Since Na^+ ingestion is increased while

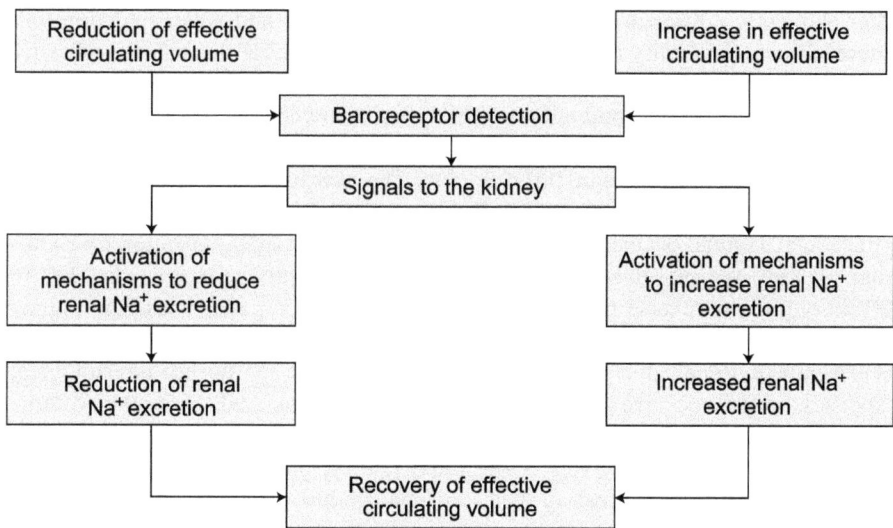

Fig. 7.6 Crucial role of the kidney in the regulation of the effective circulating volume. Changes in extracellular effective circulating volume modify urinary Na^+ excretion

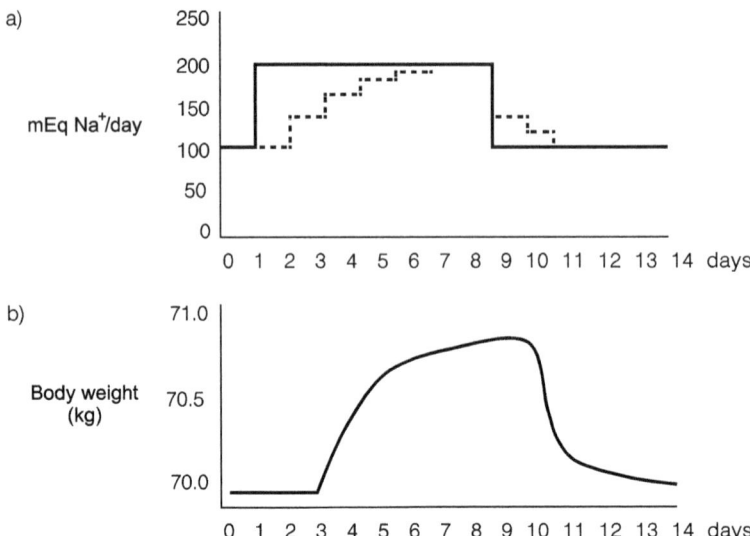

Fig. 7.7 Effect of changes in Na$^+$ ingestion in urinary Na$^+$ excretion and body weight. (**a**) Changes in Na$^+$ ingestion (solid line) produce changes in urinary Na$^+$ excretion (dotted line). (**b**) Effect of Na$^+$ ingestion in body weight; changes in Na$^+$ ingestion parallel changes in body weight

excretion is maintained, the organism attains a positive balance manifested as an increase in body weight. The increase in sodium intake translates into an increase in plasma Na$^+$ concentration and thus in plasma osmolality. Through the mechanism of thirst and the vasopressin-dependent water reabsorption, the subject will drink more water and excrete hyperosmotic urine. Both water intake and water retention will restore plasma osmolality but, at the same time, will increase the volume of extra-cellular fluid. The latter is reflected as an increase in body weight. The increase in sodium excretion occurs gradually and becomes equivalent to the intake between days 7 and 8. It should be noted that restoring the balance does not imply maintaining a Na$^+$ excretion of 100 mEq/day. The new balance setting corresponds with a renal sodium excretion of 200 mEq/day, at the same time that an expansion of extracellular volume occurred, which has been evidenced by the increase of 1 kilo-gram of body weight. If sodium intake is now drastically reduced from 200 to 100 mEq/day, the reverse homeostatic response will occur, and, while the kidney adjusts renal excretion to 100 mEq/day, the body is in negative balance as excretion exceeds intake. Renal adjustments to sodium excretion are appropriate physiological responses, which occurred in response to signals generated from the volume receptors mentioned above.

If the glomerular filtration rate is 180 L/day and the plasma Na$^+$ concentration is 140 mEq/L, the filtered load of Na$^+$ will be 25,200 mEq/day. Approximately 126 mEq/day is excreted in the urine, giving a fractional excretion of Na$^+$ (FENa$^+$)

of less than 1% of the filtered load. The Na^+ reabsorption along the tubular segments is the process that accounts for the difference between the filtered and excreted load. It should be kept in mind that there is considerable nephron heterogeneity, for example, the length of the proximal tubule and thick ascending limb varies from cortical to juxtamedullary nephrons. Thus, the fractional reabsorption in each tubular segment is subject to variations. Under euvolemic conditions, the segments of the distal nephron (distal convoluted tubule, connecting tubule, and the cortical collecting duct) are the main sites of adjustment between the Na^+ intake and excretion. In a euvolemic subject, 67% of the filtered Na^+ (17,000 mEq/day) is reabsorbed in the proximal tubule, while 25% is reabsorbed in the thick ascending loop of Henle. The distal nephron reabsorbs 5–7% of the filtered load and less than 3% is reabsorbed in the medullary collecting duct. The Na^+ reabsorption in the proximal tubule and thick ascending limb allows a constant Na^+ delivery to the distal nephron. This is important for two reasons: first, the distal nephron is a site of fine adjustments of Na^+ reabsorption to maintain the extracellular volume and Na^+ balance. Second, the distal nephron has a low reabsorption capacity and an excess of Na^+ delivery can result in an increase of Na^+ excretion. There are at least two key mechanisms responsible for maintaining constant Na^+ delivery to the cortical collecting duct: autoregulation of renal blood flow and glomerulotubular balance. Autoregulation of renal blood flow maintains constant glomerular filtration rate and therefore the filtered load of Na^+ in the presence of fluctuations in renal blood pressure. Despite this mechanism, there may be small increases in glomerular filtration rate, resulting in increase in Na^+ filtered load. If this change is not compensated, it would represent an increased Na^+ delivery to the distal nephron, resulting in increased urinary excretion of Na^+ and the consequent loss of extracellular volume. The glomerulotubular balance allows the adjustment of the tubular Na^+ reabsorption to the Na^+ filtered load. This mechanism operates mainly in the proximal tubule. The physiological role of the glomerulotubular balance is exemplified with the following numbers from a euvolemic subject. First, consider a 2% increase in the filtered load of Na^+, due to an increase in the glomerular filtration rate (from 25,200 to 25,704 mEq of Na^+/day). If Na^+ reabsorption in the proximal tubule, loop of Henle, and distal convoluted tubule does not increase proportionally, the delivery of Na^+ to the cortical collecting duct would be 1512 instead of 1029 mEq/day, resulting in an excreted Na^+ of 654 instead of 150 mEq/day. The increase in the Na^+ excretion in euvolemic conditions would sharply reduce the effective circulating volume. The physiological role of the glomerulotubular balance is to adjust the tubular reabsorption of Na^+ to the quantity delivered to a given segment. This mechanism maintains a constant Na^+ delivery to the distal nephron, reducing Na^+ urinary loss (Table 7.1).

Besides the autoregulation of renal blood flow and the glomerulotubular balance, aldosterone is the third important factor influencing Na^+ reabsorption in the distal nephron. Its mechanism of action was discussed in Chap. 4.

Table 7.1 Physiological role of glomerulotubular balance in maintaining constancy in Na^+ delivery to the cortical collecting duct, when Na^+ filtered load is increased by 2% in GFR

Variable	%	With glomerulotubular balance (mEq/day)	Without glomerulotubular balance (mEq/day)
Normal Na^+ filtered load		25,200	25,200
Increased Na^+ filtered load	2	25,704	25,704
Proximal tubule	67	17,221	16,884
Thick ascending limb	25	6426	6300
Distal convoluted tubule	4	1028	1512
Cortical collecting duct	3	771	756
Medullary collecting duct	0.4	102	102
Excreted Na^+ load		156	654

7.6   Control of Renal Sodium Excretion in Hypovolemia

The decrease in effective circulation volume (hypovolemia) is detected mainly by arterial baroreceptors. In these conditions the kidney receives signals that stimulate Na^+ reabsorption and reduce Na^+ and water (Fig. 7.8). The signals sent to the kidney are:

(a) Increased renal sympathetic discharge, resulting in increased norepinephrine secretion.
(b) Increased secretion of renin, which increases circulating levels of angiotensin II and aldosterone.
(c) Stimulation of vasopressin secretion.
(d) Inhibition of atriopeptin secretion.

The consequences of the signals that the kidney receives can be summarized in the following points:

(a) The activity of the renal sympathetic nerves and the action of angiotensin II reduce the glomerular filtration rate and therefore the filtered sodium load. There is a significant reduction in hydrostatic pressure in the glomerular capillaries, which will affect resorption in the proximal tubule.
(b) Norepinephrine, through $\alpha1$ adrenergic receptors, stimulates the activity of the proximal tubule NHE3 exchanger. The net effect is an increase in the reabsorption of $NaHCO_3$, $NaCl$, and water. On the other hand, the decrease in hydrostatic pressure in the peritubular capillaries favors volume reabsorption.

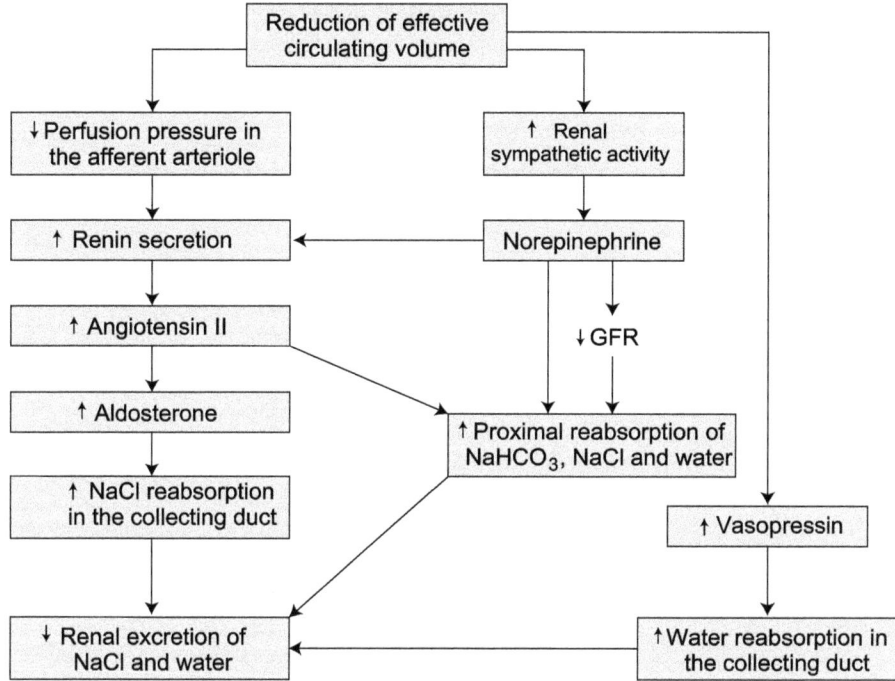

Fig. 7.8 Sequence of events generated by a reduction in circulating volume. A decrease in effective circulating volume leads to a reduction in afferent arteriole perfusion pressure and increased sympathetic renal activity. Both cascades are interconnected at the level of renin secretion, leading to the stimulation of proximal and distal Na^+ and water reabsorption

(c) Inhibition of atriopeptin reduces renal excretion of Na^+.
(d) When the blood volume decreases by 10–15%, AVP secretion is stimulated. Vasopressin has two actions. First, it stimulates AQP-2-dependent distal water reabsorption. Second, it has a synergistic action with aldosterone by increasing the abundance of preexisting ENaC in the apical membrane of the connecting and principal cells of the distal nephron. These effects are mediated by V_2 receptor binding followed by activation of the cAMP-protein kinase A signal cascade.
(e) Renin-angiotensin-aldosterone system in hypovolemia.
   Hypovolemia stimulates the renin-angiotensin-aldosterone system and transforms the distal nephron into a site where tubular Na^+ reabsorption is stimulated with minimal loss of K^+. As mentioned in Chap. 4, WNK4 kinase has a key role in adapting the function of the distal nephron to an increased Na^+ reabsorption mode. In the distal convoluted tubule, angiotensin II increases NCC activity in the apical membrane through an increase in membrane NCC abundance and phosphorylation. Both actions are mediated by WNK4. In the connecting and principal cells, angiotensin II and aldosterone increase the abundance of ENaC in the apical membrane. Aldosterone, through SGK1

kinase, suppresses the inhibition of WNK4 on ENaC. In the distal nephron, WNK4 inhibits ROMK activity. The net result is an epithelium capable of increasing tubular Na^+ reabsorption with minimal loss of urinary K^+.

7.7 Control of Renal Sodium Excretion Under Conditions of Hypervolemia

The renal response to the expansion of effective circulating volume is opposite to that observed when there is a reduction in circulating volume. The kidney receives signals from baroreceptors, which are intended to increase renal excretion of Na^+ and water (Fig. 7.9). The most important signals are:

(a) Decreased sympathetic discharge to the kidney.
(b) Decrease in renin secretion and, therefore, in the formation of angiotensin II and the synthesis and release of aldosterone in the adrenal cortex.
(c) Increased secretion of atriopeptin, which inhibits renin secretion, aldosterone synthesis, and vasopressin secretion.
(d) Inhibition of vasopressin secretion.

The consequences of the signals the kidney receives can be summarized in the following points:

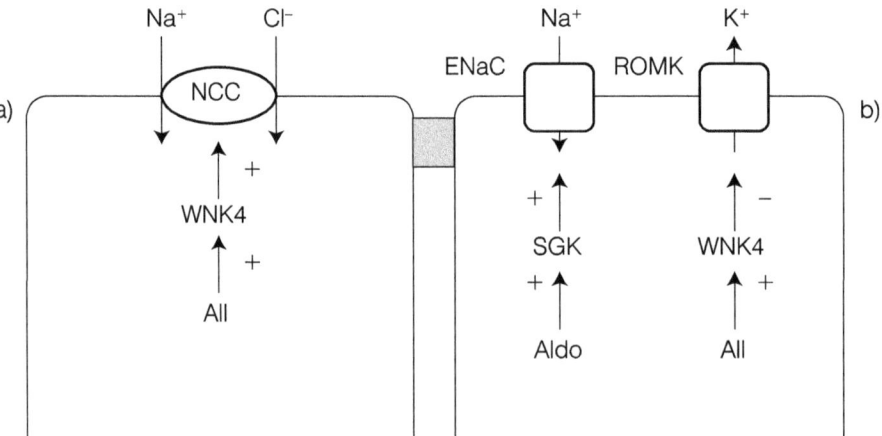

Fig. 7.9 Stimulation of NaCl reabsorption by angiotensin II and aldosterone in the distal nephron. (**a**) In distal convoluted tubule cells, angiotensin II stimulates NaCl reabsorption through a WNK4 kinase-dependent pathway that increases NCC phosphorylation and activity. (**b**) In connecting and principal cells, aldosterone increases ENaC abundance in the apical membrane through an SGK-1 kinase-dependent pathway, while angiotensin II activates WNK4 kinase that inhibits ROMK potassium channel at the apical membrane. The net effect is the stimulation of Na^+ reabsorption and the reduction in tubular K^+ secretion

(a) The glomerular filtration rate increases as a result of decreased sympathetic innervation activity and circulating levels of angiotensin II. Consistent with the above, it increases the filtered Na^+.
(b) Decreased proximal tubule reabsorption. This is mainly due to the lack of stimulating neurohormonal signals, such as norepinephrine and angiotensin II. In addition, increased glomerular filtration increases the hydrostatic pressure in the peritubular capillaries that accompany the proximal tubules in the cortical labyrinths, reducing the reabsorption of solutes and water.
(c) Reduced plasma aldosterone levels decrease the reabsorption of Na^+ in the aldosterone-sensitive distal nephron. Atriopeptin contributes in a minority way to a direct inhibition of Na^+ in the medullary collecting duct. Another factor affecting the decrease in Na^+ in the collecting tube is the increased delivery of Na^+ to this segment, resulting from a higher glomerular filtration rate (remember that this segment has low reabsorption capacity).
(d) Inhibition of vasopressin secretion decreases the water reabsorption in the connecting tubule and collecting duct and thus increases water excretion.

The net result is that the absence of factors stimulating reabsorption in different segments, as well as the increase in inhibiting factors (atriopeptin), favors the renal excretion of Na^+ and water and therefore contributes to reducing the effective circulation volume.

7.8 Edema

Edema is the accumulation of fluid in the interstitial compartment. Edema generation can result from alterations in the balance of Starling forces that determine hemodynamics at the capillary level and renal retention of NaCl and water. The kidney plays an important role in the generation of edema, since this condition is only observed when the interstitial fluid volume has increased by 2.5 to 3 L. If this volume were derived from the plasma volume (3 L) alone, the subject would develop an increased hematocrit as a product of fluid movement from the plasma compartment to the interstitial. Edema is generated because the kidney is stimulated to retain NaCl, due to the activation of the renin-angiotensin-aldosterone system.

The contribution of Starling's forces becomes evident, if we consider the equation that determines the ultrafiltration of fluid in the capillaries:

$$PUF = LpS[(PHcap - PHint) - (\pi int - \pi cap)]$$

where PUF is the net ultrafiltration pressure, Lp is the hydraulic conductivity, and S is the area available for exchange.

PH corresponds to hydrostatic pressure and π to oncotic pressure.

The amount of interstitial fluid will be determined by the balance between the Starling's forces and also by the absorption of fluid by the lymph capillaries. For example, an increase in hydrostatic pressure at the venous end of the capillary or a blockage of the lymph capillaries will result in decreased resorption and thus accumulation of interstitial fluid.

7.9   Conclusions

The sodium amount in the extracellular fluid determines the volume of the extracellular volume, whereas the sodium concentration in the extracellular volume determines the effective plasma osmolality. Therefore, changes in the amount of sodium in the extracellular volume produce changes in the extracellular volume.

The effective circulating volume is the fraction of the extracellular volume that is contained in the arterial system. This volume is subject to the pressure exerted by the left ventricle and effectively perfuses the organs of the systemic circulation.

In a euvolemic subject, the tubuloglomerular feedback and the glomerulotubular balance are the major mechanisms for reducing an excess of NaCl delivery to the distal nephron.

In a hypovolemic subject, the sympathetic system stimulates proximal tubule reabsorption and renin secretion. The latter stimulates systemic and intrarenal angiotensin II formation. Intrarenal and systemic angiotensin II stimulates proximal tubule and distal convoluted tubule NaCl reabsorption. Systemic angiotensin II stimulates aldosterone synthesis. Aldosterone stimulates NaCl reabsorption in segments of the distal nephron. Hypovolemia also stimulates vasopressin secretion, which increases water reabsorption and NaCl reabsorption in the thick ascending limb and distal nephron. Therefore, the action of the sympathetic system, angiotensin II, aldosterone, and vasopressin contributes to the reduction of NaCl urinary excretion.

In a hypervolemic subject, the renal sympathetic activity, renin secretion, angiotensin II formation, and aldosterone synthesis are all reduced. There is also an inhibition of vasopressin secretion. The reduced activity of these neurohormonal systems increases NaCl and water urinary excretion.

Review Questions

1. A healthy subject weighing 70 kg ingests 100 mEq of Na^+ daily for 3 days. At the beginning of day 4, the daily intake increases to 200 mEq/day, for a period of 9 days, returning to the intake of 100 mEq/day subsequently. On the fifth day of ingesting 200 mEq/day, his body weight had increased by 1 kg. How can you explain the weight increase experienced by this subject?
2. A subject suffers an acute episode of vomiting and diarrhea, losing 3 kg of body weight within 24 h. The plasma Na^+ concentration was 145 mEq/L, in the normal range.
   Which would be the state of the following physiological variables in this setting?
   (a) Plasma vasopressin levels.
   (b) Extracellular fluid volume.
   (c) Plasma osmolality.

(d) Urinary osmolality.
(e) Plasma renin activity.
(f) Plasma aldosterone levels.
(g) What is the physiological role of Na^+ (ENaC) in volume control?

Bibliography

Arroyo JP, Ronzaud C, Lagnaz D, Staub O, Gamba G (2011) Aldosterone paradox: differential regulation of ion transport in distal nephron. Physiology 26:115–123

Johns EJ, Kopp U (2013) Neural control of renal function. In: Seldin D, Giebisch G (eds) The kidney: physiology and pathophysiology, vol I, 5th edn. Academic Press, San Diego, pp 451–479

Melo LG, Pang S, Ackermann U (2000) Atrial natriuretic peptide: regulator of chronic arterial blood pressure. Physiology 15:143–149

Rose BD, Post T (2001a) The total body water and plasma sodium concentration. In: Clinical physiology of acid-base and electrolyte disorders, 5th edn. McGraw-Hill, New York, pp 241–257

Rose BD, Post T (2001b) Regulation of the effective circulating volume. In: Clinical physiology of acid-base and electrolyte disorders, 5th edn. McGraw-Hill, New York, pp 258–284

Renal Regulation of Acid-Base Balance

Learning Objectives
- To understand the generation of volatile and fixed acid by the metabolism.
- To describe the process of removing volatile acid.
- To describe the neutralization of fixed acids in the body and the role of the CO_2/HCO_3^- buffer system in this process.
- To describe the process of tubular bicarbonate reabsorption along the nephron.
- To describe the process of bicarbonate regeneration and its relationship to urinary buffer acidification.
- To understand primary acid-base balance disorders and their compensations.

8.1   Body Acid-Base Balance

Plasma pH is a highly regulated physiological variable with very narrow limits (7.38–7.42), although the pH range compatible with life is considerably wider (6.80–7.80; 160–16 nM, respectively). Normal arterial plasma pH is 7.40, equivalent to a H^+ concentration of 40 nM, which is one million times less than the concentrations of other cations in the plasma, such as Na^+, K^+, and Ca^{++} all in the millimolar range. The H^+ has a very small hydrated molecular radius (10^{-5} Å), which facilitates their interaction with protein. Proton binding affects protein activity. This is particularly important for enzymes and other regulatory proteins.

The organism functions under conditions of acid-base balance; that is, the amount of acid and base ingested and produced is matched by their excretion. Under normal insulin secretion and action, the metabolism of carbohydrates and lipids generates about 15,000 mmol/day of CO_2 (capable of forming 15,000 mEq/day of H^+). Normal insulin secretion should be understood as the normal response of the β pancreatic cells to various stimuli (the main one being glucose). Normal insulin sensitivity is the normal response in insulin-sensitive cells and is triggered when the hormone binds to the receptor and activates the intracellular transduction system.

The CO_2 derives primarily from oxidative cycles such as the Krebs cycle and is eliminated by the lungs (Fig. 8.1). If the CO_2 would not be removed, it would react

© Springer Nature Switzerland AG 2022
P. A. Gallardo, C. P. Vio, *Renal Physiology and Hydrosaline Metabolism*,
https://doi.org/10.1007/978-3-031-10256-1_8

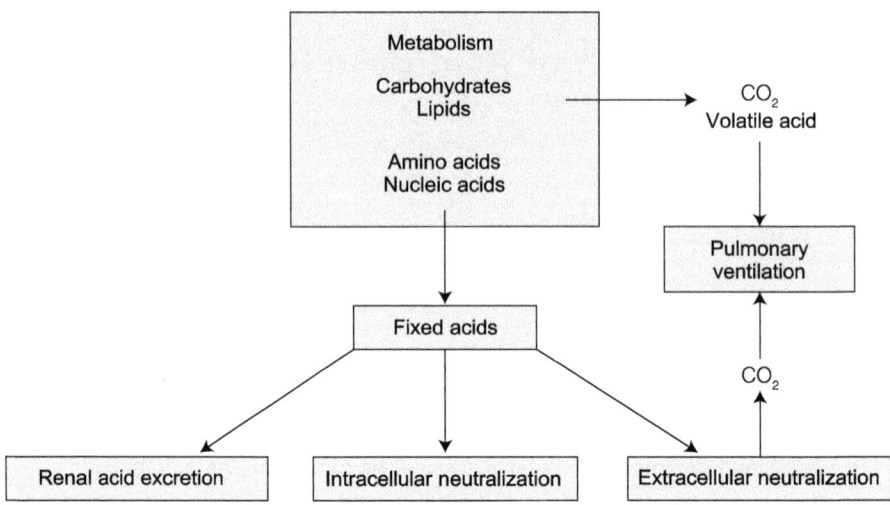

Fig. 8.1 Body acid production. Carbohydrate and fatty acid oxidation produce CO_2 that is eliminated through pulmonary ventilation; these are the main source of volatile acid production. Fixed acids derived from the catabolism of certain amino acids are neutralized in the extracellular and intracellular fluids. Through renal acid excretion, the kidneys regenerate an amount of bicarbonate equal to that spent in fixed acid neutralization

with water to form carbonic acid (H_2CO_3) according to the reaction stated in equation 8.1:

$$CO_2 + H_2O \rightarrow H_2CO_3 \tag{8.1}$$

According to the definition of an acid given by Bronsted and Lowry, CO_2 would not be an acid because it would not yield H^+ in solutions. However, equation 8.1 shows that CO_2 is in fact a potential source of H^+ formed from a volatile acid. Therefore, the concept of a volatile acid refers to the fact that it can be eliminated through pulmonary ventilation.

Amino acid metabolism is also a source of acid and base. In either case, they cannot be eliminated through pulmonary ventilation and are therefore called fixed acids or bases (Fig. 8.1). Fixed acids and bases have to be neutralized before being excreted from the organism. In a Western diet, catabolism of sulfur amino acids, such as methionine and cysteine, is the main source of fixed acid, due to the generation of sulfuric acid according to reactions 8.2 and 8.3.

$$\text{Methionine o Cysteine} \rightarrow \text{Glucose} + \text{Urea} + H_2SO_4 \tag{8.2}$$

$$H_2SO_4 \rightarrow 2H^+ + SO_4^{-2} \tag{8.3}$$

The catabolism of cationic amino acids, such as arginine and lysine, generates hydrochloric acid, according to reactions 8.4 and 8.5:

$$\text{Arginine} + \text{Cl}^- \rightarrow \text{Glucose} + \text{Urea} + \text{Cl}^- + \text{H}^+ \qquad (8.4)$$

$$\text{Lysine} + \text{Cl}^- \rightarrow \text{CO}_2 + \text{Urea} + \text{H}^+ + \text{Cl}^- \qquad (8.5)$$

This acid production is partly neutralized by the production of base. The catabolism of some organic anions like citrate generates CO_2 and water, which in turn form bicarbonate according to reaction 8.6. The production of bicarbonate is equivalent to the consumption of protons, according to reaction 8.6:

$$\text{Citrate}^{-3} + 3\text{H}^+ \rightarrow \text{CO}_2 + \text{H}_2\text{O} \rightarrow \text{HCO}_3^- \qquad (8.6)$$

In a Western diet rich in animal protein, catabolism provides an endogenous acid production that, on average, is 50 to 100 mEq/day of H^+ (1 mEq/day/kg H^+), bearing in mind that this value is highly dependent on the diet. Thus, a subject who eats a diet rich in meat will have a higher endogenous fixed acid production. In turn, in a vegetarian subject predominates the production of salts of organic acids that provide base. Therefore, to maintain the acid-base balance, the body must excrete a fixed amount of acid per day equivalent to the endogenous fixed acid production per day. This amount is known as "net acid excretion" and corresponds to 50–100 mEq/day.

8.2 Buffers in the Body

Buffers are chemicals capable of neutralizing the change in pH of a solution, when a base or an acid has been added. Therefore, a buffer solution regulates the pH of the solution where it is present. Before going into detail about buffers, it is necessary to remember the concepts of "acid" and "base"; according to Bronsted and Lowry, an acid is a substance that releases H^+ in solution, while a base picks up H^+ in solution.

Useful species as buffers are weak acids or weak bases, because they establish chemical equilibrium between the acid and a base and the products of dissociation. In contrast, a strong base or acid dissociates completely and does not establish chemical equilibrium.

Let us consider a buffer pair formed by a weak acid (HA) and the products of dissociation: a conjugated base (A^-) and proton (H^+), according to reaction 8.7:

$$\text{HA} \leftrightarrow \text{A}^- + \text{H}^+ \qquad (8.7)$$

Assume a vessel containing a fixed amount of HA at a given concentration. If you add H^+ to the solution, these will be neutralized by the conjugated base, forming HA. In other terms, adding H^+ to the solution will shift the reaction 8.7 to the left. The net result is that the concentration of H^+ (pH) would not change, even though a certain amount of acid was added. On the other hand, if a base is added, it will capture H^+ and reaction 8.7 will be shifted to the right.

The buffer capacity of the weak acid is directly related to three factors: the pH to be regulated, the pK of the buffer, and the amount available of the buffer. Ideally, the pK of the buffer should coincide with the pH of the solution to be regulated, since under these conditions there is a 1:1 ratio between acid and base. In other words, the buffer has the same capacity to buffer an added acid or base. The body has several extra- and intracellular buffer systems, whose function is to neutralize possible changes in pH against the generation of fixed acids or bases. The main intracellular buffers are phosphates, proteins, bicarbonate, and hemoglobin present in red blood cells. Extracellular buffers include bicarbonate, phosphates, and proteins. Quantitatively, the most important buffer species in the extracellular fluid is bicarbonate, which contributes half of the buffer mass in the blood.

8.3   Physiology of the CO_2-Bicarbonate Buffer

This system is the main buffer of the extracellular medium. The pK of this system is 6.1, which is a little far from the extracellular pH of 7.4 to be regulated. The main advantage of the system, which compensates for the difference between pK and pH, is the double regulation to which it is subjected: first, the partial pressure of CO_2 is regulated by pulmonary ventilation, and second, the plasma bicarbonate concentration is regulated by the kidney. The equation that describes this buffer system is:

$$CO_2 + H_2O \overset{CA}{\leftrightarrow} H_2CO_3 \leftrightarrow H^+ + HCO_3^- \qquad (8.8)$$

The reaction between CO_2 and H_2O to form carbonic acid is very slow, but in epithelial cells of the renal tubule, gastrointestinal tract, and erythrocytes, it is catalyzed by the enzyme carbonic anhydrase (CA). The reaction from carbonic acid to the formation of protons and bicarbonate is very fast, so most of the carbonic acid is converted to bicarbonate and protons. According to the above, Eq. 8.8 can be simplified to:

$$CO_2 + H_2O \overset{CA}{\leftrightarrow} H^+ + HCO_3^- \qquad (8.9)$$

The CO_2 that react with those of H_2O to form carbonic acid are molecules that are solubilized from the gas phase to the water phase. The number of gas molecules passing from the gas phase to the aqueous phase is proportional to the arterial partial pressure of the gas ($PaCO_2 = 40$ mmHg) and the solubility coefficient (α) of the gas in the plasma (0.03 mmol/L-mmHg). Therefore, the concentration of CO_2 dissolved, equivalent to the concentration of carbonic acid, can be expressed as:

$$[CO_2] \text{ dissolved} = PCO_2 \cdot \alpha = 40 \text{ mmHg} \cdot 0.03 \frac{mmol}{L} mmHg = 1.2 mmol/L \quad (8.10)$$

Equation 8.10 can be used to derive the equation of the equilibrium constant of the reaction:

$$K_a = \frac{[H^+] \cdot [HCO_3^-]}{[H_2CO_3]} \tag{8.11}$$

The value of K_a is 2.72×10^{-4}. If the $[H^+]$ concentration is 40×10^{-9} M, it is possible to calculate the concentration ratio between the acid species (H_2CO_3) and the base (HCO_3^-) like:

$$2.72 \cdot 10^{-4} = \frac{40 \cdot 10^{-9} \cdot [HCO_3^-]}{[H_2CO_3]} \tag{8.12}$$

Solving Eq. 8.12 results in a base/acid concentration ratio of 6800, i.e., there are 6800 bicarbonate molecules for every carbonic acid molecule, or in other words, the equilibrium is shifted to the right.

Equation 8.12 can be expressed as a function of pH by taking the logarithm on both sides of the equation and corresponds to the Henderson-Hasselbalch equation for a buffer system:

$$pH = pK + Log\frac{[HCO_3^-]}{PCO_2 \cdot 0.03} \tag{8.13}$$

If the concentration of $[HCO_3^-]$ in arterial plasma is 24 mEq/L, $PaCO_2 = 40$ mmHg, and pK $= 6.1$, then the calculated arterial plasma is 7.4.

8.4   Renal Regulation of Acid-Base Balance

The kidney has two essential tasks in relation to maintaining acid-base balance. The first task is the reabsorption of 100% of the filtered bicarbonate, since a loss of bicarbonate in the urine implies a gain of acid. As mentioned above, the catabolism of amino acids, carbohydrates, and lipids is the main determinant of endogenous acid production. A typical Western diet generates between 50 and 100 mEq/day of fixed acid. The H^+ acid derivatives, such as H_2SO_4 and HCl, must be neutralized immediately in order to prevent a drop in the pH of the extracellular fluid. These H^+ are buffered in the extracellular medium mainly with bicarbonate. The CO_2 generated is eliminated by pulmonary ventilation and the salt formed by the anion is excreted through the urinary tract.

$$H_2SO_4 \rightarrow 2H^+ + SO_4^{-2} + 2HCO_3^- \rightarrow 2CO_2 + H_2O \tag{8.14}$$

$$HCl \rightarrow Cl^- + H^+ + HCO_3^- \rightarrow CO_2 + H_2O \tag{8.15}$$

Equations 8.14 and 8.15 show that in the process of neutralizing H^+ there is consumption of extracellular bicarbonate. The second task of the kidney is the generation or replacement of an amount of bicarbonate equivalent to that consumed in fixed acid neutralization. The processes involved in these two tasks will be analyzed below.

8.4.1 Bicarbonate Reabsorption Along the Nephron

The bicarbonate reabsorption along the nephron is shown in Fig. 8.2. The kidney must reabsorb 100% of the filtered load of HCO_3^-. If the glomerular filtration rate is 180 L/day and the plasma $[HCO_3^-]$ is 24 mEq/L, then the filtered load of HCO_3^- is 4320 mEq/day and the excreted HCO_3^- is close to zero.

80% of the filtered load (3456 mEq/day) of HCO_3^- is reabsorbed mostly in the proximal convoluted tubule (Fig. 8.3a). The process involves apical H^+ secretion, mainly carried out by the electroneutral Na^+/H^+ exchanger (NHE3); a lesser contribution is made by the H^+-ATPase. The energy for NHE3 activity is derived from the sodium electrochemical potential gradient maintained by basolateral Na^+, K^+-ATPase. The H^+ secreted acidifies the filtered HCO_3^-, forming H_2CO_3 which is a substrate for brush border carbonic anhydrase (isoform IV), generating CO_2 and H_2O. Cytosolic carbonic anhydrase (isoform II) catalyzes the reverse reaction to produce H^+ and HCO_3^-. Cytosolic H^+ is secreted into the lumen via NHE3 and

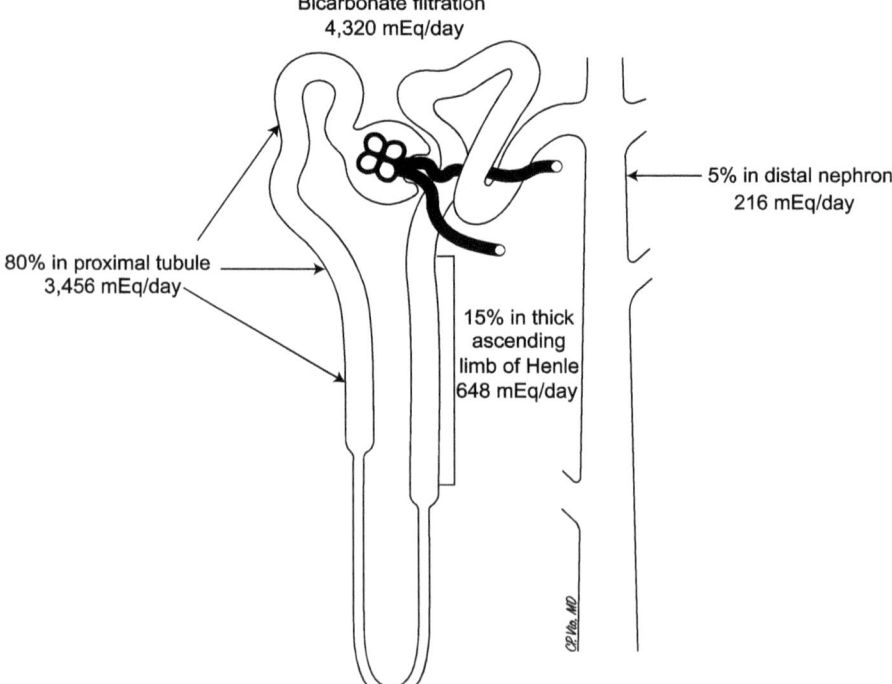

Fig. 8.2 Bicarbonate reabsorption along the nephron. Bicarbonate filtered load was calculated assuming a GFR of 180 L/day and a plasma bicarbonate concentration of 24 mEq/L

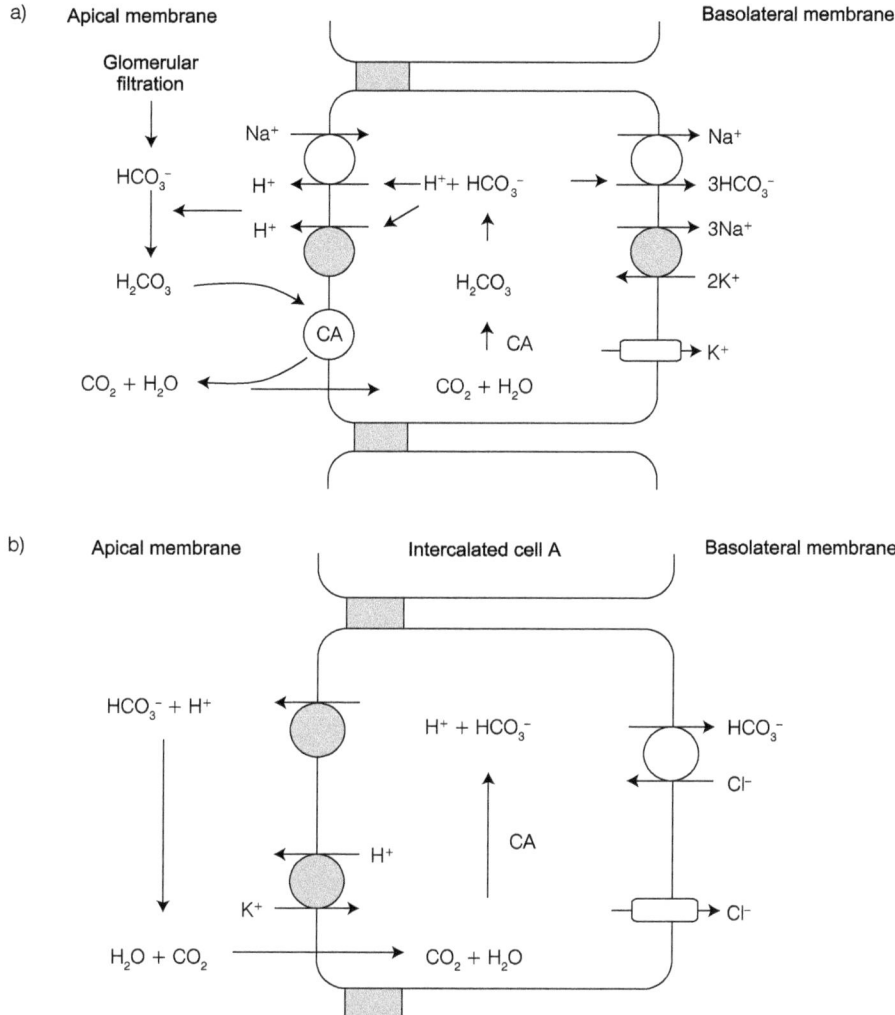

Fig. 8.3 Mechanism of tubular bicarbonate reabsorption. (**a**) Proximal tubule bicarbonate reabsorption. The mechanism depends on the activity of apical Na^+/H^+ (NHE3) exchanger, carbonic anhydrase (CA II and IV) isoforms, and basolateral $Na^+/3HCO_3^-$ (NBC) cotransporter that mediates HCO_3^- exit. (**b**) Bicarbonate reabsorption in type A intercalated cells of the distal nephron; the major apical acidification mechanism is the H^+-ATPase; under physiological conditions (e.g., normal or elevated K^+ ingestion), the H^+, K^+-ATPase plays a minor role. Basolateral bicarbonate exit is mediated by a Cl^-/HCO_3^- (AE1) exchanger

HCO_3^- is transported through the basolateral membrane, mainly by the $1Na^+$-$3HCO_3^-$ (NBC) cotransporter, whose energy is also derived from the activity of the sodium pump. Therefore, for every molecule of HCO_3^- that is acidified in the tubular lumen, an equivalent is generated in the cytosol and then transported to the plasma through the basolateral membrane. The massive reabsorption of bicarbonate in the proximal tubule reduces the bicarbonate concentration from 24 mEq/L to 10 mEq/L in the tubular fluid leaving the proximal convoluted tubule. The thick ascending limb of Henle reabsorbs about 15% of the filtered load by the same mechanism as in the proximal tubule (Fig. 8.3a). About 5% of the filtered load is reabsorbed by the type A intercalated cells of the connecting tubule and cortical collecting duct of the distal nephron (Fig. 8.3b). In these cells the H^+ is mostly secreted by the apical H^+-ATPase and less by a H^+, K^+-ATPase that is very similar to that present in the parietal cells of the gastric glands. The CO_2 which is formed by the acidification of bicarbonate molecules diffuses into cells and through the cytosolic carbonic anhydrase II is converted into H^+ and bicarbonate. HCO_3^- ions are transported through the basolateral membrane by the Cl^--HCO_3^- anion exchanger (AE1). For every molecule of bicarbonate acidified in the lumen, one molecule of bicarbonate is generated in the cytosol of the type A intercalated cell and is reabsorbed into the plasma.

In summary, 95% of the filtered load of bicarbonate is reabsorbed between the proximal tubule and the thick ascending loop of Henle; the remaining 5% is reabsorbed in the distal nephron (Fig. 8.2). In all the tubular segments involved, the key process is the tubular H^+ secretion and the activity of carbonic anhydrase enzyme: in the proximal tubule and thick ascending limb of Henle, there are the luminal and cytosolic isoforms of carbonic anhydrase. In contrast, only the cytosolic isoform exists in the collecting tube.

8.4.2   Bicarbonate Regeneration

The second task of the kidney is to replace an amount of bicarbonate equivalent to that spent on neutralizing the H^+ derived from fixed acids generated in the metabolism. The process basically consists of the urinary excretion of H^+ as acid and the retention of bicarbonate, bearing in mind that both H^+ and HCO_3^- are generated from the same reaction catalyzed by the carbonic anhydrase. The H^+ secreted is excreted as acid, thanks to the presence of urinary buffers in the tubular lumen (Fig. 8.4). The urinary buffers are bases present in the glomerular filtrate or synthesized intrarenally. The urinary buffers accept H^+ secreted by the type A intercalated cells, forming an acid that is excreted in the urine. The bicarbonate ion formed together with the H^+ by the activity of carbonic anhydrase is transported to the plasma (Fig. 8.5). The most abundant urinary buffers in the tubular fluid are the phosphate system formed by the monoprotic or alkaline component (HPO_4^{-2}) and the diprotic or acid ($H_2PO_4^-$) and the ammonia/ammonium system (NH_3/NH_4^+).

The phosphate system has a total plasma concentration of 1 mM and its pK is 6.8. Considering the physiological pH, it is possible to calculate the concentration ratio for the base/acid pair in the plasma using the Henderson-Hasselbalch equation:

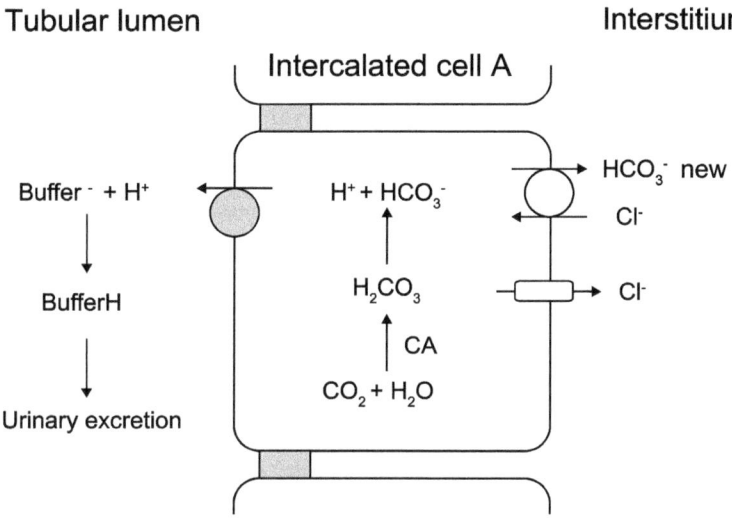

Fig. 8.4 General mechanism of acidification of urinary buffers and bicarbonate regeneration. H^+ secreted by vacuolar type H^+-ATPase acidifies a buffer molecule in the tubular lumen, generating an acid that is excreted in the urine as part of the net acid excretion. The H^+ and the HCO_3^- originate from the reaction catalyzed by cytosolic carbonic anhydrase (CA). Note that the H^+ is secreted into the lumen and the "new" HCO_3^- is transported to the plasma through a basolateral anion exchanger (AE1)

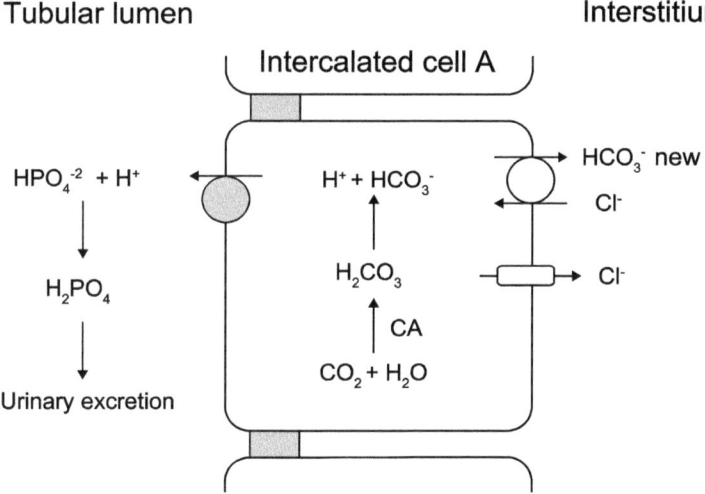

Fig. 8.5 Alkaline phosphate (HPO_4^{-2}) as a urinary buffer. H^+ secreted by type A intercalated cells acidifies alkaline phosphate to acid phosphate, which is excreted in urine and is the major component of tritable acidity in urine

$$7.4 = 6.8 + \mathrm{Log}\, \frac{\left[\mathrm{HPO_4^{-2}}\right]}{\left[\mathrm{H_2PO_4^-}\right]} \tag{8.16}$$

The value of the ratio is 4, and, considering the total plasma concentration of 1 mM, the concentration of the acidic and basic species in the plasma is 0.2 and 0.8 mM, respectively. The physiological meaning of this ratio is that in plasma, $\mathrm{HPO_4^{-2}}$ is four times more abundant than the acid component $\mathrm{H_2PO_4^-}$. About 80% of the total phosphate filters into the glomeruli, and the remaining 20% is bound to plasma proteins and therefore does not filter. About 90% of the filtered load is reabsorbed in the proximal tubule, leaving a limited amount of base available to function as a urinary buffer (Fig. 8.5). At the end of the proximal tubule, the pH of the tubular fluid is 6.8 and therefore the base/acid ratio is 1:1.

It is important to note that most of the acidification of urinary buffers occurs in the distal nephron. At this level, most of the filtered bicarbonate load has already been reabsorbed. Otherwise, the $\mathrm{H^+}$ secreted through the apical membrane would titrate filtered bicarbonate molecules, resulting in reabsorption of bicarbonate from the filtrate but not in the regeneration of new bicarbonate molecules.

The ammonia/ammonium system is intrarenal. All tubular segments are capable of ammonia synthesis. However, only the proximal convoluted tubule is able to increase ammoniagenesis in response to metabolic acidosis. Ammoniagenesis in the proximal tubule is linked to glutamine catabolism. In proximal tubule cells, glutamine uptake can occur through apical $\mathrm{Na^+}$-dependent cotransporter or a basolateral carrier.

Glutamine is synthesized in the liver from glutamate, a process that requires $\mathrm{NH_4^+}$ derived from hepatic amino acid catabolism (Fig. 8.6). In the proximal tubule cells, glutamine is deaminated by phosphate-dependent glutaminase (PDG), resulting in glutamate and the first molecule of $\mathrm{NH_4^+}$. The reaction occurs in the mitochondria. The action of glutamate dehydrogenase (GDH) converts glutamate to α-ketoglutarate and a second molecule of $\mathrm{NH_4^+}$. Therefore, for every molecule of glutamine that is catabolized in the proximal tubule cell, two molecules of $\mathrm{NH_4^+}$ and one from α-ketoglutarate are produced. The subsequent metabolism in the Krebs cycle or in gluconeogenesis generates two molecules, of which $\mathrm{HCO_3^-}$ which is transported to the plasma (Fig. 8.7). The pKa of the $\mathrm{NH_3/NH_4^+}$ is 9.2. At plasma pH of 7.4, only 1.7% of total ammonia is present as $\mathrm{NH_3}$. This implies that at an intracellular pH of 7.1 in the proximal tubule cell by far predominant species is $\mathrm{NH_4^+}$. The $\mathrm{NH_4^+}$ is mainly secreted into the tubular lumen by the NHE3 exchanger, where the $\mathrm{NH_4^+}$ binds to the $\mathrm{H^+}$ transport site in the cytosolic side of the transporter. A smaller fraction diffuses across the apical membrane as $\mathrm{NH_3}$ and is protonated to $\mathrm{NH_4^+}$ in the tubular lumen. The $\mathrm{NH_4^+}$ secreted into the lumen of the proximal tubule is reabsorbed in the thick ascending limb (Fig. 8.8). The main reabsorption pathway is the apical NKCC2 cotransporter; $\mathrm{NH_4^+}$ binds to the $\mathrm{K^+}$ site in NKCC2. $\mathrm{NH_4^+}$ is also reabsorbed paracellularly, driven by the positive transepithelial potential difference in the tubular

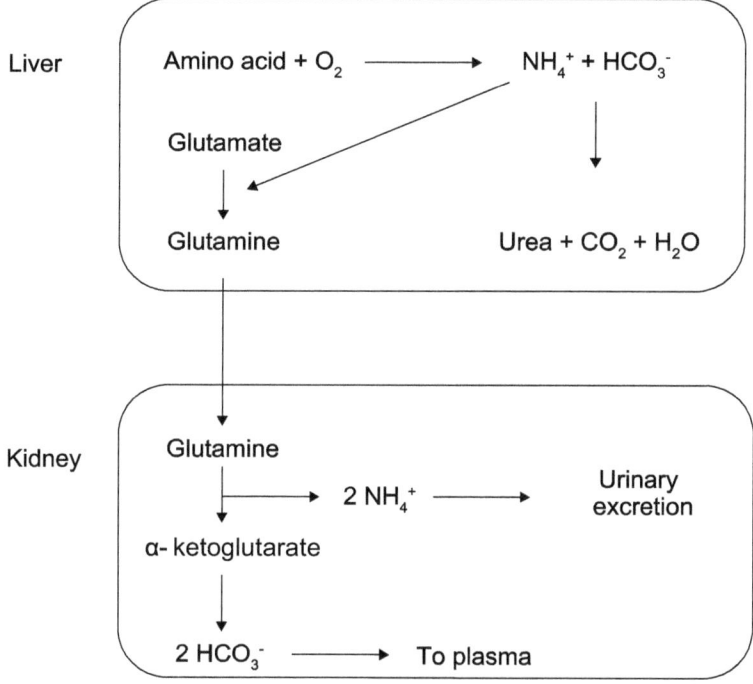

Fig. 8.6 Physiological relationship between the liver and kidney in ammonium production and handling. Hepatic ammonium synthesis derives from amino acid catabolism to ammonium and then to glutamine. In the kidney, hepatic glutamine is catabolized to $2NH_4^+$ and $2HCO_3^-$. Ammonium ions are excreted in the urine and HCO_3^- is transported to plasma

lumen. The basolateral exit mechanism of NH_3/NH_4^+ is not clearly established. In the renal medulla a fraction of the NH_4^+ is in equilibrium with NH_3 and this can diffuse into the lumen of the medullary collecting duct. Thus, an important function of NH_4^+ reabsorption in the thick ascending limb of Henle is NH_3 delivery to the collecting duct. In the collecting duct, type A intercalated cells play a key role in ammonia transport and bicarbonate regeneration. Basolateral NH_3 uptake is mediated by specific NH_3 transporters like RhBG and RhCG, while apical NHE_3 exit is mediated by RhCG. In the lumen, NH_3 molecules are acidified by H^+ secreted mainly by apical H^+-ATPase and secondarily the H^+, K^+-ATPase. H^+ together with HCO_3^- is derived from the reaction mediated by carbonic anhydrase. H^+ is secreted to the lumen, while HCO_3^- is transported through the basolateral anion exchanger 1 Cl^-/HCO_3^- (AE1). The NH_4^+ molecule formed in the lumen is excreted as part of the net acid excretion (Fig. 8.9). Therefore, the synthesis of ammonium in the proximal tubule and its subsequent tubular handling are processes that generate new HCO_3^-. The NH_3/NH_4^+ has two major advantages compared to the phosphate system:

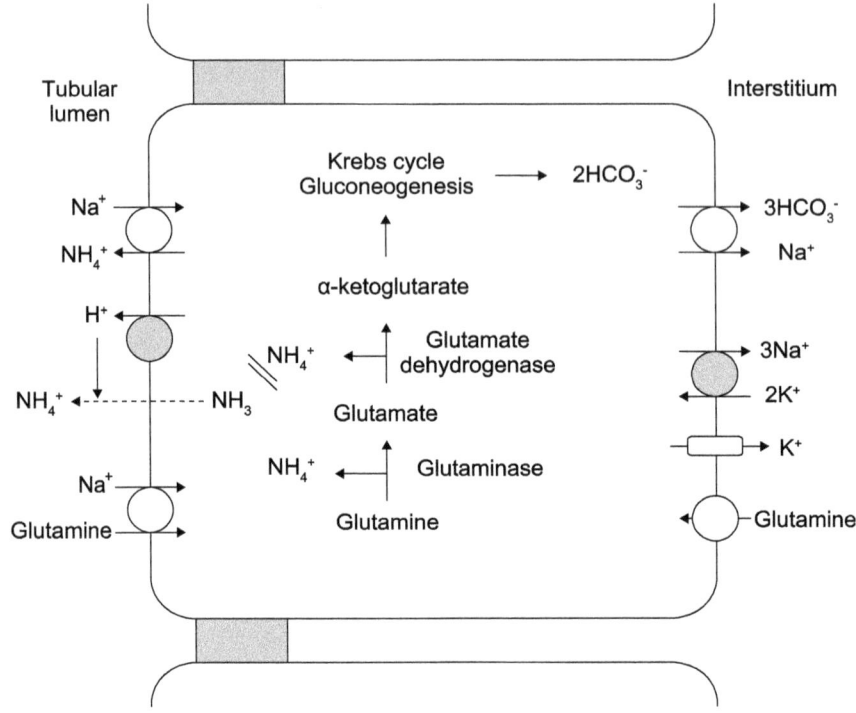

Fig. 8.7 NH_4^+ production in the proximal tubule. Glutamine enters the proximal cell either through the apical or basolateral membrane. Glutamine is the first substrate for glutaminase generating glutamate which is a substrate for glutamate dehydrogenase rendering α-ketoglutarate that enters gluconeogenesis and finally generating $2HCO_3^-$ that is transported to the plasma through the $Na^+/3HCO_3^-$ cotransporter. Glutamine catabolism also produces two molecules of NH_4^+ that should be secreted to the tubular lumen mainly through the apical NHE3 exchanger

(a) Liver synthesis of glutamine increases in conditions of acidosis (arterial pH < 7.4).

(b) The NH_3/NH_4^+ in the proximal tubule is stimulated in states of acidosis, mainly by increased synthesis of enzymes involved in glutamine catabolism.

Finally, bicarbonate regeneration is a key process because it allows the net excretion of acid, according to the requirements of the body's acid-base state. Net acid excretion is expressed as:

$$\text{Net acid excretion}\left(\text{mEq}/_{\text{day}}\right) = \left[H_2PO_4^-\right] \cdot V + \left[NH_4^+\right] \cdot V - \left[HCO_3^-\right] \cdot V \quad (8.17)$$

where V is the urinary flow (L/day); the urinary concentrations of acid phosphate and ammonium are expressed in mEq/L. In turn, the urinary excretion of acid phosphate is the main component of titratable acidity. This corresponds to the quantity of base that must be added to a volume of urine to bring its pH to the value of the arterial pH.

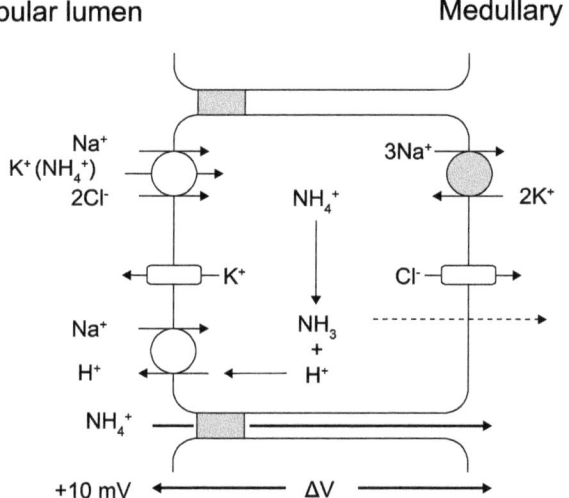

Fig. 8.8 Ammonium reabsorption in the thick ascending limb. NH_4^+ secreted in the proximal tubule is reabsorbed through the transcellular and paracellular pathways. In the first one, NH_4^+ ions substitute K^+ in the NKCC2 cotransporter. The lumen-positive transepithelial potential difference drives paracellular NH_4^+ reabsorption

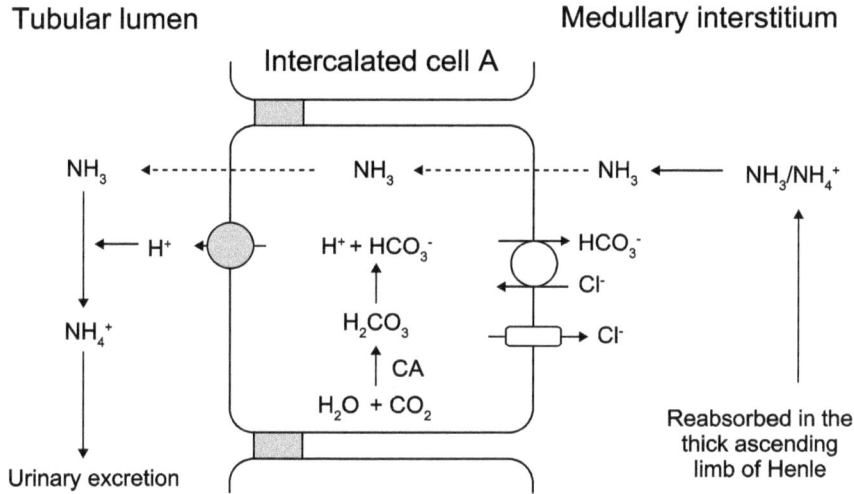

Fig. 8.9 NH_3 acidification in the collecting duct. NH_3 is transported across the basolateral membrane by RhCG and RhBG transporters. NH_3 exit across the apical membrane is mediated by RhCG. In the collecting duct tubular lumen, NH_3 is acidified by H^+ secreted by H^+-ATPase, while the HCO_3^- is transported to the extracellular fluid. Most of the ammonia secreted into the tubular lumen was reabsorbed in the thick ascending limb

8.5   Regulation of Acid-Base Transport by the State of the Acid-Base Balance

The state of the acid-base balance and some hormones modifies the speed of transport of H^+ and HCO_3^- in the renal tubule.

A state of extracellular acidosis is reflected as a drop in intracellular pH, which increases the activity of the proximal tubule NHE3 exchanger. This is because intracellular H^+ binds to an allosteric site in the cytosolic side of the transporter, increasing NHE3 activity. Under the same conditions, there is a greater insertion of vacuolar H^+-ATPase in type A intercalated cell (Fig. 8.10). Both effects lead to a higher rate of H^+ secretion. Alkalosis has the opposite effect.

As mentioned before, the NHE3 exchanger is a key element in proximal bicarbonate reabsorption and NH_4^+ secretion. Active NHE3 is located in the apical plasma membrane forming the brush border. Angiotensin II is a powerful stimulator of the

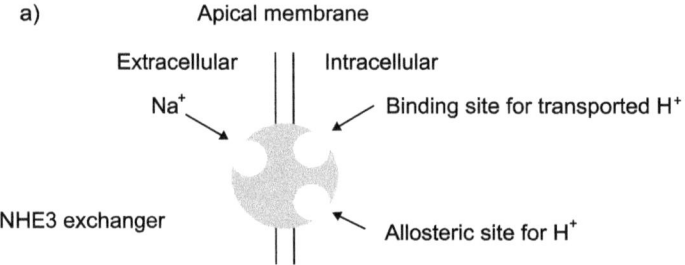

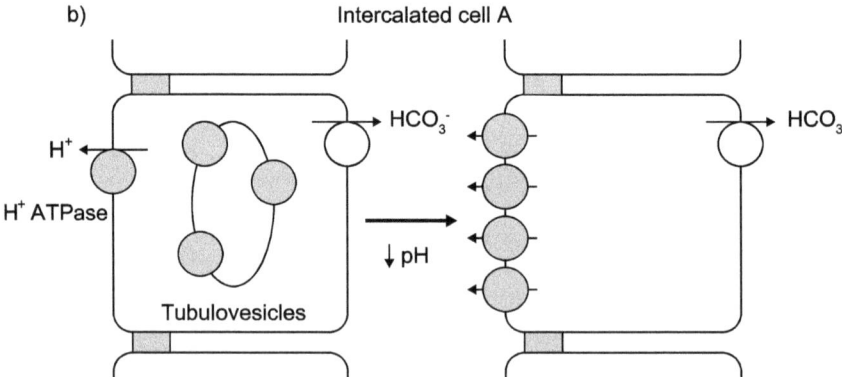

Fig. 8.10 Regulation of tubular H^+ secretion by intracellular pH. (**a**) Allosteric regulation of NHE3 exchanger in the proximal tubule. Intracellular H^+ binds to an allosteric site in cytosolic domains of the protein; the net effect is the increase in NHE3 activity. (**b**) In type A intercalated cells of the distal nephron, an acid cytosolic pH induces the translocation of H^+-ATPase from subapical tubulovesicles to the apical membrane, increasing H^+ secretion

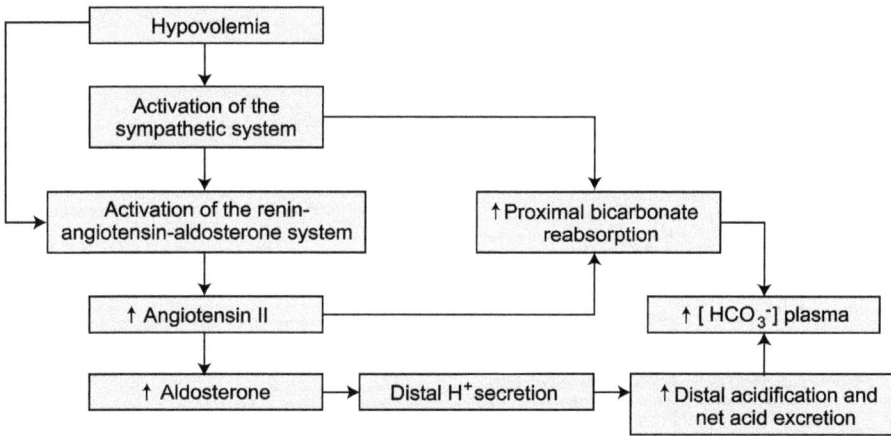

Fig. 8.11 Relationship between hypovolemia and the generation of metabolic alkalosis. Both norepinephrine and angiotensin II increase proximal tubular $NaHCO_3$ reabsorption. In the distal nephron, aldosterone stimulates Na^+ electrogenic reabsorption, which in turn increases the lumen-negative transepithelial potential difference. The latter stimulates H^+ secretion by type A intercalated cells, increasing distal acidification and net acid excretion

proximal NHE3 exchanger. This activation in the context of volume depletion results in $NaHCO_3$, NaCl, and water reabsorption. Parathyroid hormone (PTH) is a powerful inhibitor of NHE3 abundance in the microvilli plasma membrane, resulting in a reduced H^+ secretion. In the connecting tubule and cortical collecting duct, aldosterone stimulates electrogenic transport through apical ENaC. The increased abundance of ENaC results in a more lumen-negative transepithelial potential that secondarily stimulates H^+ secretion by the type A intercalated cells. In terms of acid-base balance, this mineralocorticoid action favors HCO_3^- regeneration and thus metabolic acidosis. Both angiotensin II and aldosterone are hormones whose actions prevail in conditions of hypovolemia, which explains why metabolic alkalosis can also occur in association with this disorder (Fig. 8.11).

8.6   Primary Acid-Base Balance Disorders

Primary alterations in acid-base balance are reflected in the blood pH measured in an arterial blood sample. In a state of acidosis, the arterial pH is <7.4; in a state of alkalosis, the arterial pH is >7.4 (Fig. 8.12).

If the acid-base primary alteration results from a change in arterial plasma [HCO_3^-], then the alteration is metabolic. If the alteration is associated primarily with a change in arterial partial CO_2 pressure ($PaCO_2$), then the alteration is respiratory.

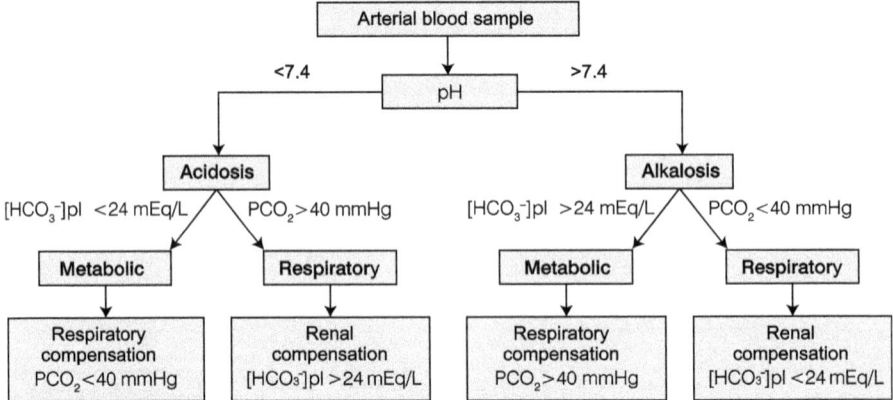

Fig. 8.12 Primary alterations in acid-base balance and their compensations. The pH of an arterial blood sample is the starting point for the analysis of acid-base alteration. Evaluation of plasma bicarbonate concentration and PCO_2 in arterial sample is considered to define the initial acid-base disturbance as metabolic or respiratory. Metabolic disturbances have a respiratory compensation; respiratory disturbances have a renal compensatory response

8.6.1 Metabolic Acidosis

Metabolic acidosis is characterized by an arterial pH < 7.4; plasma bicarbonate concentration is less than 24 mEq/L; $PaCO_2$ is less than 40 mmHg. This disorder can have multiple origins:

(a) Increased urinary bicarbonate excretion. Renal causes for metabolic acidosis would be due to a deficiency in H^+ tubular secretion or carbonic anhydrase inhibition.
(b) Gastrointestinal origin for metabolic acidosis would be the loss of HCO_3^- through an intestinal fistula. At this point, the reader should remember that the pancreatic secretion is a HCO_3^--rich solution (120 mEq/L). A second cause is diarrhea. Therefore, the urinary or gastrointestinal loss of HCO_3^- is equivalent to the addition of acid to the body.
(c) Increased metabolic production of fixed acids or the addition to the body of a compound whose subsequent metabolism generates a fixed acid. For example, in uncontrolled diabetes mellitus there is an increased production of fixed acids like β-hydroxybutyric acid and acetoacetic acid.

The first homeostatic response of the body to metabolic acidosis is the fall in $PaCO_2$. This respiratory compensation is caused by the stimulation of central and peripheral chemoreceptors and the response is an increase in alveolar ventilation. Hyperventilation causes a drop in $PaCO_2$ of 1.25 mmHg for every 1 mEq/L of drop in plasma HCO_3^- concentration. The respiratory compensation is limited and alveolar ventilation would not increase more than 7–8 L/min. The second homeostatic response comes from the kidney with an increase in net acid excretion. This is given by an increase in urinary excretion of tritable acidity (mainly acid phosphate) and a very pronounced increase in ammonia excretion. The anion gap is the difference between measured plasma cations (Na^+ and K^+) and anions (Cl^- and HCO_3^-); this simple calculation is useful to know the origin of the acid. The anion gap is calculated as follows:

$$\text{Anion gap} = ([Na^+] + [K^+]) - ([Cl^-] + [HCO_3^-]) \qquad (8.18)$$

The normal range of the anion gap (with K^+) is 16 to 20 mEq/L. The plasma K^+ concentration is low; thus, this variable can be omitted from Eq. 8.18. In this case, the normal range of the anion gap is 12 to 16 mEq/L. The usefulness of the anion gap can be understood by considering two cases of metabolic acidosis.

Case 1. Generation or intake of a mineral acid such as hydrochloric acid (HCl). This strong acid will be completely dissociated into H^+ and Cl^- in the stoichiometry of 1:1. The H^+ will be buffered with HCO_3^- ; for every mole of acid that is neutralized, 1 mole of HCO_3^- and 1 mole of CO_2 are produced. The stoichiometry between the added H^+ and consumed plasma HCO_3^- is 1:1. Consistent with the above, plasma HCO_3^- concentration will decrease in the same proportion as the increase in plasma Cl^- concentration. In this setting, the anion gap value will remain within the normal range. This is because the decrease in one measured anion (HCO_3^-) will be compensated by a proportional increase in the other measured anion (Cl^-). The proportional change is related to the 1:1 stoichiometry in hydrochloric acid dissociation and the 1:1 ratio in H^+ buffering with HCO_3^-. In this case, metabolic acidosis is hyperchloremic with normal anion gap, because of the increased Cl^- extracellular concentration. That is, the decrease in the plasma concentration of bicarbonate is accompanied by an equivalent increase in the plasma concentration of Cl^-. The reaction for hydrochloric acid dissociation is described in Eq. 8.19:

$$HCl \rightarrow H^+ + Cl^- \qquad (8.19)$$

Equation 8.20 describes the reaction for H^+ buffering with HCO_3^- as:

$$H^+ + HCO_3^- \rightarrow CO_2 + H_2O \qquad (8.20)$$

$$\text{Anion gap} = ([Na^+] + [K^+]) - (\uparrow [Cl^-] + \downarrow [HCO_3^-]) \qquad (8.21)$$

Equation 8.21 describes the anion gap applied to this case. Anion gap remains constant because the decrease in bicarbonate concentration is compensated by an equivalent increase in chloride concentration.

Case 2. Increased production of an endogenous acid (HA). The anion of this acid is not measured and not considered in the anion gap formula. As in the previous case, the H^+ will be neutralized with bicarbonate. Consequently, their plasma concentration is reduced. Equation 8.22 describes the dissociation of the endogenous acid, generating H^+ and anion (A^-); Eq. 8.23 represents H^+ buffering with HCO_3^-:

$$HA \rightarrow A^- + H^+ \tag{8.22}$$

$$H^+ + HCO_3^- \rightarrow CO_2 + H_2O \tag{8.23}$$

H^+ buffering with HCO_3^- will produce a fall in bicarbonate concentration; this decrease will not be accompanied with change in plasma Cl^- concentration. Equation 8.24 describes the anion gap applied for this setting. Plasma chloride concentration will remain unchanged, while plasma bicarbonate will decrease due to H^+ buffering; thus, the anion gap of this case will be increased.

$$\text{Anion gap} = ([Na^+] + [K^+]) - ([Cl^-] + \downarrow [HCO_3^-]) \tag{8.24}$$

In summary, the anion gap provides information that is useful in the investigation of the cause of metabolic acidosis. The most frequent types of metabolic acidosis are diabetic ketoacidosis, lactic acidosis, and various poisonings by compounds that in their catabolism generate organic acids.

8.6.2 Metabolic Alkalosis

This disorder is characterized by an arterial pH higher than 7.4 and plasma HCO_3^- concentration greater than 24 mEq/L; $PaCO_2$ is greater than 40 mmHg. Possible origins for metabolic alkalosis could be:

(a) Addition of a nonvolatile base to the body, as with the intake of antacids.
(b) Secondary to hypovolemia (Fig. 8.11). In this setting, it is necessary to remember that in this state the sympathetic system and the renin-angiotensin-aldosterone system are activated. In the proximal tubule, norepinephrine and angiotensin II stimulate bicarbonate reabsorption. Aldosterone increases electrogenic Na^+ reabsorption in the distal nephron, generating a more negative transepithelial potential in the tubular lumen that favors the tubular secretion of H^+ in the type A intercalated cells A. One cause of metabolic alkalosis and hypovolemia is the loss of fixed acids that occurs in prolonged vomiting. The loss of gastric secretion corresponds to a simultaneous loss of acid and extracellular fluid. The loss of fixed acid contributes to the development of metabolic alkalosis. The loss of volume (gastric secretion) activates the renin-angiotensin-

aldosterone system which through its actions is also responsible for alkalosis. Once the extracellular volume is recovered, renal correction of the acid-base disorder will occur, consisting of increased bicarbonate excretion.

(c) Primary hyperaldosteronism. Mineralocorticoids, such as aldosterone, stimulate the tubular secretion of H^+ in the intercalated cells of the distal nephron. As mentioned above, the H^+ and the HCO_3^- are the products of carbon dioxide activity, H^+ is excreted in the urine as acid, and bicarbonate passes into the plasma, causing the pH to become alkaline. Metabolic alkalosis can also be secondary to syndromes like apparent excess of mineralocorticoids and Liddle, Bartter, and Gitelman syndromes.

8.6.3 Respiratory Acidosis

It is characterized by arterial pH lower than 7.4 and $PaCO_2$ greater than 40 mmHg. This disorder results from inadequate gas exchange in the alveoli, resulting from poor ventilation or decreased gas diffusion. Poor ventilation may be caused by a depression in the activity of the brainstem respiratory center, destruction of the airways, or neuromuscular disorders affecting the rib cage. Neutralization in respiratory acidosis occurs in the intracellular medium. The compensatory response is renal and consists of an increase in the net excretion of acid by increased ammonium synthesis and bicarbonate reabsorption. However, this response is late and takes several days to complete. Therefore, in an acute phase the neutralization is exclusively intracellular and in a later stage the renal response allows the correction of the disorder. In the acute phase, the concentration of plasma bicarbonate increases by 1 mEq/L for every 10 mmHg rise in $PaCO_2$.

8.6.4 Respiratory Alkalosis

The PCO_2 is less than 40 mmHg and the arterial pH is greater than 7.4. It may be due to an increase in gas exchange due, for example, to a stimulation of the respiratory centers. Hyperventilation is also a typical response in states of anxiety and fear. The fall in $PaCO_2$ and the elevated pH inhibit the reabsorption of bicarbonate in the proximal tubule and distal nephron. Later, urinary ammonium excretion will be reduced, which, in addition to the decrease in bicarbonate reabsorption, will reduce net acid excretion. This kidney compensation reduces the plasma bicarbonate concentration by 5 mEq/L for every 10 mmHg drop in $PaCO_2$.

8.7 Conclusions

In a Western diet, carbohydrate and lipid metabolism produce a considerable amount of CO_2 that is eliminated through pulmonary ventilation. The same diet also produces daily between 70 and 100 mEq of acid, derived from protein catabolism.

These acids cannot be eliminated through pulmonary ventilation. They have to be neutralized mostly with bicarbonate in the intracellular and extracellular fluid. Thus, acid neutralization consumes bicarbonate.

Regarding acid and base metabolism, the kidneys have two crucial roles. First, they must reabsorb 100% of the bicarbonate filtered load. The most important segment for this task is the proximal tubule followed by the thick ascending limb of Henle. Second, the kidney must replace an amount of bicarbonate to replace that used to neutralize fixed acids from protein catabolism. This function is accomplished by the distal nephron and requires the acidification of urinary buffers like filtered phosphate and ammonia derived from glutamine catabolism in the proximal tubule.

Therefore, under a Western diet and to accomplish acid-base balance the organism must first, eliminate 15,000 mmol of CO_2 through pulmonary ventilation; and second, reabsorb all the bicarbonate filtered load; and third, excrete 70–100 mEq of acid in the urine.

Review Questions
1. If the arterial pH is 7.4, $PaCO_2$ is 40 mmHg, the plasma bicarbonate concentration is 24 mEq/L, and the pK is 6.1, calculate the concentration ratio between the base and acid of the CO_2/HCO_3^-. What is the physiological significance of the calculated value?
2. What is the difference between reabsorption and regeneration of bicarbonate? What are the physiological roles of these processes?
3. An artificial system is available that contains 1 L of plasma with a bicarbonate concentration of 24 mEq/L and PCO_2 of 40 mmHg.
   (a) Assume that the container is hermetically sealed and therefore there is no exchange of molecules with the outside. Calculate the millimoles of HCl that you should add under these conditions to bring the pH to 7.1.
   (b) Now assume that the same airtight container has a mechanism to maintain PCO_2 by 40 mmHg. How many millimoles of acid would need to be added to cause the same pH change?
   (c) Now the same system allows the PCO_2 go down to 20 mmHg. Calculate the millimoles of acid to be added to obtain the pH of 7.1.
   (d) Identify the difference between the three conditions and relate it to the result you obtained in each case. If you had to standardize one of the situations with what happens in the body, which one would you choose?
4. Acetazolamide is a carbonic anhydrase inhibitor. What will happen to the net excretion of acid if a normal subject is given this drug?
5. What will happen to the renal excretion of Na^+ on the subject of the previous problem?
6. The following data table contains parameters measured in arterial blood from patients with various disorders.

Normal arterial blood values:

pH = 7.37–7.42
PCO_2 = 37–43 mmHg
$[HCO_3^-]$ = 23–25 mM
Anion gap = not considered $[K^+]$ plasma = 12–16; 16–20 mEq/L considering $[K^+]$
plasma

Alteration	pH	PCO_2 (mmHg)	$[HCO_3^-]$ plasma (mEq/L)	Disorder acid-base
Prolonged vomiting	7.55	46		
Intake of NH_4Cl		28	10	
Uncompensated diabetes mellitus		28	12	
Anxiety attack	7.57		21	
Chronic bronchitis	7.33	68		

(a) Calculate the missing variables for each case and define the acid-base disorder.
(b) The intake of NH_4Cl is equivalent to the addition of HCl by the following reactions:

$$2NH_4Cl \rightarrow 2NH_4^+ + 2Cl^-$$

$$2NH_4^+ + CO_2 \rightarrow Urea + 2H^+ + H_2O$$

Compare the value of the anion gap in this case with that of decompensated diabetes mellitus and explain the origin of the difference found. Plasma electrolytes (mEq/L); intake of NH_4Cl^-: Na^+ = 140, Cl^- = 115; uncompensated diabetes: Na^+ = 148; Cl^- = 98.

(c) How do you explain the bicarbonate concentration in the subject with vomiting?
(d) What will happen to the net excretion of acid in the first three cases?

Bibliography

Aronson PS, Nee J, Suhm MA (1982) Modifier role of internal H^+ in activating the Na^+-H^+ exchanger in renal microvillus membrane vesicles. Nature 299:161–163

Curthoys NP, Moe OW (2014) Proximal tubule function and response to acidosis. Clin J Am Soc Nephrol 9:1627–1638

Geibel J, Giebisch G, Boron WF (1990) Angiotensin II stimulates both the Na^+-H^+ exchange and Na^+/HCO_3^- cotransport in the rabbit proximal tubule. Proc Natl Acad Sci U S A 87:7917–7920

Good DW (1994) Ammonium transport by the thick ascending limb of the Henle's loop. Annu Rev Physiol 56:623–647

Hamm LL, Nakhoul N, Hering-Smith KS (2015) Acid-base homeostasis. Clin J Am Soc Nephrol 10:2232–2242

Rose BD, Post T (2001) "Acid-base physiology", "regulation of acid-base balance". In: Clinical physiology of acid-base and electrolyte disorders, 5th edn. McGraw-Hill, New York, pp 299–371

Weiner ID, Verlander JW (2016) Renal acidification mechanisms. In: Brenner BM (ed) The kidney, vol I, 10th edn. Elsevier, Philadelphia, pp 234–257

Potassium Balance Regulation

9

Learning Objectives
- To describe the distribution of potassium in the body.
- To understand the importance of transmembrane potassium distribution and its relationship to the membrane potential of excitable cells.
- To describe the function of the internal potassium balance and its regulation by physicochemical and hormonal factors.
- To describe the function of the external potassium balance and its regulation by plasma potassium and aldosterone concentration.

The regulation of K^+ concentration of the extracellular fluid is critical for normal nerve and muscle function. The extracellular K^+ concentration and the high permeability of the plasma membrane to K^+ are the main determinants of the membrane potential. Potassium plays an important role in several cellular processes such as mitosis and glycogenesis and is necessary for the normal activity of several enzymes. For example, a decrease in the concentration of plasma K^+ affects the renal capacity to concentrate urine; this effect is directly related to a loss of connecting and principal cells capacity to respond to AVP. The result is a marked polyuria, which is reversed when the extracellular and intracellular concentration of K^+ is reestablished.

The body regulates the total K^+ content as well as the plasma K^+ concentration. The regulation of body K^+ contemplates two mechanisms: the internal and the external balance, which will be discussed later.

9.1 Distribution of the K^+ in the Body

Figure 9.1 shows the distribution of K^+ in the body. A normal subject weighing 70 kg has an approximate total K^+ content of 3500 mEq. Roughly, 98% of total body potassium resides in the intracellular compartment, while only 2% is in the extracellular medium. The most important deposit of body K^+ is the skeletal muscle (2635 mEq). Under physiological conditions, the body functions in K^+ balance. This balance is basically defined as a constant equality between potassium intake

© Springer Nature Switzerland AG 2022   177
P. A. Gallardo, C. P. Vio, *Renal Physiology and Hydrosaline Metabolism*,
https://doi.org/10.1007/978-3-031-10256-1_9

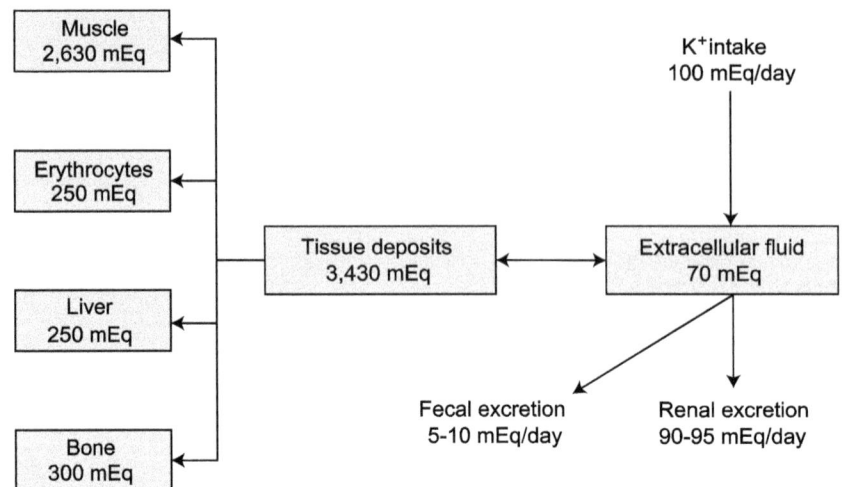

Fig. 9.1 Body potassium distribution. K^+ is the major cation in the intracellular fluid. Skeletal muscle is the most important tissue as an intracellular store. Extracellular potassium is only 70 mEq and is maintained by an equality between potassium ingestion and excretion mainly in urine

(100 mEq/day) and K^+ output, mainly represented by K^+ urinary excretion (90–95 mEq/day), plus a small contribution of fecal excretion (5–10 mEq/day) (Fig. 9.1). Under pathophysiological conditions, such as renal failure, or in animal models adapted to a high intake of K^+, K^+ secretion in the distal colon plays an important role in maintaining the external balance and may constitute up to 30% of the total excretion.

9.2   Transmembrane Distribution of K^+

The transmembrane distribution of potassium is asymmetric (Fig. 9.2). As mentioned in the previous section, 98% of the body potassium is located in the intracellular compartment at a concentration ranging from 120 to 140 mEq/L. The normal range of plasma K^+ concentration (kalemia) is between 3.5 and 5.0 mEq/L. The concentration difference of K^+ between intracellular and extracellular compartments creates a chemical potential gradient that favors the diffusion of K^+ through ion channels. In other words, this chemical potential favors K^+ exit through K^+ channels that are open under resting conditions. Besides, all cells maintain an electrical potential difference across the plasma membrane that is negative in the intracellular side with respect to the extracellular side. This electrical gradient counteracts the chemical gradient of K^+ from the cells. In sum, the chemical gradient given by the concentration difference and the electrical gradient given by the electric membrane potential constitute the K^+ electrochemical potential gradient ($\Delta\mu K^+$) (Eqs. 9.1 and 9.2). The algebraic addition of both gradients favors the exit of K^+ from the cells.

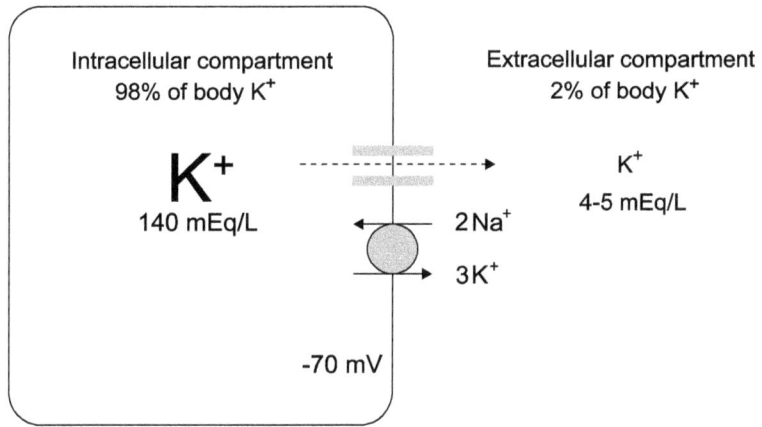

Fig. 9.2 Transmembrane distribution of K⁺. Nearly 98% of the total body potassium resides in the intracellular compartment with an average concentration of 140 mEq/L. Only 2% of total body potassium resides in the extracellular compartment with an average concentration of 4 mEq/L. The plasma membrane is very permeable to K⁺; Na⁺, K⁺-ATPase activity plays a key role in maintaining the electrochemical potential gradient for K⁺

$$\Delta\mu K^+ = \text{Chemical gradient} + \text{Electrical gradient} \tag{9.1}$$

$$\Delta\mu K^+ = RTLn\frac{[K^+]e}{[K^+]i} - zF(\Delta Vm) \tag{9.2}$$

In Eq. 9.2, the first term is the chemical potential gradient that favors K⁺ exit from the cells; the second term is the electrical gradient that is determined by the negative electrical potential inside the cells and does not favor K⁺ exit. Quantitatively, Eq. 9.2 is dominated by the chemical gradient favoring the output of K⁺. The passive exit of K⁺ through ion channels is counteracted with active K⁺ uptake across the plasma membrane. The most important active transport mechanism for K⁺ uptake is the Na⁺, K⁺-ATPase, which exchanges 3Na⁺ from the inside to the outside of the cell in exchange for 2K⁺ moving in the opposite direction. The function of the Na⁺, K⁺-ATPase allows the transmembrane asymmetric distribution of the K⁺. The Na⁺, K⁺-ATPase plays a key role both in the internal and external balance.

The maintenance of the K⁺ electrochemical potential gradient is key to the excitability of all cells, especially muscle fibers and neurons (Fig. 9.3). Under resting conditions, the plasma membrane has a high permeability to K⁺. Therefore, the transmembrane distribution of K⁺ is close to thermodynamic equilibrium. This is reflected in the fact that the value of the membrane potential is very close to the value of the K⁺ equilibrium potential of K⁺ (Nernst potential, −88 mV). An increase in the plasma K⁺ concentration (hyperkalemia) will reduce the chemical gradient across the plasma membrane. Thus, it will depolarize the membrane potential, reducing the excitability of muscle fibers and nerves. This depolarized membrane potential will

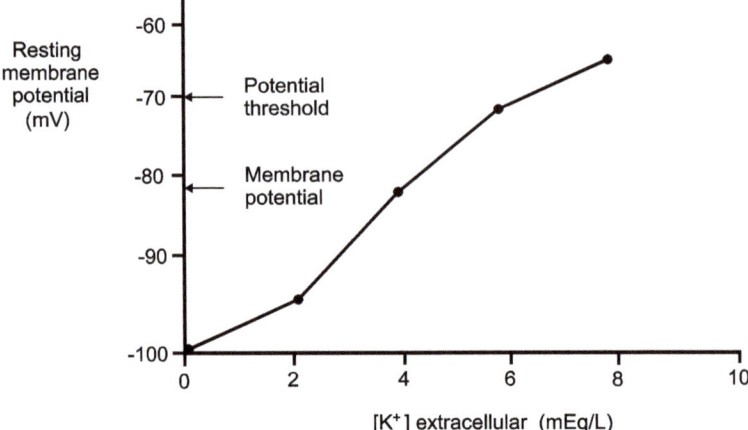

Fig. 9.3 Effect of extracellular K^+ concentration on resting membrane potential. With a normal plasma K^+ concentration, the resting membrane potential is between 80 and 85 mV. An increase in plasma K^+ concentration depolarizes the resting membrane towards the threshold potential. If plasma K^+ concentration is lower than 4 mEq/L, the resting membrane potential is hyperpolarized. The effects of changes in plasma K^+ concentration are related to high plasma membrane permeability to K^+ in resting conditions

inactivate voltage-dependent Na^+ channels that play a key role in the depolarization phase of the action potential. A reduction in plasma K^+ concentration (hypokalemia) increases the chemical gradient, favoring K^+ exit from the cell. The latter will hyperpolarize the membrane potential, moving it away from the threshold value that it must reach to trigger an action potential. Therefore, hypokalemia reduces excitability by hyperpolarizing the membrane potential. Alterations in the concentration of plasma K^+ alter the electrophysiology of the cell. Such disturbances can be observed in electrocardiograms of patients with altered plasma K^+ concentration.

9.3 Internal Balance

The internal balance is key for maintaining the constancy of plasma K^+ concentration within a very narrow range. When a normal subject is submitted to a K^+ overload, nearly 80% of the potassium is translocated to the intracellular compartment within the first hour after the load. This transport of K^+ to the intracellular medium mitigates changes in kalemia. The ingestion of most foods adds significant amounts of K^+ to the extracellular medium. A typical meal contains about 50 mEq of K^+ which is rapidly absorbed in the small intestine. If this amount of K^+ will remain in the extracellular fluid (14 L in a 70 kg subject), the plasma K^+ concentration would rise by 3.6 mEq/L, reaching a dangerous value of 8 mEq/L. This value constitutes a state of severe hyperkalemia. As mentioned before, hyperkalemia depolarizes the membrane potential which translates into alterations in electrocardiogram and heart function. However, this is not the case because the internal

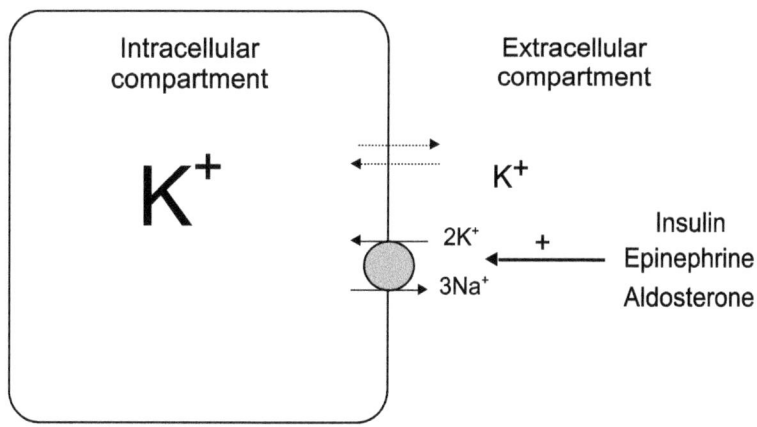

Fig. 9.4 Internal K⁺ balance. The internal balance is the translocation of K^+ between the extracellular and intracellular compartments. Insulin, epinephrine, and aldosterone stimulate K^+ translocation from the extracellular to intracellular compartment. This process occurs mainly in the skeletal muscle, adipose, and hepatic tissue. This movement of K^+ into the cell is active and mediated through the Na⁺, K⁺-ATPase

balance allows a shift of K^+ from the extracellular to the intracellular medium, thus preventing the development of hyperkalemia. In other words, internal balance acts as a buffering mechanism preventing postprandial hyperkalemia. The translocation of K^+ occurs against the K^+ electrochemical potential gradient, and thus active transport is mediated by Na⁺, K⁺-ATPase, stimulated by several hormones (Fig. 9.4).

The internal balance of K^+ is a rapid mechanism and is subject to hormonal regulation. Food intake and most products of intestinal digestion, such as glucose and amino acids, stimulate insulin secretion. Insulin increases active K^+ uptake especially in muscle fibers, fat tissue, and liver. This insulinic effect is independent of the stimulation of GLUT4 insertion in the plasma membrane that increases glucose transport that occurs in muscle and fat tissue. At the cellular and molecular level, the insulin effect on muscle and fat tissue is based on the stimulation of the α1 isoform of Na⁺, K⁺-ATPase and recruitment to the membrane of the α2 isoform of Na⁺, K⁺-ATPase. The latter isoform is expressed primarily in muscle and fat tissue (Fig. 9.5). In the hepatocytes, insulin increases the activity of the Na⁺-H⁺ exchanger, which secondarily activates the Na⁺, K⁺-ATPase. Epinephrine induces the translocation of K^+ to intracellular space. The effect is mediated by β₂-adrenergic receptors that activate the cAMP-protein kinase A signaling cascade, which in turn stimulates the activity of Na⁺, K⁺-ATPase into the muscle. This effect is physiologically important in conditions like exercise (Fig. 9.5). Aldosterone also participates in the internal balance, stimulating the translocation of K^+ from the extracellular to the intracellular medium; the mechanism involved is unknown. The net result is an increase in the capacity of these tissues to carry out active transport of K^+.

Some pathological states are capable of altering the internal balance, associated with an alteration in the function or abundance of Na⁺, K⁺-ATPase. Two examples

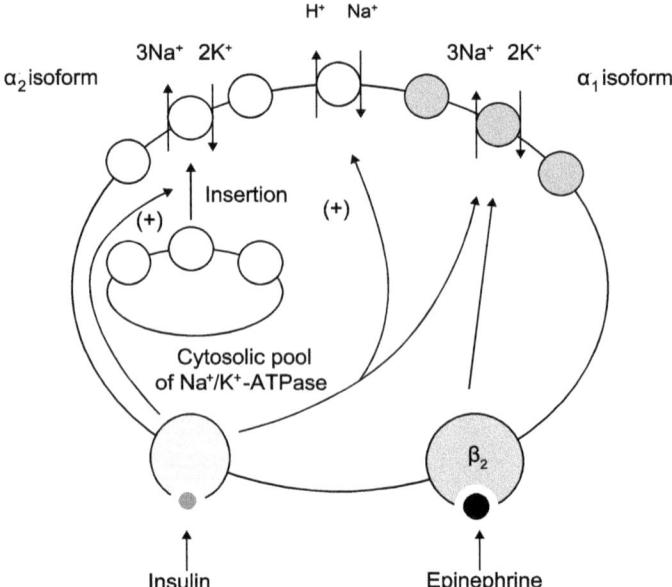

Fig. 9.5 Physiological role of insulin, epinephrine, and skeletal muscle in internal balance. In unstimulated conditions, the plasma membrane pool of Na^+, K^+-ATPase is composed of α-1 isoform; insulin stimulates the insertion of a population of α-2 isoform of Na^+, K^+-ATPase in the plasma membrane. This insulin effect is important for preventing postprandial hyperkalemia. Insulin also stimulates the Na^+/H^+ exchanger. Epinephrine binding to $\beta 2$ adrenergic receptors also stimulates Na^+, K^+-ATPase activity; this effect is important during exercise

are hypothyroidism and chronic renal failure. In hypothyroidism, the translocation of K^+ to the intracellular medium is diminished because the thyroid hormones stimulate the synthesis of Na^+, K^+-ATPase, thus maintaining the number of pumps in the cell membrane. In chronic renal failure it has been shown that, in addition to altered renal excretion of K^+, there is a deficient translocation of a K^+ load from the extracellular to the intracellular compartment. This is explained by a reduction in the Na^+, K^+-ATPase activity (Fig. 9.6) in muscle and adipose tissue and by changes in the relative abundance of the pump isoforms. Both findings help to understand why there is K^+ intolerance in chronic renal failure.

Several physicochemical factors are capable of modifying the internal K^+ balance. The most important are plasma osmolality, acid-base status, and chronic intake of K^+. Changes in plasma osmolality cause changes in water distribution between the intracellular and extracellular compartments. These variations can modify the intracellular K^+ concentration and therefore the chemical gradient. For example, an increase in plasma osmolality produces a net water flux from the cells to the extracellular fluid, increasing the intracellular K^+ concentration. The latter increases the chemical gradient that favors K^+ exit from the cell.

The effect of the acid-base balance status depends on whether it is acidosis or alkalosis. In metabolic acidosis, the movement of H^+ from the extracellular fluid to

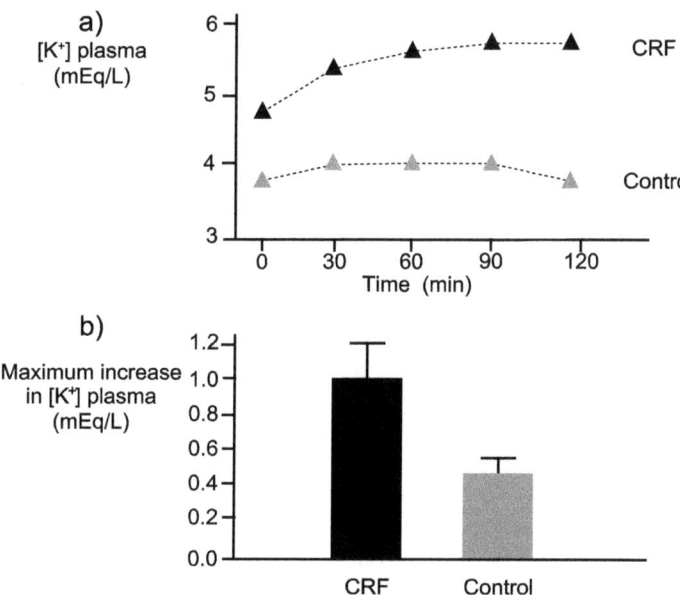

Fig. 9.6 Role of chronic renal failure in plasma K$^+$ concentration. (**a**) Temporal course of plasma K$^+$ concentration during an oral glucose and potassium load in control and patients with chronic renal failure (CRF). (**b**) Maximal increment of plasma potassium concentration attained during a glucose and potassium load in the same patients (mean $\pm$ SE, $n = 8$)

Fig. 9.7 Effect of acidosis in K$^+$ internal balance. Metabolic acidosis favors the development of hyperkalemia; the magnitude of the disturbance depends if the conjugated base is permeant or impermeant in the plasma membrane. (**a**) Permeant conjugated base has a lesser impact on plasma potassium concentration. (**b**) Impermeant conjugated base has a greater impact on plasma K$^+$ concentration

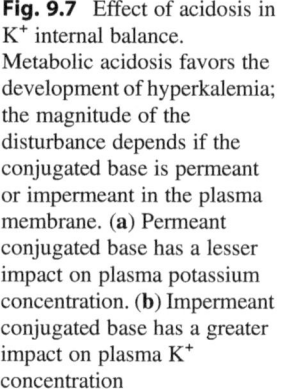
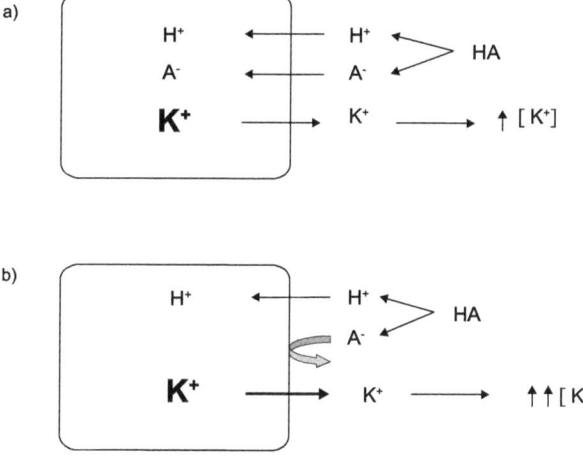

the intracellular fluid is compensated by K$^+$ exit, generating a variable degree of hyperkalemia. The magnitude of hyperkalemia depends on the nature of the anion generated from acid dissociation. If the anion is permeant in the plasma membrane, it will enter the cell with H$^+$. In this setting, acidosis will have little impact in plasma K$^+$ concentration (Fig. 9.7a). If the anion is impermeant, it will remain in the extracellular fluid and will not enter the cell. In this setting, the H^+ influx to the cell will be compensated by K^+ efflux, which will have a greater impact in plasma K^+; the subsequent hyperkalemia will be quantitatively more important (Fig. 9.7b). Conversely, alkalosis promotes the release of H^+ to the extracellular medium, which is compensated by a movement of K^+ to the cytoplasm. In summary, while acidosis tends to cause hyperkalemia, alkalosis induces hypokalemia. The ingestion of a chronic high-K^+ diet produces several adaptive responses in the organism; one of these is a faster uptake of K^+ to the intracellular compartment. This response would be mediated by an increase in Na^+, K^+-ATPase.

Cell lysis produces a release of K^+ from the cytoplasm that contributes to increasing the plasma K^+ concentration of K^+. This process is associated with necrosis, infarction, or muscular polytraumatism.

9.4   External Balance

The internal balance allows the body to regulate the plasma K^+ concentration in acute form, but is not directly involved in maintaining constancy of the total K^+ of the organism. In contrast, the external balance allows the maintenance of total body potassium. Through the external balance, about 90–95% of daily ingested K^+ is excreted in the urine, while the remaining 5% is eliminated in the stool. In terms of time course, the external balance is slower than the internal balance. For example, a K^+ load will have been completely excreted within 4 h after ingestion.

The underlying process in external balance is renal K^+ excretion; the cellular process responsible of K^+ excretion is tubular K^+ secretion. This corresponds to the transcellular movement of K^+ from blood to the tubular lumen. Tubular K^+ secretion is confined to the segments of the distal nephron: connecting tubule and cortical collecting duct. Within these segments, the connecting cells and principal cells are responsible for K^+ secretion.

It is worth to note that the connecting tubule cell is the place of synthesis of kallikrein, the main enzyme of the kallikrein-kinin system whose effector hormone bradykinin is natriuretic and diuretic (*more information on Chaps. 11 and 12*).

Figure 9.8 shows the tubular handling of the K^+. Daily, 720 mEq/day of K^+ is filtered freely through the glomerular barrier. About 67% of the filtered load is passively reabsorbed in the proximal tubule in relation to the movements of Na^+ and Cl^-; between 20% and 25% is reabsorbed in the thick ascending limb of Henle. In the latter, K^+ reabsorption occurs through the transcellular and paracellular pathways. The paracellular reabsorption of K^+ depends on the positive transepithelial potential in the lumen, which in turn depends on the activity of the NKCC2 cotransporter and K^+ channel ROMK, both expressed in the apical membrane. Therefore, only 5–10% of the filtered load reaches the distal nephron (Fig. 9.8). With normal or high K^+ intake, the connecting and principal cells of the distal nephron secrete K^+, which can reach up to 80% of the filtered load (Fig. 9.8). With a low-K^+ diet type A intercalated cells of the distal nephron reabsorb K^+; this

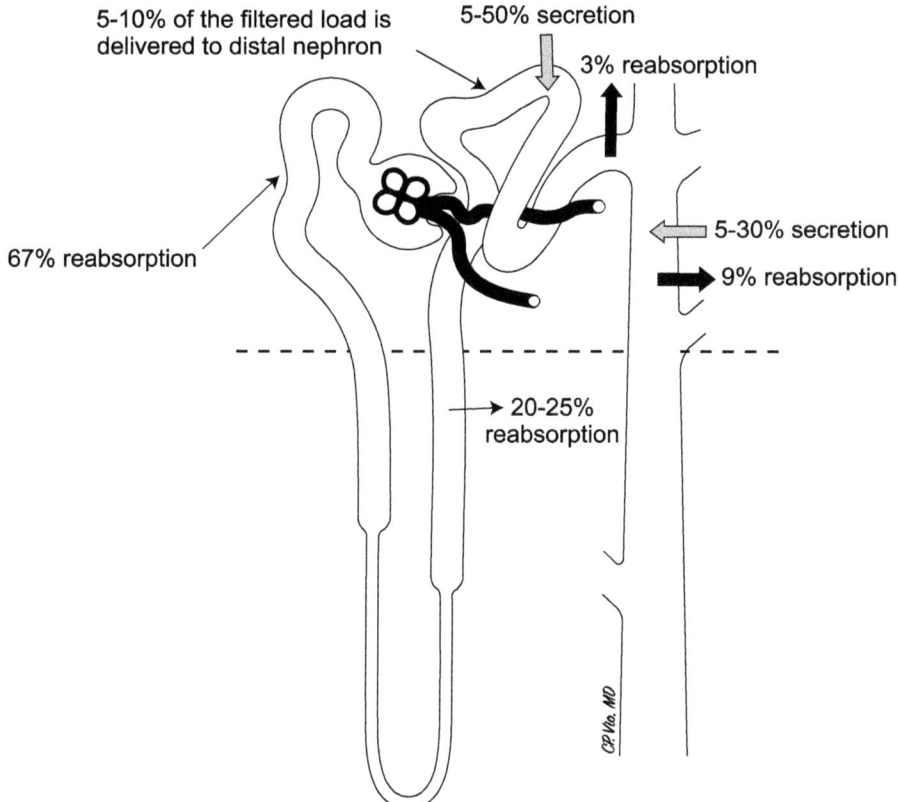

Fig. 9.8 Renal K$^+$ handling. Gray arrows show renal K$^+$ handling in a subject in a normal or high-K$^+$ diet; the black arrows show the renal handling in low-K$^+$ diet. In both cases, the proximal tubule and thick ascending limb reabsorb 67% and 20–25% of the filtered load, respectively. With normal or high-K$^+$ diet, the connecting tubule and the cortical collecting duct secrete K$^+$ between 10% and 80% of the filtered load. In a low-K$^+$ diet, the same segments reabsorb about 12% of the filtered load

process is mediated by the H$^+$, K$^+$-ATPase, expressed in the apical membrane (Fig. 9.9).

As mentioned in the previous paragraph, the distal nephron secretes K$^+$ in the presence of a normal or high K$^+$. Tubular K$^+$ secretion is a two-step process (Fig. 9.10). In the first step, K$^+$ is transported across the basolateral membrane by the Na$^+$, K$^+$-ATPase from the extracellular fluid to the cytosol of the connecting and principal cells. In the second step, K$^+$ diffuses across the apical membrane into the tubular lumen through K$^+$ channels. At least two types of K$^+$ channels are expressed in the apical membrane. The most important channel mediating K$^+$ secretion is ROMK. Also expressed in the apical membranes is the BK or Ca^{++}-dependent K$^+$ channel. The channel is activated by an increase in cytosolic Ca^{++} (Fig. 9.11). The driving force for K$^+$ secretion is given by the electrochemical K$^+$ gradient between the cytosol and the tubular lumen. This force is composed of the chemical gradient

Tubular lumen   Cortical interstitium

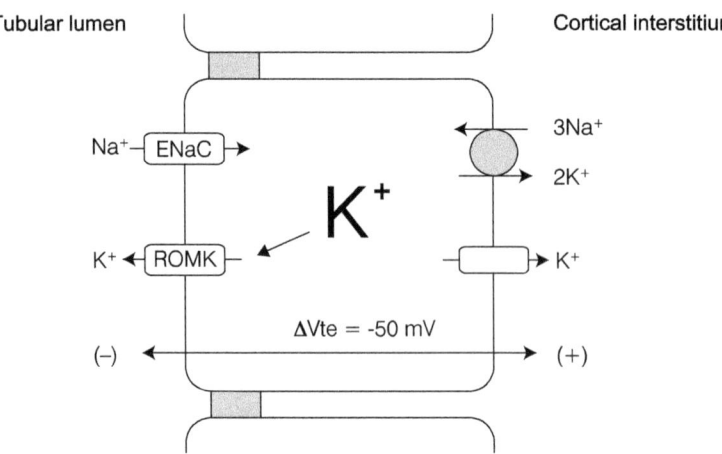

Fig. 9.9 Cell model for tubular K$^+$ secretion in the distal nephron. The connecting and principal cells secrete K$^+$. The most important apical K$^+$-secreting channel is ROMK followed by the Ca^{++}-dependent BK channel (not shown in the figure). The gradient favoring tubular K$^+$ secretion derives from the chemical K$^+$ gradient across the apical membrane and the lumen-negative transepithelial potential difference generated by the electrogenic Na$^+$ transport through ENaC

Tubular lumen   Interstitium

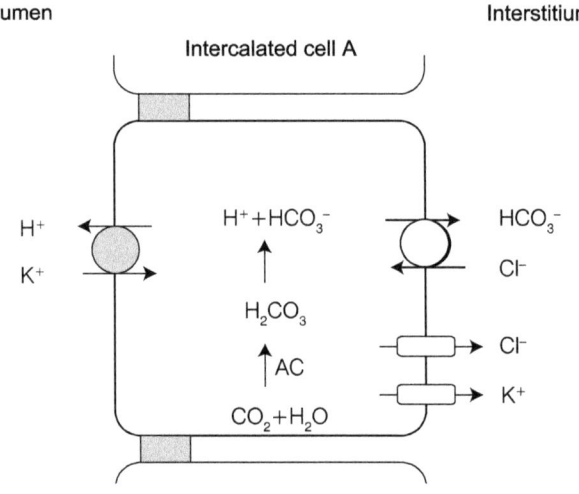

Fig. 9.10 Tubular K$^+$ reabsorption in the distal nephron. Under conditions of low K$^+$ ingestion, type A intercalated cells carry out active K$^+$ reabsorption through apical H$^+$-K$^+$-ATPase

across the apical membrane and the electrical gradient. The K$^+$ chemical gradient is maintained by the activity of the Na$^+$, K$^+$-ATPase and the electrical gradient is due to the negative transepithelial potential difference that exists in the tubular lumen, a direct consequence of the electrogenic transport of Na$^+$ via ENaC (Fig. 9.10).

The tubular secretion of K$^+$ is regulated by luminal and peritubular factors (Fig. 9.11). Within the luminal factors are:

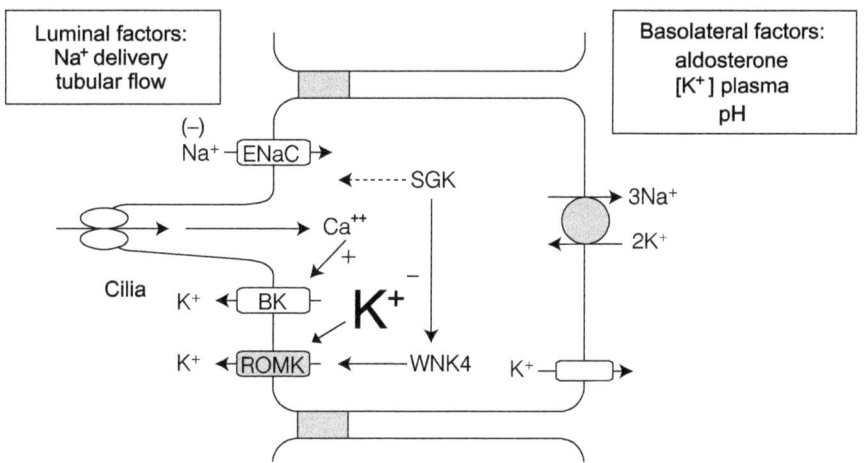

Fig. 9.11 Stimulation of K^+ secretion by luminal and basolateral factors. Tubular fluid flow activates mechanosensitive TRP channels mediating Ca^{++} entry to cytosol; the subsequent increase in cell Ca^{++} activates Ca^{++}-dependent K^+ channels (BK channels). Aldosterone stimulates K^+ secretion through the effect of SGK kinase on ROMK channels

(a) Na^+ concentration in the lumen of the distal nephron
(b) Na^+ delivery to the distal nephron
(c) Tubular flow in the distal nephron

The apical entry of Na^+ via ENaC is the limiting stage in the Na^+ electrogenic reabsorption in the connecting tubule and cortical collecting duct. Therefore, a higher concentration of Na^+ in the tubular fluid reaching the distal nephron, or a greater delivery of Na^+ to the distal nephron, will increase the apical transport of Na^+. Secondarily, this process will increase the electrochemical potential gradient that favors the tubular secretion of K^+. This leads to the conclusion that any factor that inhibits the electrogenic reabsorption of Na^+ will also reduce K^+ tubular secretion of K^+, conducing to hyperkalemia. The tubular flow also affects the secretion of K^+. The principal cells have a primary cilium, whose membrane expresses the Ca^{++} channel PKD1/PKD2 (TRPP1/2), belonging to the TRP channel superfamily. This complex of polycystins functions as a flow sensor. The increase in flow activates the channel, increasing the Ca^{++} entry into the cell and cytosolic Ca^{++}. This in turn activates the apical Ca^{++}-activated K^+ channel (BK channel) (Fig. 9.11).

The main peritubular factors are:

(a) Plasma concentration of aldosterone
(b) Plasma K^+ concentration
(c) Acid-base status

An increase on plasma K^+ concentration stimulates aldosterone synthesis and release in glomerulosa cells of the adrenal cortex. Aldosterone stimulates K^+

secretion in the distal nephron by binding to mineralocorticoid receptors. The first protein induced by aldosterone is SGK protein kinase, which has two important actions. First, it inhibits the endocytosis of ENaC from the apical membrane, increasing the density of ENaC. This action increases the electrogenic transport of Na^+ which creates in part the driving force for the tubular secretion of K^+. SGK also phosphorylates and inactivates the protein kinase WNK4, relieving the inhibition of WNK4 on the ROMK K^+ channel (Fig. 9.11).

An increase in plasma K^+ concentration stimulates tubular K^+ secretion. Multiple mechanisms are responsible for the increased secretion. First, an increase in plasma K^+ concentration inhibits Kir4.1/5.1 K^+ channel in the basolateral membrane of distal convoluted tubule cells. The net effect is a reduction in apical NCC activity and increased Na^+ delivery to connecting tubule and cortical collecting duct, which increase K^+ secretion. Second is the stimulation of aldosterone synthesis described in the previous paragraph.

The acid-base balance also affects the secretion of K^+. Both metabolic and respiratory alkalosis promote tubular K^+ secretion. This is related to K^+ translocation to the intracellular medium that occurs in this condition, with the consequent elevation of intracellular K^+ that favors K^+ secretion. On the other hand, an increase in bicarbonate concentration in the fluid reaching the distal nephron also favors tubular secretion of K^+. This effect is due to the fact that the distal nephron has little capacity to reabsorb bicarbonate; it remains in the lumen behaving as an impermeant anion. The latter increases the transepithelial potential difference, which increases the electrical gradient that allows K^+ secretion. Acidosis reduces tubular secretion, which is directly related to the output of K^+ from the intracellular medium to the extracellular medium through the basolateral membrane, which decreases the electrochemical gradient of K^+. On the other hand, acidosis reduces the activity of Na^+, K^+-ATPase.

9.5   Adaptation to Potassium

As noted above, the connecting and principal cells of the distal nephron have the ability to secrete K^+ under normal or increased K^+ intake. With a low-K^+ diet, the type A intercalated cells of the distal nephron reabsorb K^+. Therefore, the direction of the net K^+ transport in the distal nephron correlates with adjustment in external K^+ balance to meet K^+ homeostasis.

Tubular K^+ secretion can also be stimulated by parenteral administration of a K^+ load. Under these conditions, hyperkalemia secondary to acute load is the main factor responsible for the increased tubular secretion of K^+. On the other hand, hyperkalemia stimulates the release of aldosterone. The mechanism of action and effects of aldosterone were described previously. In this case, the increased tubular secretion of K^+ manifests itself several hours after the acute load is administered.

Studies on rats, rabbits, and dogs have shown that eating a diet rich in K^+ for several days (2–3 weeks) results in the development tolerance to K^+ by the body. This tolerance or adaptation to K^+ has an extrarenal and a renal component.

The extrarenal component had been studied in animals chronically fed with a high-K^+ diet. In this case there is an increased translocation of K^+ from extracellular to intracellular fluid in response to a K^+ load. The mechanisms underlying this response include increased activity and density of Na^+, K^+-ATPase especially in the skeletal muscle and adipose tissue. On the other hand, the distal colon of animals adapted to K^+ has an increased secretory capacity for K^+ secretion. When the intake of K^+ is normal, the distal colonic contribution to K^+ excretion is modest. However, in pathologies such as chronic renal failure there is an increase in the secretory capacity of K^+ of the distal colon. Although this increased colonic secretion does not compensate for the renal excretion, it does contribute to the maintenance of K^+ homeostasis in this setting. The renal component corresponds to an increase in the secretory capacity of the connecting and principal cells in the distal nephron. The increased secretory capacity has a morphological and a biochemical background. The morphological background consists in the amplification of the basolateral membrane: infoldings of the basolateral membrane are more abundant and penetrate deeper in the cytoplasm. The biochemical adaptation is related to an increased abundance of Na^+, K^+-ATPase in this membrane domain. Similar adaptations occur in the superficial epithelium of the distal colon. The full cellular adaptation to increased K^+ secretion has two components: one that is aldosterone-dependent and one aldosterone-independent. These two components came clear from studies carried out in animal models. Adrenalectomized animals that maintained with basal levels of aldosterone and submitted to increased intake of K^+ developed, although less pronounced, an increased K^+ secretory capacity that was more prominent in the connecting cells than principal cells. Increasing aldosterone stimulated the distal nephron K^+ secretion. Thus, K^+ itself is capable of stimulating its own secretion. Very recent studies strongly suggest that K^+ activates mTORC2 which phosphorylates SGK1 kinase in an aldosterone-independent manner.

It is worth to note that the connecting tubule cell is the place of synthesis of kallikrein, the main enzyme of the kallikrein-kinin system whose effector hormone bradykinin is natriuretic and diuretic (*more information on Chaps. 11 and 12*).

9.6   Conclusions

Potassium is the most abundant intracellular cation. Nearly 98% of total body potassium is in the intracellular compartment. Only 2% of body potassium is in the extracellular fluid. The potassium electrochemical gradient is maintained by the Na^+, K^+-ATPase that actively exchanges 2 K^+ into the cell for $3Na^+$ out of the cell.

Transmembrane potassium distribution and the high membrane permeability to potassium are determinants of the resting membrane electrical potential, especially in nerve and muscle cells. Thus, disturbances in plasma potassium concentration alter cellular electrical activity especially in excitable cells like myocardiocytes.

Internal balance is the shift of potassium from the extracellular to intracellular fluid. This is an active movement carried out by the Na^+, K^+-ATPase. Insulin and epinephrine stimulate cell potassium uptake. The insulinic effect is important for preventing postprandial hyperkalemia. The effect occurs in skeletal muscle and adipose tissue. β2 adrenergic activation in skeletal muscle prevents exercise-induced hyperkalemia.

Potassium excretion is the major process in external balance and allows the body to eliminate an amount of potassium equal to that ingested. Urinary potassium excretion is the main route for potassium output, whereas the distal colon plays a small role in potassium excretion. Under normal or high-potassium intake, the cellular process in urinary excretion is potassium secretion by cells of the distal nephron.

Review Questions
1. A subject with insulin deficiency is given the same oral load of K^+ than a normal subject. What do you expect to happen, as a function of time, in both subjects with the plasma K^+?
2. A subject has an autosomal genetic disease, consisting of a mutation with gain in function of the epithelial Na^+ (ENaC). What will happen to plasma K^+ concentration? What change would you expect in the extracellular volume?
3. How do you expect to find the concentration of K^+ under the following conditions?
   (a) Subject ingesting amiloride (ENaC blocker)
   (b) Subject with hypoaldosteronism
   (c) Subject who suffered a hemorrhage with a loss of 1 L of blood
4. Hyperkalemia may initially produce an increase in excitability, but later there is a reduction in excitability. How can this phenomenon be explained?

Bibliography

Arroyo JP, Ronzaud C, Lagnaz D, Staub O, Gamba G (2011) Aldosterone paradox: differential regulation of ion transport in distal nephron. Physiology 26:115–123

Clausen T (1996) Long- and short-term regulation of the Na^+-K^+-pump in skeletal muscle. News Physiol Sci [currently Physiology] 11:24–30

Giebisch G, Krapf R, Wagner C (2007) Renal and extrarenal regulation of potassium. Kidney Int 72:397–410

Palmer BF (2015) Regulation of potassium homeostasis. Clin Am J Soc Nephrol 10:1050–1060

Patel A, Honoré E (2010) Polycystins and renovascular mechanosensory transduction. Nat Rev Nephrol 6:530–538

Rosa RM, Epstein FH (2000) Extrarenal potassium metabolism. In: Seldin D, Giebisch G (eds) The kidney: physiology and pathophysiology, vol II, 3rd edn. Lippincott Williams and Wilkins, Philadelphia, pp 1551–1574

Rose BD, Post T (2001) Potassium homeostasis. In: Clinical physiology of acid-base and electrolyte disorders, 5th edn. McGraw-Hill, New York, pp 372–402

Rieg T, Vallon V, Sausbier M, Sausbier U, Kaissling B, Ruth P, Osswald H (2007) The role of the BK channel in potassium homeostasis and flow-induced renal potassium excretion. Kidney Int 72:566–573

Wang WH (2004) Regulation of renal K transport by dietary K intake. Annu Rev Physiol 66:547–569

Welling PA, Ho K (2009) A comprehensive guide to the ROMK potassium channel: form and function in health and disease. Am J Physiol Renal Physiol 297:F849–F863

Wingo CS, Cain BD (1993) The renal H-K-ATPase: physiological significance and role in potassium homeostasis. Annu Rev Physiol 55:323–347

Woudenberg-Vrenken TE, Bindels RJM, Hoenderop JGJ (2009) The role of transient potential channels in kidney disease. Nat Rev Nephrol 5:441–449

Youn JH, McDonough AA (2009) Recent advances in understanding integrative control of potassium homeostasis. Annu Rev Physiol 71:381–401

Tubular Transport of Calcium, Phosphate, and Magnesium

10

Learning Objectives
- To understand the physiological roles of Ca^{++} and phosphate in the body.
- To understand the tubular handling of Ca^{++} and phosphate.
- To understand endocrine regulation of Ca^{++} and phosphate.

10.1 Role of the Ca^{++} and Phosphate in the Body

Calcium plays a key role in multiple processes, such as the secretion of neurotransmitters and peptide hormones, a cofactor for numerous enzymatic reactions, and is important in the excitation-contraction coupling mechanism in muscle. It also participates in signal transduction systems initiated by the activation of membrane receptors. It is part of the bone matrix and dental tissue and finally has an important role in blood coagulation.

The major features of calcium homeostasis are shown in Fig. 10.1. The only source of calcium is the diet and the major pathway for calcium excretion is urine. The bone matrix constitutes the major source of available calcium. PTH and calcitriol are the most important hormones to maintain calcium homeostasis.

Phosphate is also very important in the body: first, because it is part of the structure of several molecules, such as nucleotides, nucleosides, phospholipids, coenzymes, and second messengers like cAMP. Phosphate is also used to change the activity of numerous proteins through phosphorylation mediated by various protein kinases. It is also part of the body's tissues, such as the bone matrix and dental tissue. Figure 10.2 shows phosphate homeostasis. All body phosphate comes from the diet; the bone matrix is the major source of phosphate. Urine is the main pathway for phosphate excretion.

© Springer Nature Switzerland AG 2022

P. A. Gallardo, C. P. Vio, *Renal Physiology and Hydrosaline Metabolism*,
https://doi.org/10.1007/978-3-031-10256-1_10

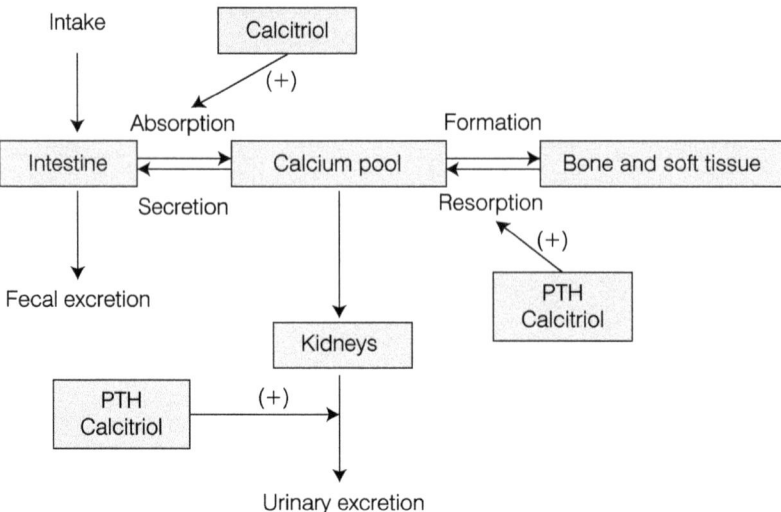

Fig. 10.1 Calcium balance. Calcium ingested in food is absorbed in the small intestine. Intestinal calcium absorption contributes to the extracellular calcium pool. The main calcium deposit in the body is the bone matrix. Calcium is fixed in the bone matrix and is also removed by the process of resorption. The major pathway for calcium output is urinary excretion. Intestinal calcium absorption, renal reabsorption, and bone resorption are under hormonal control

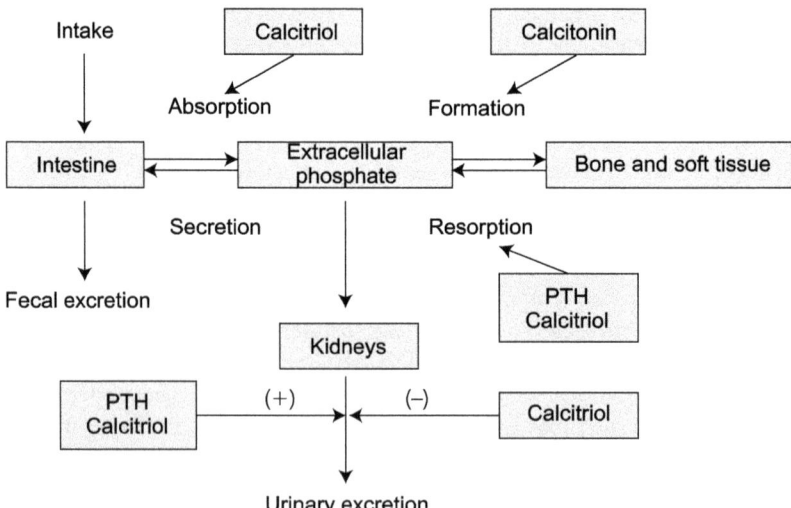

Fig. 10.2 Phosphate balance. Phosphate intake occurs with food ingestion; urinary excretion is the major pathway for phosphate excretion. As with calcium, bone matrix is the main phosphate body deposit. Phosphate renal reabsorption and bone resorption are under hormonal control

10.2   Components of Total Calcium and Phosphate in Plasma

Ca^{++} exists in plasma in several forms. Approximately 40% of the total Ca^{++} is bound to plasma proteins, such as albumin. This is a form of Ca^{++} that is not diffusible as it cannot pass through the endothelial pores. Nearly 45% of the total Ca^{++} is ionized and therefore diffusible. About 15% of the total calcium concentration is forming diffusible complexes, where the Ca^{++} is bound with anions, such as citrate, phosphate, etc. Protein-bound calcium transport in plasma can be affected by several factors like extracellular pH and albumin concentration. Calcium binding to albumin is affected by the extracellular pH. Acidosis reduces the Ca^{++} bound to albumin and increases the concentration of free Ca^{++}; alkalosis has the opposite effect. A second factor that can alter the dynamics of Ca^{++} in plasma is the concentration of plasma proteins, mainly albumin. An increase in albumin concentration will reduce the concentration of free ionic Ca^{++} and vice versa. The components of calcium transport in plasma are summarized in Table 10.1.

Phosphate circulates in the plasma in various forms. About 50% of total phosphate concentration is in ionized form, 40% is forming complex with diffusible cations, and 10% is bound to proteins. Plasma phosphate is found as acid phosphate $(H_2PO_4^-)$ and alkaline phosphate (HPO_4^{-2}); the pK of this system is 6.8. At an arterial pH of 7.4, the alkaline form predominates over the acid form. Table 10.2 summarizes the components of phosphate transport in plasma.

Table 10.1 Forms of calcium transport in plasma

Transport form	mg/dL	mM	% of the total
Ionized	4.7	1.2	45
Bound to diffusible anions	1.4	0.3	15
Protein bound (non-diffusible)	3.9	1.0	40
Total	10	2.5	100

Diffusible calcium is transported as ionized calcium and bound to diffusible anions. Non-diffusible calcium is bound to plasma proteins

Table 10.2 Forms of phosphate transport in plasma

Transport form	mg/dL	mM	% of the total
Ionized ($H_2PO_4^- + HPO_4^{-2}$)	2.1	0.7	50
Bound to diffusible cations	1.5	0.5	40
Protein bound (non-diffusible)	0.6	0.2	10
Total	4.2	1.4	100

With arterial plasma pH of 7.4, most of the ionized phosphate is in the alkaline form

10.3   Tubular Transport of Calcium

Nearly 60% of the total calcium in plasma is available for filtration. The kidney reabsorbs about 99% of the filtered Ca^{++}. The major reabsorption site is the proximal tubule, followed by the thick ascending limb of Henle and the distal convoluted tubule (Fig. 10.3). The proximal tubule reabsorbs 50–60% of the filtered Ca^{++}which is not under endocrine regulation. Proximal resorption of Ca^{++} occurs through the paracellular pathway and depends on the electrochemical potential gradient for Ca^{++}. A second factor affecting proximal reabsorption is the state of the effective circulating volume. A reduction of the effective circulating volume stimulates the proximal Na^+ and fluid reabsorption, favoring the paracellular reabsorption of Ca^{++}.

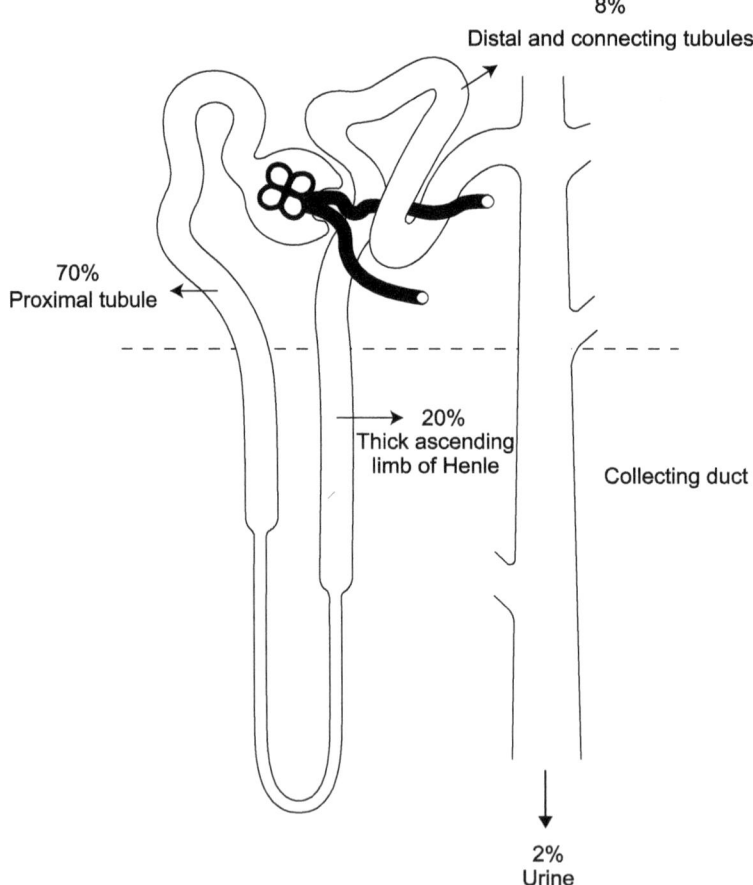

Fig. 10.3 Renal calcium handling. 70% of the filtered calcium load is reabsorbed passively in the proximal tubule. The thick ascending limb of Henle reabsorbs 20% of the filtered load. The distal convoluted tubule and connecting tubule reabsorb 8% of the filtered load. Calcium urinary excretion is nearly 2% of the filtered load

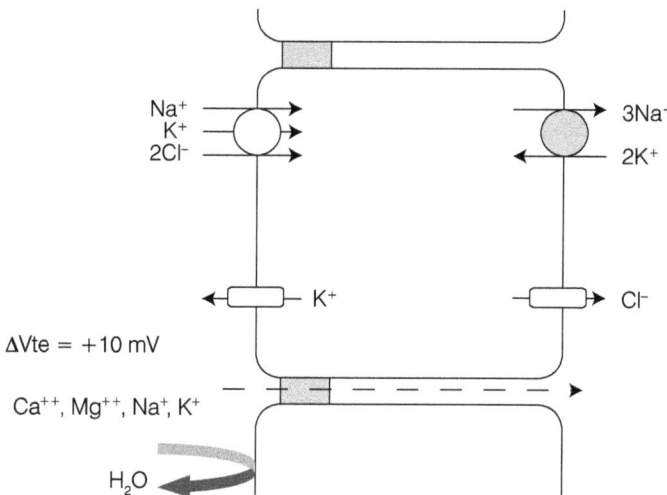

Fig. 10.4 Mechanism of calcium reabsorption in the thick ascending limb. Calcium reabsorption occurs through the cation permeable paracellular pathway. NaCl reabsorption through NKCC2 generates a lumen-positive transepithelial potential difference that acts as driving force for paracellular cation reabsorption

An increase in extracellular volume reduces the proximal reabsorption of Na^+ and therefore Ca^{++} reabsorption.

Nearly 15% of the filtered load of Ca^{++} is reabsorbed by the thick ascending limb of Henle (Fig. 10.4). The reabsorption of Ca^{++} occurs through the paracellular pathway and depends on the reabsorption of NaCl via NKCC2. The recycling of K^+ to the tubular lumen via apical ROMK channel and basolateral Cl^- exit via CLC-K channels generates a lumen-positive transepithelial potential difference that acts as a driving force for paracellular reabsorption of Ca^{++} and Mg^{++}. The expression of claudin 16 in the tight junction of this tubular segment is crucial to allow the selective paracellular reabsorption of cations. The importance of the paracellular pathway and claudin 16 is evident in subjects who present mutations with loss of function in claudin 16. They develop hypercalciuria and hypomagnesuria syndrome.

Reabsorption is influenced by the concentration of Ca^{++} in plasma through the Ca^{++} sensing receptor (CaSR). This membrane protein belongs to the family of G-protein-coupled receptors. The union of Ca^{++} to the receptor activates phospholipase C and subsequently protein kinase C, resulting in a reduction of NKCC2 activity and thus paracellular resorption of Ca^{++} and Mg^{++}. Loop diuretics, such as furosemide, which inhibit NKCC2 activity, secondarily increase urinary excretion of Ca^{++} and Mg^{++}. The thick ascending limb is the site of action of parathyroid hormone (PTH) (*see below*).

About 9% of the filtered load of Ca^{++} is reabsorbed in the distal convoluted tubule and connecting tubule (Fig. 10.5). In these segments, the Ca^{++} reabsorption is transcellular. The apical entry of Ca^{++} occurs following an electrochemical potential gradient for Ca^{++}. The low cytosolic concentration of Ca^{++} and the negative

Fig. 10.5 Mechanism of calcium reabsorption in the distal convoluted and connecting tubules. Calcium reabsorption is transcellular. Apical calcium entry occurs through epithelial calcium channels (ECaC, TRPV5 channel). In the cytosol, calcium ions bind to calbindin protein. Basolateral calcium exit is active mediated by NCX antiporter and Ca^{++}-ATPase

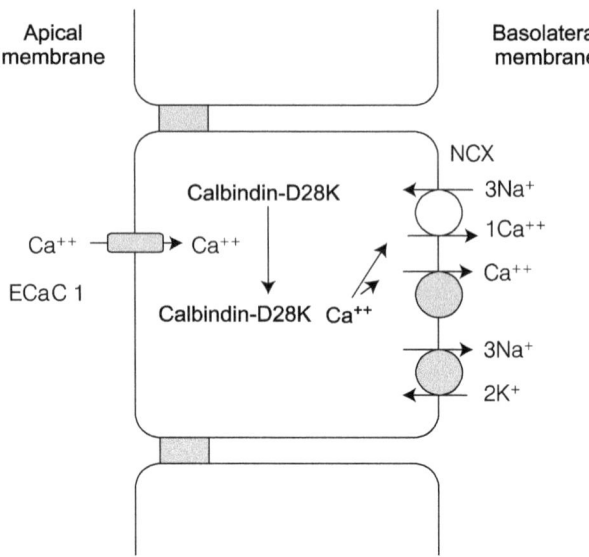

membrane potential drive the entry of Ca^{++} through the TRPV5 and 6 channels, also known as ECaC 1 and 2, respectively (ECaC: epithelial calcium channel). In the cytosol, Ca^{++} ions bind to the calbindin protein, which keeps the low cytosolic concentration of Ca^{++}. The basolateral exit of Ca^{++} occurs by active transport through the Ca^{++}-ATPase and $3Na^+/1Ca^{++}$ (NCX) exchanger.

10.4   Endocrine Regulation of Tubular Calcium Transport

Parathyroid hormone (PTH) is the most important calciotropic hormone. PTH1R receptors are located on the basolateral membrane of cells in the proximal tubule, thick ascending limb and distal and connecting tubule. The receptor is coupled to the Gs, leading to the activation of PKA. In the thick ascending limb, binding of PTH to its receptor activates the cAMP-protein kinase A cascade, leading to the activation of NKCC2. This effect will translate into an increase of the lumen-positive transepithelial potential difference that drives Ca^{++} and Mg^{++} paracellular reabsorption. In the distal convoluted and connecting tubules, PTH increases the opening time of TRPV5, thus increasing the apical entry of Ca^{++}. In the long term, PTH stimulates the expression of TRPV5, calbindin, and NCX. These late effects increase the cellular capacity for transepithelial transport of Ca^{++}.

Calcitriol is a hormone whose last stage of synthesis occurs in the proximal tubule. All the steps on calcitriol synthesis are shown in Fig. 10.6. Hydroxylation of carbon 25 rendering 25(OH) cholecalciferol occurs in the liver. This precursor binds to a carrier protein in plasma. The complex is filtered in the glomerular barrier. The apical membrane of the proximal tubule cells expresses the endocytic protein complex megalin-cubilin. The complex binds to the megalin-cubilin complex and is

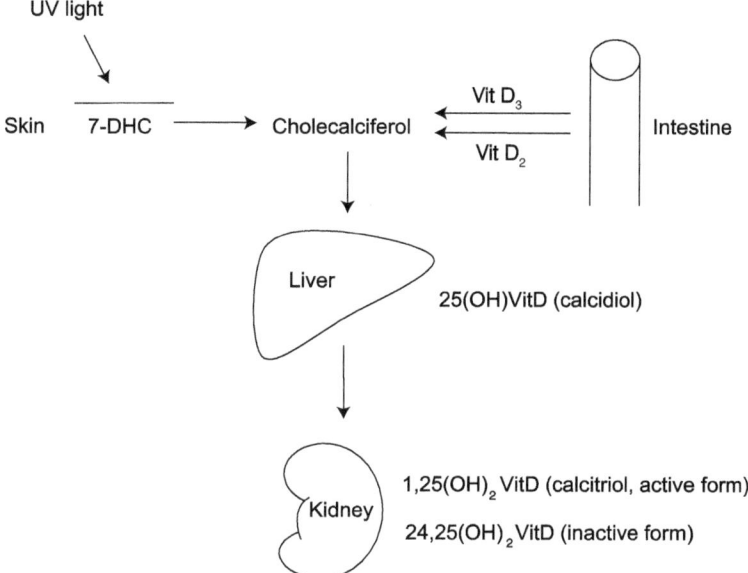

Fig. 10.6 1,25 (OH)₂ cholecalciferol (calcitriol) synthesis. Cholecalciferol can be synthesized from 7-dehydrocholesterol in the skin; the process needs ultraviolet light. Vitamins D2 and D3 from the diet are also substrate for cholecalciferol formation. Hydroxylation in carbon 25 takes place in the liver and hydroxylation in carbon 1 occurs in the proximal tubule and is mediated by 1-α-hydroxylase, which is stimulated by PTH. In a negative feedback mechanism, calcitriol stimulates the activity of 24-hydroxylase, involved in the synthesis of the inactive 24,25(OH)₂-cholecalciferol

endocytosed. In the proximal tubule cell, 25(OH) cholecalciferol is hydroxylated in carbon 1, rendering 1,25(OH)₂ cholecalciferol or calcitriol. This last hydroxylation is directly related to free calcium concentration in plasma. Through PTH, hypocalcaemia stimulates the activity and abundance of 1α-hydroxylase. In a negative feedback mechanism, calcitriol inhibits 1α hydroxylase expression and stimulates the transcription of 24-hydroxylase. The net effect is a reduction in 1,25 (OH)₂ cholecalciferol and an increase in 24,25(OH)₂ cholecalciferol; the latter is inactive as a hormone. Calcitriol binds to a nuclear receptor and stimulates transcription of ECaC channels, calbindin, and NCX (Fig. 10.5). A similar mechanism is activated by calcitriol in duodenal enterocytes. It is necessary to remember that the duodenum is the only site of regulated absorption of Ca⁺⁺ in the small intestine.

Vasopressin (AVP) stimulates Ca⁺⁺ transport in the thick ascending limb. An increase in plasma osmolality leads to an increase in the plasma concentration of AVP. Through V₂ receptors, vasopressin activates the cAMP/PKA transduction system and stimulates NaCl reabsorption via NKCC2. The lumen-positive transepithelial potential difference generated drives paracellular reabsorption of Ca⁺⁺ and Mg⁺⁺.

On the other hand, hypercalcemia reduces the kidney's ability to concentrate and dilute urine. In this case, the increase in plasma Ca⁺⁺ activates CaSR in the basolateral membrane of the cells of the thick ascending limb, which results in a reduction of the activity of the NKCC2 cotransporter. The reduction in NaCl reabsorption reduces the hypertonicity of the medullary interstitium and thus the osmotic gradient for water reabsorption in the collecting duct.

10.5 Phosphate Tubular Transport

Ionized phosphate and phosphate that is complexed with anions are the two forms that can filter through the glomerular barrier. This corresponds to a concentration of between 0.7 and 1.3 mM. As mentioned above, at a physiological pH of 7.4, 80% of phosphate is in the alkaline form (HPO_4^{-2}) and 20% is found as acid phosphate ($H_2PO_4^-$). Assuming a glomerular filtration rate of 180 L/day and a phosphate concentration of 4.2 mg/dL, the kidneys filter about 7000 mg of phosphate daily. According to Table 10.2, the concentration of filterable phosphate is 1 mM. At this plasma concentration, phosphate reabsorption is very close to its maximum transport rate. Figure 10.7 shows the renal handling of phosphate. About 80–95% of the filtered phosphate is reabsorbed in the proximal tubule by apical Na^+-dependent cotransporters 3NaPiIIa and NaPiIIc.

Between 5% and 20% of the filtered phosphate escapes reabsorption and is excreted in the urine. This excretion occurs at physiological plasma phosphate concentrations and is due to the fact that the plasma phosphate concentration is very close to the maximum phosphate transport rate. This excretion of phosphate is of great physiological relevance in the context of the acid-base balance. Phosphate that is not reabsorbed in the proximal tubule behaves like a urinary buffer, and its acidification in the distal nephron is a mechanism for excreting H^+ in the form of acid and contributes to the regeneration of bicarbonate.

10.6 Endocrine Regulation of Phosphate Tubular Reabsorption

PTH is the most important hormone in the regulation of phosphate tubular transport. PTH significantly reduces proximal phosphate resorption. This effect is mediated by PTH1R receptors coupled to Gs protein; subsequent activation of protein kinase A results in endocytosis of the cotransporter. This acute reduction in the abundance of apical NaPi cotransporters results in a reduction in proximal phosphate resorption.

Phosphatonins are hormones that inhibit proximal phosphate reabsorption. These proteins were identified in diseases that produce loss of urinary phosphate. One of the best characterized is FGF23. This growth factor is secreted by osteocytes and osteoblasts, in response to hyperphosphatemia. The functional receptor for FGF23 consists of FGFR1 and the Klotho receptor. The functional receptor has tyrosine kinase activity and reduces the expression of NaPi cotransporter, resulting in hyperphosphaturia. FGF23 also reduces the expression of 1α hydroxylase and thus calcitriol synthesis in the proximal tubule.

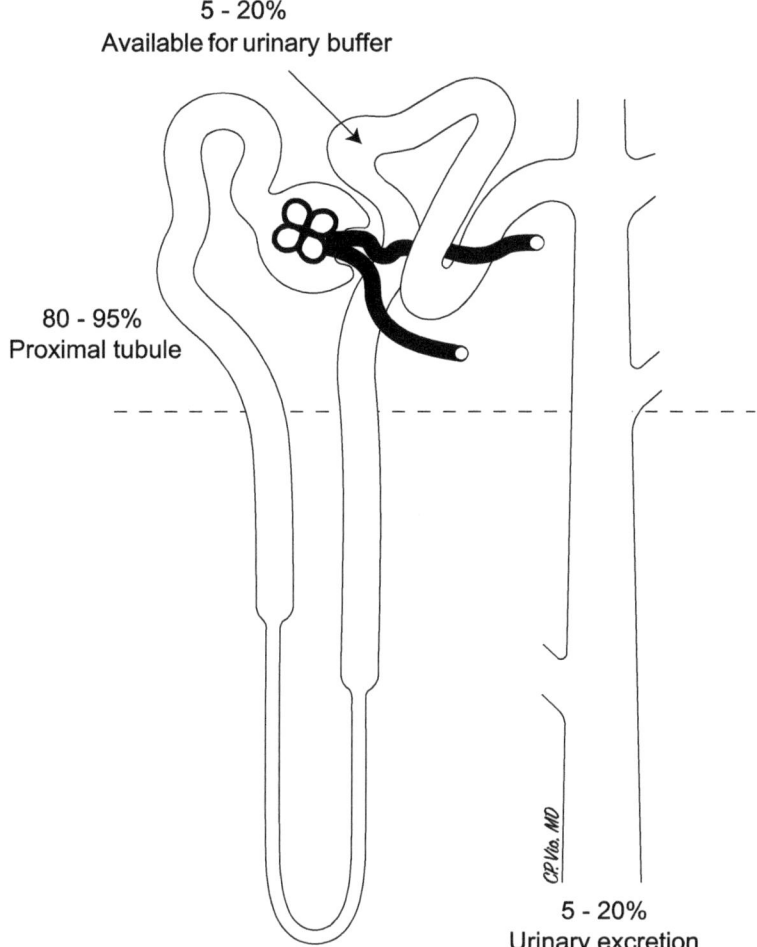

Fig. 10.7 Renal phosphate handling. 80–95% of the filtered load is reabsorbed in the proximal tubule through NPT cotransporters. 5–20% of the filtered load is used as urinary buffer and is excreted in the urine. PTH inhibits phosphate reabsorption in the proximal tubule

10.7 Conclusions

The kidney is the major organ for calcium excretion, thus maintaining calcium homeostasis. Filtered calcium is reabsorbed in the proximal tubule, thick ascending limb, distal convoluted, and connecting tubules. Tubular reabsorption is stimulated by PTH. Urinary phosphate excretion is also important for acid-base balance.

In the kidneyα occurs the last step in calcitriol synthesis. 1α hydroxylase in proximal tubule cells is the key enzyme in calcitriol formation. This hormone stimulates intestinal calcium and phosphate absorption.

Review Questions

1. How is the reabsorption of NaCl in the thick ascending limb related to calcium reabsorption in this segment?
2. Describe the mechanism of PTH stimulation of calcium reabsorption and inhibition of phosphate reabsorption.
3. Describe the physiological relevance of the proximal tubule for calcium homeostasis.
4. Describe the integrated response of the organism to hypocalcemia.

Bibliography

Berndt TJ, Thompson JR, Kumar R (2016) The regulation of calcium, magnesium and phosphate excretion by the kidney. In: Brenner BM, Rector FC (eds) The kidney, vol I, 10th edn. Elsevier, Philadelphia, pp 185–203

Blaine J, Chonchol M, Levi M (2015) Renal control of calcium, phosphate, and magnesium homeostasis. Clin J Am Soc Nephrol 10:1257–1272

Gamba G, Friedman P (2009) Thick ascending limb: the Na^+:K^+:$2Cl^-$ co-transporter, NKCC2, and the calcium-sensing receptor, CaSR. Pflugers Arch 458:61–76

Hoenderop JGJ, Nilius B, Bindels RJM (2005) Calcium absorption across epithelia. Physiol Rev 85:373–422

Murer H, Hernando N, Foster I, Biber J (2003) Regulation of Na/Pi transporter in the proximal tubule. Annu Rev Physiol 65:531–542

Kidney Hormones and Their Action

11

Learning Objectives
- To describe the components of the renin-angiotensin system in the adult kidney and during renal development.
- To describe the components of the kallikrein-kinin system in the adult kidney and during renal development.
- To describe the renal distribution of the synthesis pathways for prostaglandins.
- To know the renal distribution of the synthesis pathways of nitric oxide in adult kidney and during renal development.

In this chapter we will refer to the function of the kidney as an endocrine organ and consider those hormones whose synthesis or transformation takes place in the kidney and whose actions can be renal or extrarenal.

11.1 Renin-Angiotensin System

The renin-angiotensin system (RAS) is one of the main vasoactive systems responsible for body fluid volume, electrolyte balance, and blood pressure regulation. This system can be divided into two pathways—the first known as the "canonical" or conventional, formed by the enzymes renin and angiotensin I-converting enzyme (ACE); the substrate angiotensinogen; the hormones angiotensin I (Ang I), angiotensin II (Ang II), and angiotensin III (Ang III); and the angiotensin type 1 receptor (AT_1) and the other pathway known as "non-canonical" or unconventional that is formed by angiotensin I-converting enzyme 2 (ACE2), the hormones angiotensin 1–9 and angiotensin 1–7, and the AT_2 and Mas receptors. In addition, another pathway is composed of pro-renin, pro-renin receptor (PRR).

© Springer Nature Switzerland AG 2022
P. A. Gallardo, C. P. Vio, *Renal Physiology and Hydrosaline Metabolism*,
https://doi.org/10.1007/978-3-031-10256-1_11

11.1.1 Function

The known functions of the conventional RAS pathway in the kidney are mediated by the action of Ang II acting on its AT_1 receptor. These include increased tubular reabsorption of sodium, bicarbonate, and water in the proximal tubule and vasoconstriction of the efferent arterioles. On this point, it is important to note that the physiological action of Ang II is observed on the efferent artery, while its action on the afferent artery is observed at very high (non-physiological) concentrations of Ang II. In addition, Ang II induces contraction of the glomerular mesangial cells and the straight vessels (*vasa recta*) derived from the efferent arterioles of the juxtamedullary nephrons which provide blood supply to renal medulla. This vasoconstriction results in a decreased renal plasma flow, papillary flow, changes in glomerular filtration rate, and sodium excretion. In addition to the direct effects of Ang II on blood vessels and kidney tubules, this hormone can stimulate the release of vasoactive factors from blood vessel endothelial and smooth muscle cells, mesangial cells, interstitial cells, and other cells located in the kidney.

The non-canonical RAS pathways mediated by the AT_2 and Mas receptors have opposite functions to those of the canonical system; thus, the activation of these receptors by their Ang (1–7) agonist produces vasodilation and the excretion of sodium and water.

11.1.2 Renal Location of RAS Components

The RAS consists of the substrate (the angiotensinogen), the enzymes (renin, angiotensin I-converting enzyme (ACE) and 2 (ACE2), and neutral endopeptidase 24.11 (NEP 24.11)), the products (angiotensins I, II, III, IV, angiotensin 1–7, 1–9), and the receptors (AT_1, AT_2, AT_4, and Mas). The kidney is the main site of synthesis and origin of the components of the renin system, although angiotensinogen is mainly synthesized in the liver, and ACE is abundant in the lung endothelium—but the kidney has five times more ACE than the lung (Fig. 11.1).

11.1.3 Renin

Renin is an aspartyl-protease encoded, in most species, by a single gene. In the adult kidney, renin is stored in granules in a small number of granular cells of the afferent artery, and under stimuli such as a reduction in renal perfusion pressure or pharmacological inhibition with ACE inhibitors or AT_1 receptor antagonists, a recruitment of cells that synthesize and secrete renin is induced, spreading along a large number of vascular muscle cells.

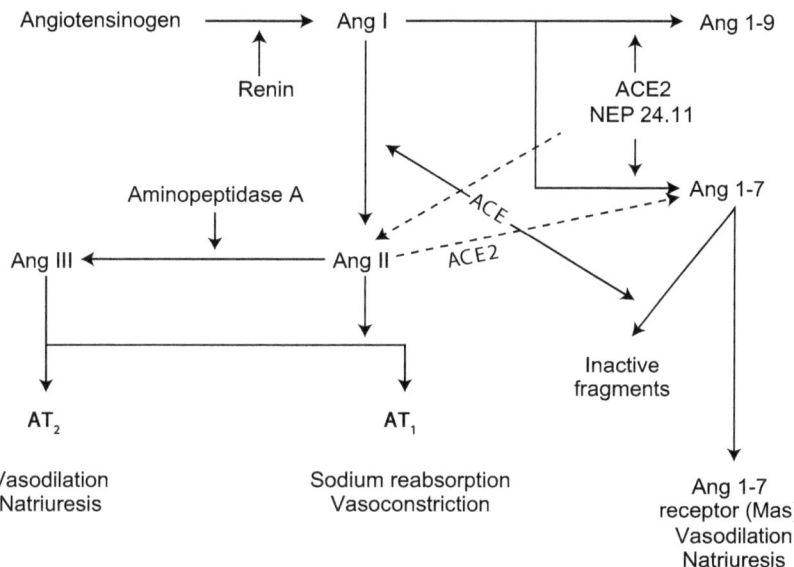

Fig. 11.1 Renin-angiotensin system. Scheme of the renin-angiotensin system where the main enzymes (renin) are highlighted, those that process hormones, angiotensin I-converting enzyme (ACE), ACE2, and aminopeptidase A. The AT_1 and AT_2 receptors appear together with the main physiological effects

11.1.4 Angiotensin I-Converting Enzyme (ACE), ACE2, and NEP 24.11

ACE is a dipeptidyl carboxypeptidase that cuts the last two amino acids (aa) of angiotensin I and bradykinin. It then transforms Ang I into Ang II (10 to 8 aa), and the bradykinin (9 aa) degrades it to BK 1–7 (7 aa). This enzyme is located in the endothelium of various organs including the lung. Because of the large endothelium content of the lung, it has been described that the largest amount of ACE is found in the lung, but the kidney has 5–10 times ($\times$ mg tissue) more ACE than the lung. In the kidney it is located mainly in the brush border of the S2 segment of the proximal tubule and in much less quantity in the glomerular endothelial cells. ACE is induced at sites of kidney damage in the tubulo-interstitium and increases in proximal tubules in hypertension, diabetes, and chronic kidney damage, conditions that have local increase of Ang II. ACE2 is an exopeptidase, carboxypeptidase, that cuts the last amino acid and transforms Ang I to Ang 1–9, or Ang II to Ang 1–7. This enzyme has been described in similar places than ACE in the proximal and distal tubules. However, they have differential distribution in the renal corpuscle, while ACE is found in the glomerular endothelium, ACE2 is described in the podocytes and mesangial cells. The expression of this enzyme is increased by the administration of angiotensin AT_1 receptor antagonists and ACE inhibitors. In addition, this enzyme is diminished in pathologies such as diabetes or kidney failure. NEP 24.11 or neprilysin was originally described in renal tissue—as was ACE in the brush border of the proximal tubule—and is an enzyme that degrades BK and Ang 1–7 form and also degrades atrial natriuretic hormone (ANP). The effects on BK and ANP have led to the development of drugs that inhibit NEP 24.11 and thus increase the effects of BK and ANP by removing sodium and water. Although neprilysin was originally discovered in the kidney, we now know that it is widely distributed in the brain, lung, and other tissues, where it metabolizes endothelin, encephalins, and substance P.

11.1.5 Angiotensinogen

In the rat nephron the mRNA for angiotensinogen has been located in the straight and convoluted segments of the proximal tubule and, in less intensity, in the glomerulus, while its protein has been located in granules located in the subapical region of the cells. The action of renin on angiotensinogen generates Ang I; once formed, Ang I is processed by several peptidases. The action of ACE generates Ang II, while the action of NEP 24.11 generates Ang 1–7. An alternative way to generate Ang 1–7 is by the action of ACE2 which is highly efficient in converting Ang II to Ang 1–7. The actions of Ang 1–7 generally oppose those of Ang II; therefore, while Ang II stimulates contraction, Ang 1–7 induces vasodilation through the production of NO, vasodilating prostaglandins and enhancing the effects of bradykinin.

11.1.6 AT_1 and AT_2 Receptors

The main biological actions of Ang I and Ang II are mediated by two receptors AT_1 and AT_2. In the adult, the AT_1 receptor is widely distributed, and, in the kidney, it is located in the vascular smooth muscle cells of the efferent arterioles, in the mesangial cells and podocytes of the renal corpuscle, and in the basolateral and apical membranes of the cells of the proximal tubule in the thick ascending loop of Henle, the macula densa, the distal convoluted tubule, and the cortical collecting duct. The AT_2 receptor, on the other hand, has its maximum expression in the mesenchymal tissue of the fetus, decreasing its expression to almost undetectable levels a few days after birth. In the adult kidney AT_2 is expressed in very low quantity, being located in epithelial cells of the renal corpuscle, in the most cortical segments of the nephron, and in interstitial cells. Its expression is regulated by the intake of sodium, increasing in animals with low-sodium diets. Methodological difficulties derived from the structure of this receptor have prevented its precise location until now. In adults, AT_2 receptor distribution is very limited but it is particularly high in the adrenal medulla where it could contribute to the regulation of the response to stress stimuli.

Both receptors share several similarities; they belong to the family of receptors with seven transmembrane domains coupled to G-protein and bind Ang II with different affinity. There is only a 30% homology between the two receptors, and the intracellular signals activated by each receptor are different. While AT_1 activates mitotic signals and induces vasoconstriction, AT_2 activates anti-mitotic and vasodilatory signals. These differences are mainly due to the difference in the cellular location of the receptors and in the intracellular signaling mechanisms.

Through binding tests, it has been shown that Ang 1–7 binds to a receptor that corresponds to the protein encoded by the proto-oncogene *Mas*.

11.1.7 Mas Receptor

In 2003 the proto-oncogene *Mas* was identified as the receptor for Ang 1–7. This receptor is expressed mainly in the brain and testicles but has also been detected in the kidney, heart, and blood vessels. In the kidney this receptor is expressed mostly in the proximal tubules although its definitive location is pending as there are only few studies addressing its cellular distribution. The protein belongs to the family of receptors with seven transmembrane domains; however, to date the mechanisms of signal translation that are activated after ligand binding remain controversial. Among the signals that have been activated is the phosphorylation of AKT and the production of arachidonic acid. Among the most relevant effects are the reduction in oxidative stress and fibrosis.

The activation by Ang 1–7 produces opposite effects than angiotensin II; thus, Ang 1–7 induces vasodilation, natriuresis, and other actions under study.

11.1.8 Pro-Renin Receptor

The pro-renin receptor was discovered in 2002 as a membrane protein of 350 amino acids, capable of binding and activating pro-renin in a non-enzymatic way. This receptor is part of a protein complex corresponding to the H^+ vacuolar ATPase present in the intercalated cells. This receptor appears to be located in the intracellular vesicles in many cell types of the body and its contribution to a physiological function in RAS is under study. However, at in vitro studies, it has been shown that by binding to pro-renin, this receptor activates the intracellular ERK-2 enzyme by stimulating cell proliferation and the production of TGF-β. In the human kidney, the PRR has been located mainly in mesangial cells and in the subendothelium of the renal arteries. In the rat kidney, the PRR has been located in the intercalated cells of both the cortical and medullary collecting ducts.

11.1.9 Ontogenia

The expression of renin during fetal development undergoes significant changes; the first traces of this enzyme are detected as early as 15–17 days in the rat. At this stage, we observe the presence of renin widely distributed along the radial cortical and arcuate arteries and in the developing arterioles. The development of renin progressively decreases, and at day 1 postnatal renin is restricted to the arcuate arteries being located closer to the bifurcation of the arterioles. At 10 days postnatal, renin is located in afferent arterioles, in ring-shaped smooth muscle cells, and with a much more discrete immunostaining, reaching at 20 days of development an appearance similar to that of the adult. This decrease in renin coincides with the decrease in renal vascular resistance observed in the mature kidney. It is interesting to note that in the postnatal kidney, there is a decrease in renin and an increase in kallikrein. These changes favour vasodilation, which is also a feature of the adult kidney.

In adults the inhibition of ACE or AT_1 receptor antagonism induces the recruitment of vascular muscle cells expressing renin in a similar way to the postnatal and fetal period, thus pointing to the existence of an angiotensin II-mediated negative feedback regulating renin expression and secretion.

In rats, the presence of angiotensinogen is detected at day 18; this protein increases until it reaches a maximum in the newborns and then decreases towards adulthood.

The distribution of the **AT_1 and AT_2 receptors** in the rat fetus is different for each type. The mRNA for AT_1 is detected at day 13 in several organs including the kidney. AT_2 mRNA is detected at 15 embryonic days in the renal cortex. At day 19 the signal for renal AT_2 decreases in intensity; however, its distribution becomes wider, covering both the cortex and the external medulla.

The importance of RAS in kidney development is evident in animals knock out for ACE or AT_1 receptor; in these cases a thickening of the arterial walls, focal areas of interstitial fibrosis, and tubular and papillary atrophy are observed. Additionally, these animals are not able to concentrate urine properly, so this morphological defect also has an important functional correlate.

Angiotensin is important for kidney development and ACE inhibitors are teratogenic and cannot be prescribed during pregnancy.

Despite the high expression of the AT_2 receptor during the fetal period, no abnormalities in kidney development have been described in animals knock out for this receptor.

11.2   Kallikrein-Kinin System

The renal kallikrein-kinin system (KKS) participates in complex processes such as blood pressure regulation, extracellular volume regulation, regulation of sodium and water excretion, glomerular hemodynamics, and renin secretion.

11.2.1 Functions

Multiple functions have been proposed for KKS in both physiological and pathophysiological situations. In the kidney KKS participates at two levels; at the vascular level it induces dilation of afferent arterioles by regulating smooth muscle tone, and of medullary straight vessels, and at the tubular level it inhibits reabsorption of sodium and water by increasing their elimination in the urine. Kinins also have local effects on other tissues such as on the initiation of pain signals, on the transport of chlorine by the colon, on reproduction, on inflammation, and on the allergic response.

11.2.2 Renal Location of KKS Components

KKS consists of the substrate (high and low molecular weight kininogens, the enzyme kallikrein), the hormones bradykinin and lys-bradykinin, the B1 and B2 receptors, and the enzymes that degrade kinins, collectively called kininases (Fig. 11.2).

11.2.2.1 Kallikrein

The enzyme originating in tissues like the kidney (EC 3.4.21.35) is called "tissue kallikrein" and has local functions. Tissue kallikrein is a glycoprotein with a serine residue in its active site (serine-protease) that cleaves the Met-Lys and Arg-Ser bonds of low molecular weight kininogen generating bradykinin. The tissue kallikrein gene belongs to a family of genes with high homology. Its mature mRNA encodes for a 261-amino acid pre-prekallikrein, which gives rise to a protein with a molecular weight close to 45 kDa. In the kidney, kallikrein is located exclusively in the connecting cells (CNTc) of the connecting tubule (Fig. 11.3b). These CNT cells have abundant Na^+, K^+-ATPase on their basolateral side, and ROMK potassium

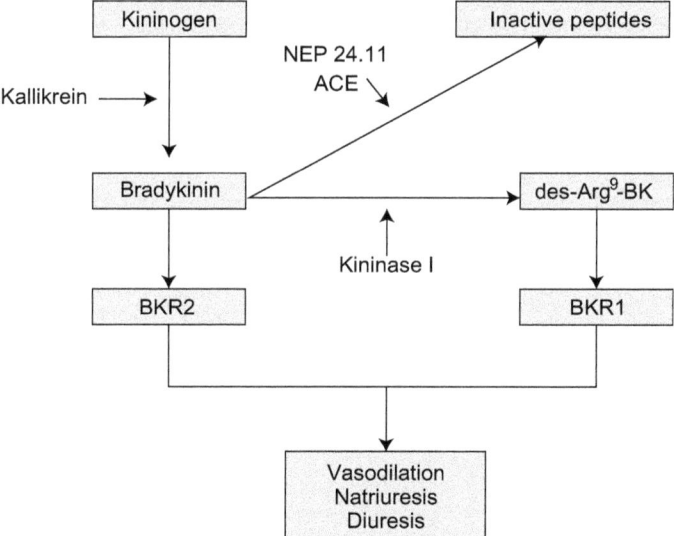

Fig. 11.2 Kallikrein-kinin system. Scheme of the kallikrein-kinin system where the main enzymes (kallikrein) stand out, those that process hormones, kininase II or angiotensin I-converting enzyme (ACE), kininase I, and neutral endopeptidase (NEP 24.11). The BKR2 and BKR1 receptors appear together with the main physiological effects, natriuresis, diuresis, and vasodilation

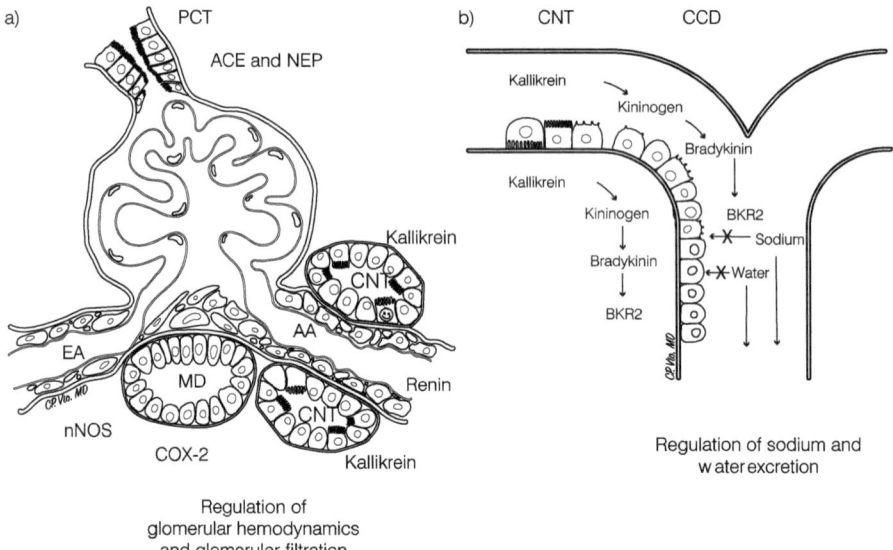

Fig. 11.3 Localization of components of the kallikrein-kinin system in the kidney. In (**a**) the location of kallikrein in the connecting tubules (CNT) is shown in close contact with the afferent arteriole (AA) and close to the macula densa (MD) and efferent arteriole (EA). This microarchitecture is close to the juxtaglomerular apparatus and has the vasoactive components renin, kallikrein, cyclooxygenase-2 (COX-2), neuronal nitric oxide synthase (nNOS), angiotensin I-converting enzyme (ACE), and neutral endopeptidase (NEP). In (**b**) the localization of kallikrein in the connecting tubule (CNT) and the bradykinin receptors in the collecting duct (CCD) is shown. In these microenvironments, vasoactive hormones participate in glomerular hemodynamics and the excretion of water and sodium

channel, and constitute a main site of potassium regulation. With a high dietary potassium, they hypertrophy, secrete more potassium, and increase the synthesis of kallikrein and production of bradykinin. This mechanism contributes to the natriuretic and hypotensive effect of the potassium-rich diet, in addition to the inhibitory effect on the activation of sodium chloride cotransporter NCC. Another relevant aspect of the CNT tubule is that it forms a second loop and comes into contact with the afferent arteriole, the site of renin synthesis. This anatomical conformation gave rise to a second mechanism of renal feedback, the connecting tubule glomerular feedback *(CTGF), which is an important regulator of glomerular function and which antagonizes and regulates the tubuloglomerular feedback of the juxtaglomerular apparatus (TGF).*

11.2.2.2 Kininogens

There are two forms of kininogens in humans: high molecular weight (HMW-Kg, PM 88–114 kDa) and low molecular weight (LMW-Kg, PM 50–68 kDa). HMW-Kg is the substrate for plasma kallikrein and participates in the coagulation process, while LMW-Kg is the substrate for tissue kallikrein. In the kidney there are sites of kininogen synthesis; these are found in the principal cells of nephron collecting ducts in anatomical proximity to cells of the macula densa and afferent arterioles (Fig. 11.3a, b). The existence of renal kininogen-producing cells anatomically close to kallikrein-producing cells allows postulating a local generation of kinins either in the interstitium or in the tubular lumen that supports its action as a paracrine hormone.

11.2.2.3 B1 and B2 Receptors

Two receptors have been identified for kinins; B2 (BKR2), which has a greater affinity for bradykinin (BK), is widely distributed in the body; its expression is constitutive and is considered the main mediator of the physiological effects of kinins. The B1 receptor (BKR1), on the other hand, has a greater affinity for the des-Arg^9-BK, which is the product of the action of kininase I on BK (Fig. 11.2). Its expression is very low or null in physiological conditions; however, it is induced in front of pathological processes such as chronic inflammation, trauma, and hyperalgesia. Both receptors belong to the family of proteins with seven transmembrane domains, whose signal transduction is coupled to the activation of a G-protein. However, the type of G-protein and the signaling activated below appear to be specific to each type of receptor. In the kidney, the B2 receptor has been identified in the glomerulus, the straight portions of the proximal tubule, and the distal tubule, collecting duct, and in the smooth muscle of radial cortical arteries, concentrating on distal convoluted tubules and collectors. Less information is available for receptor B1. The presence of its mRNA has been identified in Bowman's capsule cells and in the thin segment of Henle's loop in normal human kidneys; however, the presence of the protein in this tissue has not yet been identified.

The main enzymes that degrade kinins or kininases are kininase I; kininase II, better known as angiotensin I-converting enzyme (ACE); neutral endopeptidase 24.11 (NEP); endopeptidase 24.15; and aminopeptidase P. In the kidney the most active enzymes in the metabolism of kinins are ACE and NEP or neprilysin. Both are found in the brush border of the proximal convoluted tubule and can be found in the urine. ACE is also located in the endothelium of the renal blood vessels. The main products of bradykinin metabolism are BK 1–7 and 1–5, which are products of the action of ACE and NEP interchangeably, and des-Arg^9-BK, product of the action of kininase I.

ACE is a major kininase, and all enzyme kinetic parameters (K_m, k_{cat}/K_m) show that its preferred substrate is bradykinin (BK 1–9 and BK 1–7). The enzyme's preference for BK over Ang I is very evident and is the main form of BK degradation in humans and rodents.

11.2.3 KKS Ontogeny

Numerous studies have been conducted to determine when each component of KKS appears in kidney development. This is in order to identify their participation in the development and maturation of this body.

Kallikrein is detectable at 19 fetal days (pregnancy in the rat is 21 days) in the development of the rat kidney, located exclusively in the upper S-body loop. In the kidney of newborns, kallikrein is located in the distal regions of the more mature nephrons. As the kidney matures, the distribution of kallikrein becomes wider, from a distribution limited to the inner cortex to a wider distribution that includes the outer cortex. Once the kidney has fully matured, kallikrein is located primarily in the outermost rind. This apparent migration of kallikrein occurs because the maturation pattern of the kidney is a centrifugal pattern, where maturation occurs from the inner to the periphery (cortex). Kallikrein is always located in the CNT connecting tubule cells, its site of origin, and these tubules migrate to the periphery with the maturation of the nephron segments.

The important thing is that kallikrein increases during postnatal development, unlike renin which decreases in the same period. This predominance, in the kidney, of a vasodilator over a vasoconstrictor system explains why the kidney moves from a system with high vascular resistance to a vasodilated system with low resistance and high blood flow.

Kininase II, also known as angiotensin I-converting enzyme (ACE), begins to be detected in fetal states, where it is located in glomerular capillaries and in the proximal tubule. In the newborn it is detected at the level of mRNA and protein. The maximum expression has been observed between 15 and 20 days of postnatal life. Several ontogenetic characteristics have been observed that are shared by kallikrein and kininase II. On the one hand, the synthesis of both enzymes increases in the postnatal period, and on the other hand, the maximum expression of both enzymes coincides with the weaning period in the rat.

With regard to renal kininogen, its mRNA is detectable at 15 fetal days in the rat. During early nephrogenesis, kininogen is located in the branches of the ureteral bud. In the later stages of nephrogenesis, kininogen is located in the cortical collecting duct whose embryonic origin is the ureteral bud. The maximum expression of the protein is reached in the first day of extra-uterine life and then decreases in the adult.

The expression of the B2 receptor of bradykinin during development is interesting, since its mRNA is overexpressed exclusively in the kidney and heart, reaching levels ten times higher during this period than in adulthood. In the developing kidney, the B2 mRNA is expressed mostly in the immature glomeruli and in the distal tubules and collecting ducts. This specific expression of receptors during development suggests a significant involvement of KKS in kidney maturation.

This importance is evident in mouse models knock out for the bradykinin B2 receptor gene treated with high-sodium diets from fetal periods, where distortion of renal architecture has been observed at 16 fetal days with alterations in the formation of the most distal tubular segments of the nephron and with cyst formation. These results postulate that KKS is involved in maintaining fetal kidney structure and epithelial integrity. These non-B2-receptor mice develop high blood pressure when placed on a high-sodium diet during adulthood. This is interesting taking into consideration that our Western society is on a high-sodium diet on a daily basis (200 mmol vs. 100 mmol/day).

11.3 Prostaglandins

Prostaglandins (PGs) or eicosanoids are local lipid hormones that exert their para-crine or autocrine effects by acting on local specific receptors in multiple physiolog-ical and pathological responses (e.g., inflammation, pain, cancer). Among the physiological functions of the kidney, PGE_2 and PGI_2 participate in the vascular tone and the excretion of water and sodium by which they are natriuretic, diuretic, and vasodilator and thus counteracting the effects of Ang II; at the same time, they can potentiate the effects of BK. On the other hand, the PGF_2 and thromboxane A_2 decrease the speed of glomerular filtration and thus sodium excretion and induce vasoconstriction.

The substrate from which prostaglandins are synthesized is arachidonic acid (AA) or eicosatetraenoic acid, a polyunsaturated fatty acid of 20 carbon atoms with four double bonds, which is formed by the elongation and desaturation of its precursor linoleic acid. After its formation, AA is esterified in the phospholipids of the cell membrane. AA can be released from the plasma membrane by the enzyme phospholipase A_2 (Fig. 11.4).

When released from membranes AA is metabolized by prostaglandin G/H synthase, known also as "cyclooxygenase" (COX). Two reactions occur, one for cyclooxygenation that converts AA to PGG_2, and one for peroxidation that further transforms PGG_2 to PGH_2.

This reaction occurs in the hydrophobic channel of the enzyme's active site; subsequently, in a nearby region, peroxidation of prostaglandin G_2 and the conse-quent formation of prostaglandin H_2 occur. Prostaglandin H_2 is converted to PGE_2, PGI_2, PGF_2, and TXA_2 depending on the isomerase-specific enzymes present in each cell type where it exerts local specific functions. Thus, the TXA_2 is synthesized in the platelets, the PGI_2 in endothelial cells, and PGE_2 in the collecting duct.

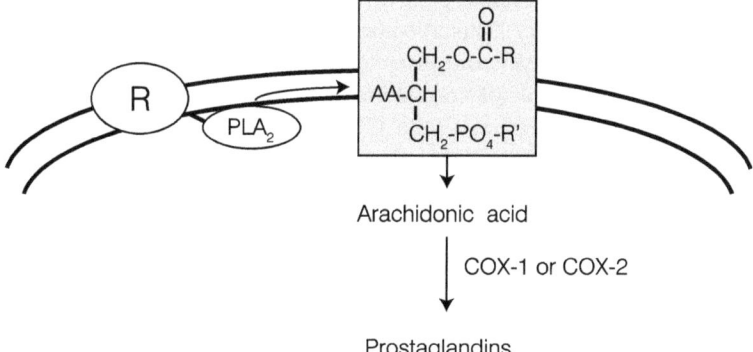

Fig. 11.4 Prostaglandin synthesis. The figure shows the synthesis of prostaglandins from arachidonic acid by the consecutive action of phospholipase A_2 (PLA_2) and cyclooxygenase 1 or 2 that gives rise to cyclic endoperoxides (PGH_2 and PGG_2), which give rise to specific tissue isomerases to prostaglandins PGI_2, PGE_2, PGD_2, PGF_{2a}, and TXA_2

Until some time ago it was believed that COX was a single enzyme; however, since the 1990s the existence of two isoforms, cyclooxygenase-1 (COX-1) and cyclooxygenase-2 (COX-2), has been demonstrated. COX-1 was identified in the 1960s and was considered a constituent enzyme that generated PGs involved in physiological functions such as vasomotor tone regulation, gastric protection, and platelet function.

Years later, in the 1990s, a second isoform was identified, which was called COX-2, and that its expression is rapidly induced by inflammatory cytokines in peritoneal or skin macrophages, and that it was thought not to be present in normal tissues so it was called the inducible enzyme, while COX-1 was the constituent. This isoform has a homology of approximately 60% with the COX-1 being the catalytic site one of the most preserved.

Currently, it is known that COX-2 is present in normal tissues, where it performs specific functions. For this reason, the function of this isoform has been re-evaluated and it has been considered more of a highly regulated constituent form. Some of the stimuli that regulate the expression of COX-2 are glucocorticoids, sodium, bradykinin, renal artery perfusion pressure, and water balance. The importance of constitutive expression of COX-2 in the kidney was observed in mice knock out for this enzyme. Two simultaneous publications showed that these mice born with normal renal function developed severe renal lesions and renal failure in the postnatal period; in addition they are infertile and have ventricular and vascular fibrosis, liver damage, and other alterations that confer them high mortality.

The hypothesis about the role of constitutive versus induced COX isozymes was that under normal homeostatic conditions, COX-1 was the constitutive enzyme present in most cells and tissues, whose activity accounts for the continuous production of physiologically important prostaglandins, whereas COX-2 was induced at the site of inflammation and produces pro-inflammatory prostaglandins. This is an elegant theory, with important therapeutical implications. Thus, enormous resources were invested in developing selective inhibitors of COX-2 based on the hypothesis that these compounds will reduce pain, fever, and inflammation without causing gastrointestinal or renal injury. This hypothesis was challenged with the demonstration of a constitutive, glucocorticoid-insensitive COX-2 in a subset of thick ascending limb of Henle (TAL) cells of normal kidneys. This pool of constitutive COX-2 observed in TAL does not coexist with COX-1 in the kidney, since the latter is present in arterial vascular endothelial cells, medullary and cortical collecting ducts, and medullary interstitial cells. The physiological role for COX-2 in TAL is the generation of prostaglandins which can contribute locally to the handling of NaCl, as the TAL segment has a crucial role in salt and water homeostasis. Renal COX-2 expression and function in vivo is regulated by physiological stimuli such as dietary sodium, angiotensin II, and glucocorticoids. We have demonstrated that COX-2 in the kidney is normally under the inhibition of endogenous corticoids, an effect that is unmasked by adrenalectomy. Thus, adrenalectomy increases expression of enzymatically active COX-2 and induces a recruitment of TAL cells that, in the basal state, do not express COX-2 protein. This recruitment occurs in the cortical TAL and proceeds in a defined pattern towards the outer medullary TAL. Moreover, evidence for a very important role of COX-2 in kidney development has emerged from mice with COX-2 KO gene disruption. These mice develop severe renal abnormalities weeks after birth; the abnormalities are not detectable at birth and are evidenced only with increasing age, indicating an important role of COX-2 during renal postnatal development. Concordant with this, we have observed high amounts of COX-2 present in TAL cells during early postnatal development, with a progressive decline to adult levels after weaning; this decrease in COX-2 is being associated with the increase in plasma levels of endogenous glucocorticoids.

The presence of this constitutive renal COX-2 in normal animals, also observed in humans, can have important therapeutical consequences since the COX-2 selective inhibitors were designed to spare the kidney and gastrointestinal tract assuming that the enzyme was absent in the normal kidney and was only induced in pathological conditions. Since that assumption was proven incorrect for the kidney and the gastrointestinal system, the therapeutical indications of selective COX-2 inhibitors required re-evaluation to assess whether they were sufficiently renal and gastrointestinal sparing to warrant their use in patients requiring long-term therapy.

11.3.1 Distribution of COX-1 in the Kidney

COX-1 is widely distributed in the kidney with greater expression in the inner medulla. Immunoreactivity has been observed in endothelial cells of the terminal portion of the afferent artery. Immunoreactivity in the renal tubules has been described in cells of the thin loop of Henle and in the collecting duct. Additionally, immunoreactivity has been detected in both cortical and medullary interstitial cells.

During postnatal development COX-1 has been located in podocytes, increasing its intensity from comma-shaped structures to the fully vascularized glomerulus.

11.3.2 Distribution of COX-2 in the Kidney

In the normal rat kidney, our studies have shown that COX-2 is located in a subtype of the thick ascending loop of Henle cells found in the cortex and the outer medulla and which under normal conditions are in the vicinity of the macula densa. This renal COX-2 is highly regulated and has a negative feedback mechanism mediated by the PGE_2 acting on the EP3 receptor. This negative feedback is similar to that of renin which is regulated by Ang II levels via the AT_1 receptor.

The function of COX-2 in TAL cells is related to the local generation of PGE_2 which in this segment inhibits the reabsorption of NaCl, an effect that is mediated in part by the inhibition of Na^+, K^+-ATPase and by the regulation of the NKCC2 cotransporter. More distal, the PGE_2 antagonizes the antidiuretic effect of ADH via the V_2 receptor by acting on its EP3 (from PGE_2); thus, the PGE_2 has a diuretic and natriuretic effect on the kidney.

The involvement of COX-2 in fetal kidney development has also been demonstrated. In the rat, large amounts of COX-2 have been observed during early stages of postnatal development with a progressive decrease as the animal reaches the adult stage. These high levels of COX-2 in the postnatal period occur because there are very low levels of glucocorticoids in this period. As these glucocorticoid levels increase, the expression of COX-2 is progressively inhibited until it reaches adult levels. Sensitivity to glucocorticoids and the importance of COX-2 in kidney development lead to caution about the use of glucocorticoids (dexamethasone) in humans born prematurely where dexamethasone is used to accelerate lung maturation.

Renin and COX-2 share regulation by negative feedback and in addition both present the phenomenon of recruitment, in which the expression of the enzyme increases by recruiting a greater number of smooth muscle cells (renin) or TAL cells (COX-2). In both cases it is by inhibition of the enzyme (ACE or COX-2) or antagonism of the receptor corresponding to the hormone (AT_1 or EP3).

This expression remains low until birth, reaching a maximum at 2 weeks postnatal and then decreasing to levels found in adults at about 6 weeks postnatal in rodents.

11.4   Nitric Oxide

Nitric oxide is a gas generated by the action of the enzyme nitric oxide synthase (NOS) on its substrate, the amino acid arginine. This signaling pathway participates in the homeostatic response of the kidney to changes in salt intake and extracellular fluid volume and in the perfusion pressure of the kidney, favoring vasodilation, natriuresis, and diuresis.

NO has a very short half-life reacting rapidly with the superoxide radical (O_2^-) to generate the peroxynitrite radical ($ONOO^-$), which subsequently oxidizes to NO_2 and NO_3. For this reason, this gas has a local action interacting with nearby thiol groups and heme protein centers. One of its main targets is guanylyl cyclase (GC) which when activated generates cyclic GMP (Fig. 11.5).

11.4.1 Distribution of NOS Isoforms in the Kidney

Three NOS isoforms have been identified, the endothelial (eNOS), the neuronal (nNOS), and the inducible (iNOS); all three isoforms are constitutively expressed in different cell types within the kidney. nNOS is mainly found in the cells of the macula densa where they participate in the function of the juxtaglomerular apparatus and in tubular-glomerular feedback and renin secretion. Additionally, it has been described in the parietal cells of Bowman's capsule and in some cells of the thick ascending loop of Henle. In the vasculature, nNOS has been identified in the efferent arterioles and there is evidence suggesting its presence in the descending straight vessels. iNOS has been located mainly in the afferent artery and intercalated cells of

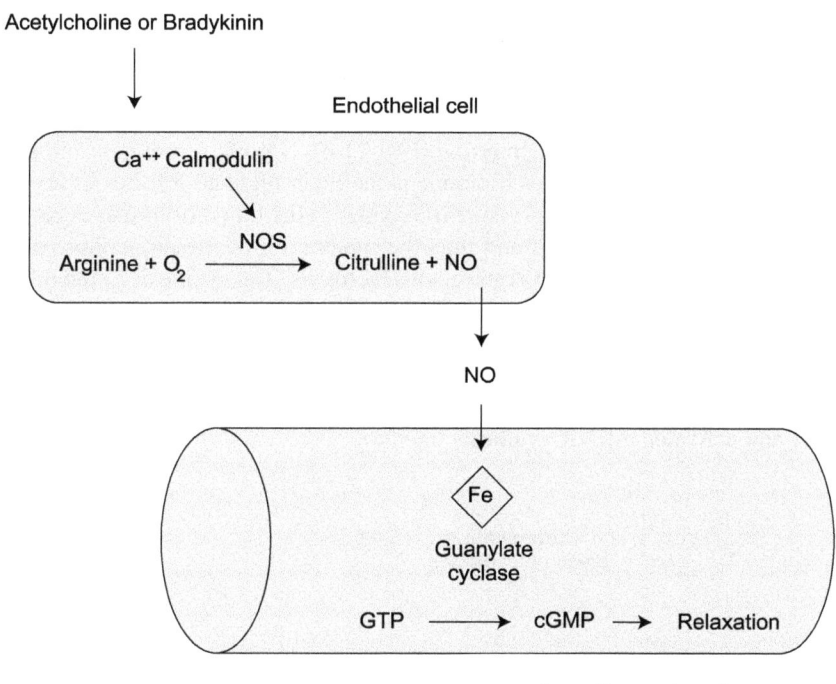

Acetylcholine or Bradykinin

Endothelial cell

Ca⁺⁺ Calmodulin

NOS

Arginine + O₂ ⟶ Citrulline + NO

NO

Fe
Guanylate cyclase

GTP ⟶ cGMP ⟶ Relaxation

Smooth muscle cell

Fig. 11.5 Synthesis and action of nitric oxide (NO). The scheme shows the signal transduction pathway activated by endothelial stimulators. The NO generated from the epithelium or in the renal vascular endothelium acts on cells that are near the synthesis site

the cortical collecting duct, but there are also reports suggesting its presence in other tubular segments. This renal isoform appears to be different from the iNOS found in other tissues, since in the kidney this isoform is constitutively expressed. The eNOS has been located in the endothelium of the renal vessels, being observed in the glomerulus, in the interlobular arteries, in the afferent arterioles, and in the efferent arterioles and its presence is also suggested in the descending straight vessels.

The location of iNOS or eNOS has not been studied in the developing kidney. With respect to nNOS, its presence has been identified in the rat by immunohistochemistry and by in situ hybridization at 2, 6, and 15 days postnatal. Its location has been identified in the developing distal tubule from the early stages of nephrogenesis, the S-bodies, and seems to be participating in the organization of the macula densa and the juxtaglomerular apparatus. Quantification of NOS gene expression by in situ hybridization and catalytic activity (NADPH-diaphorase) shows maximum expression on day 6 in contrast to renin which has its maximum expression on day 2 postnatal.

11.5   Erythropoietin

Erythropoietin (Epo) is a glycoprotein composed of 165 amino acids with an estimated molecular weight of 35 kDa, whose function is to induce the production of red blood cells from the bone marrow (Fig. 11.6). Normal serum concentrations range from 8 to 18 mU/mL. This hormone is the main regulating factor of erythropoiesis. Thus, when the partial pressure of oxygen in the blood or tissues decreases, the secretion of Epo is induced and thus the maturation of the progenitor cells is accelerated and the formation of erythrocytes increases. The synthesis of the mRNA of this protein is regulated by hypoxia and the synthesis of the protein is directly proportional to the content of the messenger, i.e., the more mRNA is transcribed, the more protein will be translated. The regulation mechanism of Epo is mainly at the messenger level. It is a protein called "hypoxia-inducible factor 1" (HIF-1) that binds to DNA and activates mRNA synthesis for Epo.

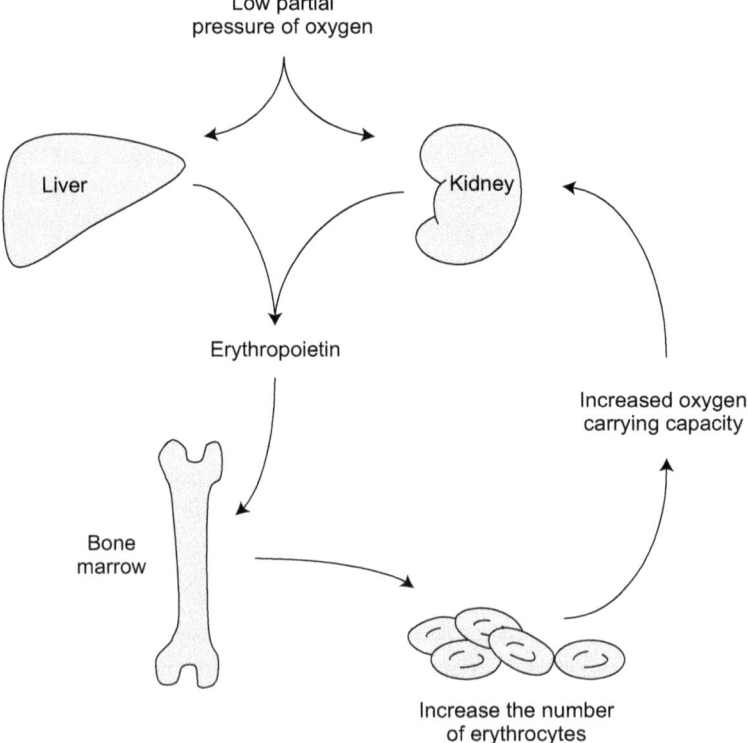

Fig. 11.6 Calcitriol synthesis. The scheme shows how 7-dehydrocholesterol (7-DHC) is transformed into cholecalciferol (Vit D_3) in the skin. This can also enter through the diet, like ergocalciferol (Vit D_2). These are transformed in the liver into 25(OH)VitD and subsequently undergo a second hydroxylation to become calcitriol (1,25(OH)$_2$VitD) or its inactive form 24,25 (OH)$_2$VitD

In contrast to other organs where the metabolic needs of the tissue determine the blood flow, in the kidney it is the blood flow that determines its oxygen requirements. The reabsorption of Na^+ is the main determinant of renal oxygen consumption, so that by inhibiting the reabsorption of Na^+, the production of Epo decreases.

11.5.1 Location

The site of Epo synthesis in the adult is mainly the kidney. The cells in charge of this function are peritubular cells of the fibroblast type that are in close contact with the renal capillaries. Under physiological conditions, Epo production is limited to the most juxtaposed region of the cortical labyrinths. However, in contrast to anemia, Epo production occurs in more superficial areas of the renal cortex. This increase in Epo synthesis is given by a recruitment of cells, which in conditions of normoxia do not synthesize Epo.

11.5.2 Ontogeny

This hormone begins to be detected very early in gestation at about 41 fetal days in the sheep kidney. Epo expression in sheep is maximum at 60 fetal days and then decreases. In the ovine fetus, Epo has been located in the fibroblasts that are found in the extratubular areas near the base of the proximal tubule cells, these being considered the sites of protein synthesis. The number of mRNA-positive fibroblasts for Epo directly determines the speed of Epo production and its serum levels.

11.6 Conclusions

The kidney contains several renal vasoactive hormones that contribute to the kidney excretory function, to renal hemodynamic, and to blood pressure regulation.

The main renal vasoactive systems are the renin-angiotensin system (RAS) and the kallikrein-kinin system (KKS) with its mediators. Among the most relevant mediators are nitric oxide (NO) and eicosanoids or prostaglandins (PGs).

Renin-angiotensin and kallikrein-kinin systems and their mediators have traditionally been described as having opposite physiological effects on sodium excretion, vascular tone regulation, blood flow distribution, glomerular hemodynamics, and blood pressure regulation.

These vasoactive components are precisely located on defined anatomical microenvironments from where they affect specific tubular and hemodynamic functions; therefore, its localization in adult kidneys and during development is of great importance to understand their contribution to function in normal and pathological conditions.

Thus, in the adult kidney renin is located in granules in a small number of granular cells of the afferent artery from where it is secreted under stimuli, whereas kallikrein is located exclusively in the connecting cells of the connecting tubule. The localization of renin in the afferent arteriole contributes to the function of the tubuloglomerular feedback (TGF) where the macula densa which contains COX-2 and neuronal nitric oxide synthase also participates. A more recent discovery is the close location of the connecting tubules containing kallikrein with the afferent arteriole. This microenvironment provides the base for the function of the connecting tubule glomerular feedback (CTGF); both feedback mechanisms contribute to regulation of renal function.

Novel integrative functions are underlined such as the contribution of CNT cells to potassium regulation. Furthermore, in a high dietary potassium, CNTc hypertrophy, secrete more potassium, and increase the synthesis of kallikrein and production of bradykinin. This mechanism contributes to the natriuretic and hypotensive effect of the potassium-rich diet, in addition to the inhibitory effect on the activation of sodium chloride cotransporter NCC.

Review Questions

1. Describe the relationship between angiotensin I, angiotensin II, ACE, and bradykinin.
2. Which is the importance of the expression of ACE and angiotensinogen in the proximal tubule?
3. Explain the relationship between the connecting tubule and the KKS.
4. If you perform an immunohistochemistry protocol for EPO, where should the signal be expected?
5. In the normal adult rat kidney, COX is expressed in a subset of cells of the thick ascending limb of Henle; what is the importance of this finding?

Bibliography

Brown AJ, Dusso A, Slatopolsky E (2000) Vitamin D. In: Seldin D, Giebisch G (eds) The kidney: physiology and pathophysiology, vol II, 3rd edn. Lippincott Williams and Wilkins, Philadelphia, pp 1047–1090

Carey RM, Siragy HM (2003) Newly recognized components of the renin-angiotensin system: potential roles in cardiovascular and renal regulation. Endocr Rev 24:261–271

Chai SY, Fernando R, Peck G, Ye SY, Mendelsohn FAO, Jenkins TA, Albiston AL (2004) The angiotensin IV/AT4 receptor. Cell Mol Life Sci 61:2728–2737

El-Dahr SS, Figueroa CD, Gonzalez CB, Müller-Esterl W (1997) Ontogeny of bradykinin B2 receptors in the rat kidney: implications for segmental nephron maturation. Kidney Int 51:739–749

Handa RK, Ferrario CM, Strandhoy JW (1996) Renal actions of angiotensin-(1-7): in vivo and in vitro studies. Am J Physiol Renal Physiol 270:F141–F147

Harris RC, Breyer MD (2004) Arachidonic acid metabolites and the kidney. In: Brenner BM (ed) The kidney, vol I, 7th edn. Saunders, Philadelphia, pp 727–773

Hewison M, Zehnder D, Bland R, Stewart PM (2000) 1alpha-hydroxylase and the action of vitamin D. J Mol Endocrinol 25:141–148

Hoenderop JGJ, Nilius B, Bindels RJM (2002) Molecular mechanisms of active Ca^{2+} reabsorption in the distal nephron. Annu Rev Physiol 64:529–549

Kurtz A, Eckardt K-U (2000) Hematopoiesis and the kidney. In: Seldin D, Giebisch G (eds) The kidney: physiology and pathophysiology, vol II, 3rd edn. Lippincott Williams and Wilkins, Philadelphia, pp 1091–1132

Maxwell PH, Ferguson DJ, Nicholls LG, Iredale JP, Pugh CW, Johnson MH, Ratcliffe PJ (1997) Sites of erythropoietin production. Kidney Int 51:393–401

Solhaug MJ, Ballèvre LD, Guignard JP, Granger JP, Adelman RD (1996) Nitric oxide in the developing kidney. Pediatr Nephrol 10:529–539

Vio CP, Cespedes C, Gallardo P, Masferrer JL (1997) Renal identification of cyclooxygenase-2 in a subset of thick ascending limb cells. Hypertension 30:687–692

Vio CP, An SJ, Cespedes C, McGiff JC, Ferreri NR (2001) Induction of cyclooxygenase-2 in thick ascending limb cells by adrenalectomy. J Am Soc Nephrol 12:649–658

Wilcox CS (2000) L-arginine-nitric oxide pathway. In: Seldin D, Giebisch G (eds) The kidney: physiology and pathophysiology, vol II, 3rd edn. Lippincott Williams and Wilkins, Philadelphia, pp 849–872

Yamaguchi S, Tamura K, Nyui N, Hibi K, Ishigami T, Kihara M, Yabana M, Sesoko S, Ishii M, Umemura S (1998) Developmental changes in expression of angiotensinogen mRNA in rat nephron segments. Hypertens Res 21:155–161

Zhang MY, Wang X, Wang JT, Compagnone NA, Mellon SH, Olson JL, Tenenhouse HS, Miller WL, Portale AA (2002) Dietary phosphorus transcriptionally regulates 25-hydroxyvitamin D-1alpha-hydroxylase gene expression in the proximal renal tubule. Endocrinology 143:587–595

Pathophysiology of Hypertension or High Blood Pressure

12

Learning Objectives
- To understand the physiological basis of arterial hypertension.
- To describe the role of the kidneys in the development of arterial hypertension.
- To understand the role of intrarenal hormonal systems in arterial hypertension.

12.1 Sodium-Sensitive Hypertension and Kidney Hormones

In the previous chapter, we described the renal vasoactive hormones and their contribution to kidney excretory function and blood pressure regulation. This chapter addresses the pathophysiological alterations of these vasoactive hormonal systems and their contribution to chronic kidney disease and salt-sensitive high blood pressure.

The main renal vasoactive systems are the renin-angiotensin system (RAS) and the kallikrein-kinin system (KKS) with its mediators. Among the most relevant mediators are nitric oxide (NO) and eicosanoids or prostaglandins (PGs). These PGs are produced by the action of cyclooxygenases (COX-1 and COX-2) on their substrate arachidonic acid or eicosatetranoic acid (AA). AA is present in mammals and gives rise to prostaglandins of series 2 (PGE_2, PGI_2). In fish, the substrate corresponds to eicosapentaenoic acid (omega-3 fatty acid derived from fish oil) which gives rise to prostaglandins of series 3. The prostaglandins of series 3 are anti-inflammatory or do not participate in inflammation; that is why the use of omega-3 oils has been recommended, which partially displaces arachidonic acid in some tissues.

RAA and KKS systems and their mediators have traditionally been described as having opposite physiological effects on sodium excretion, vascular tone regulation, blood flow distribution, glomerular hemodynamics, and blood pressure regulation. Equally important to these functional effects, RAS and KKS have opposite effects on cardiovascular and renal architecture. Thus, one of the RAS effectors, angiotensin II (AII), promotes cell proliferation, collagen synthesis, and fibrosis. On the other hand, bradykinin (BK), the hormone that affects KKS, is important in antagonizing

© Springer Nature Switzerland AG 2022
P. A. Gallardo, C. P. Vio, *Renal Physiology and Hydrosaline Metabolism*,
https://doi.org/10.1007/978-3-031-10256-1_12

the effects of angiotensin II. In this way, BK inhibits cell proliferation, stimulates collagenases, and inhibits ventricular, renal, and vascular fibrosis. A third hormone, the hemoregulatory tetrapeptide N-acetyl-seryl-aspartyl-lysyl-proline (AcSDKP), is involved in the control of hematopoietic cell proliferation and is normally degraded by the angiotensin I-converting enzyme (ACE) or kininase II. In the contribution of local levels of peptides (AII, BK, and AcSDKP), the enzymes ACE and kallikrein that generate angiotensin II and bradykinin play a critical role. Consistent with this, an important part of the beneficial effects of ACE inhibitors is explained by the triple effect of inhibiting the formation of angiotensin II and inhibiting the degradation of BK and AcSDKP, both in rats and humans.

12.2 Imbalance Between Vasoactive Hormones as a Mechanism of Renal Damage: Increase in ACE and Decrease in Kallikrein Alters the Balance between Vasoactive Systems

12.2.1 Local Induction of ACE and Progression of Kidney Disease

The RAS and KKS, with antagonistic functional effects, are closely related to each other and the resulting physiological effects depend on the balance between them. The most prominent link between both RAS and KKS systems is ACE (Fig. 12.1), an enzyme that forms angiotensin II and degrades bradykinin. ACE also degrades AcSDKP, which is involved in controlling the proliferation of hematopoietic cells by preventing their entry into the S-phase of the cell cycle. This enzyme is an

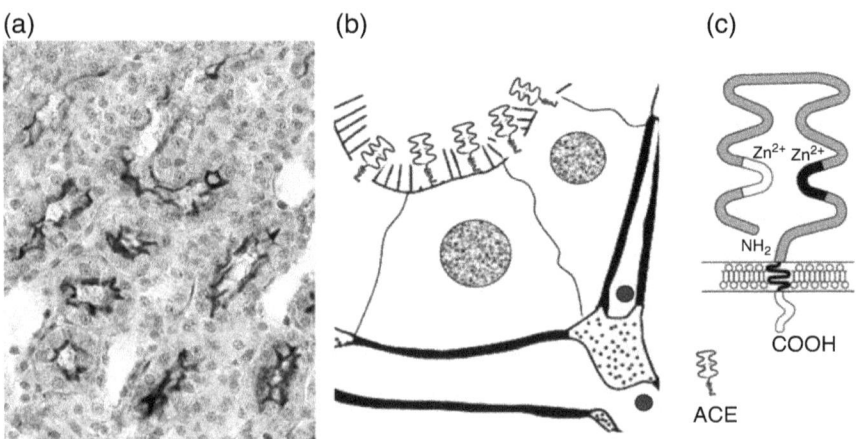

Fig. 12.1 Location and structure of the angiotensin I-converting enzyme. (**a**) Immunolocalization of the enzyme at the apical border of proximal tubules. (**b**) and (**c**) Location and general structure of the enzyme, which has an extracellular domain, a transmembrane domain, and a small C-terminal cytosolic domain

Table 12.1 Imbalance between ACE and kallikrein under clinical or experimental conditions

Clinical or experimental situation	↑ Angiotensin II	↓ Bradykinin
	ACE	Kallikrein
Arterial hypertension	Increased	Diminished
Catecholamine infusion	Increased	Diminished
Angiotensin II infusion	Increased	Diminished
Hypokalemia	Increased	Diminished
5/6 nephrectomy	Increased	Diminished
Diabetes	Increased	Diminished

ectoenzyme with two active sites on the extracellular, an intramembrane domain, and an intracellular carboxyl tail. This intracellular segment establishes communication with the intracellular segment of the bradykinin B2 receptor. There is broad agreement that local augmentation of RAS, via increased angiotensin II, contributes to tubulointerstitial injury and progression of kidney damage. This local increase in angiotensin II was originally thought to be the consequence of an increase in renin; however, in much of the human kidney diseases and in experimental models, renin is either normal or decreased. Because angiotensin II exerts a negative feedback on renin, their low levels can be explained by the increase in angiotensin II. Since renin levels are not responsible for the local increase in angiotensin II, we hypothesized that there could be a local increase in ACE to explain its increased production. In the recent past, it has been shown that there is indeed a local increase in ACE in damaged kidneys from several experimental models and in human kidney biopsies. This local increase in ACE occurs in sites where the enzyme is normally present (proximal tubule and renal vessel endothelium), but also in sites where it is not normally present, so it corresponds to a local induction of the enzyme. Local induction of ACE occurs in the damaged tubulointerstitial space where it contributes to the local production of angiotensin II, which in turn contributes to further kidney damage. As shown in Table 12.1 the increase in ACE explains the high levels of angiotensin II, and these in turn, together with the local decrease in bradykinin (and AcSDKP), explain much of the persistent kidney damage seen in chronic kidney diseases.

The imbalance between the vasoactive systems with increased ACE and diminished kallikrein produces an increase of angiotensin II and reduction in kallikrein. This imbalance has been observed in a number of pathologies and contributes to kidney damage.

12.2.1.1 Importance of Tubulointerstitial Space

The renal interstitium has received very little attention from a structural or functional point of view, its structure or composition occupying a few lines in histology texts. From the point of view of its function, only the corticomedullary osmolarity gradient and its importance in the concentration of urine are referred to.

We now know that the renal interstitium plays an integral role in normal kidney function and is also a major determinant of the evolution of diseased kidneys.

Chronic progressive kidney disease injury is unequivocally associated with both quantitative and qualitative changes in the renal interstitium, commonly known as "tubulointerstitial fibrosis."

The renal interstitium is defined as the intertubular, extraglomerular, and extra-vascular space of the renal tissue. It is bounded on all sides by tubular and vascular basement membranes and contains scarce cells, extracellular matrix, and interstitial fluid. Its distribution varies within the kidney as it represents approximately 8% of the total volume of the cortex parenchyma and up to 40% in the internal medulla.

The term "renal interstitium" is often used to refer to the peritubular interstitium (the space between tubules, glomeruli, and capillaries); periarterial connective tissue and extraglomerular mesangium are considered specialized interstitium.

It is still a matter of discussion whether the microvessels and capillaries, which are found within the peritubular space, are actually part of the renal interstitium or simply pass through it. In addition, the lymphatic vessels must be considered as a constituent interstitial part. In this text, we will use the denomination of the tubulointerstitial space to the space that exists between the tubules including the microvessels and capillaries contained in this space.

The tubular interstitium in the cortex and medulla differs with respect to cell content, composition of the extracellular matrix, relative volume, and endocrine function, which justifies considering the cortical and medullary interstitium as separate or different structures.

A central component of chronic kidney injury is cell infiltration and fibrosis that occurs in the tubulointerstitial space. Under normal conditions this space is com-posed of the capillaries and the interstitium surrounding the tubules and with the occasional presence of few macrophages, fibroblasts, and dendritic cells (Fig. 12.2). This space is of greatest physiological importance since all the substances reabsorbed by the tubular cells pass through it and it is from here that the capillaries are responsible for taking them to the systemic circulation, thus maintaining the hydrosaline balance and the constancy of the internal environment. To measure the magnitude of this process, remember that from this space about 180 l of water and 1.3 kg of NaCl daily pass to the circulation, post-tubular reabsorption, and that this is possible by an expeditious tubulointerstitial space.

Tubulointerstitial injury begins when the tubular cell is damaged by a noxa; the cell responds with synthesis de novo of chemotactic peptides, such as osteopontin (OPN) and monocyte chemoattractant protein-1 (MCP-1), which attract macrophages (ED1) and fibroblasts to the tubulointerstitial space. These infiltrative cells are activated; the fibroblasts are transformed into myofibroblasts (α-SMA) and synthesize collagen III that is deposited locally generating fibrosis (Fig. 12.2). Additionally, activated infiltrative cells synthesize inflammatory cytokines, which locally establishes a vicious circle that maintains local fibrosis. In this damaged interstitium, ACE is locally induced, which promotes the production of IIA (pro-inflammatory and pro-fibrotic), degrades bradykinin (anti-fibrotic), and degrades the peptide AcSDKP (antiproliferative), thereby increasing local cell proliferation, inflammation, and fibrosis (Fig. 12.3). This altered tubulointerstitial

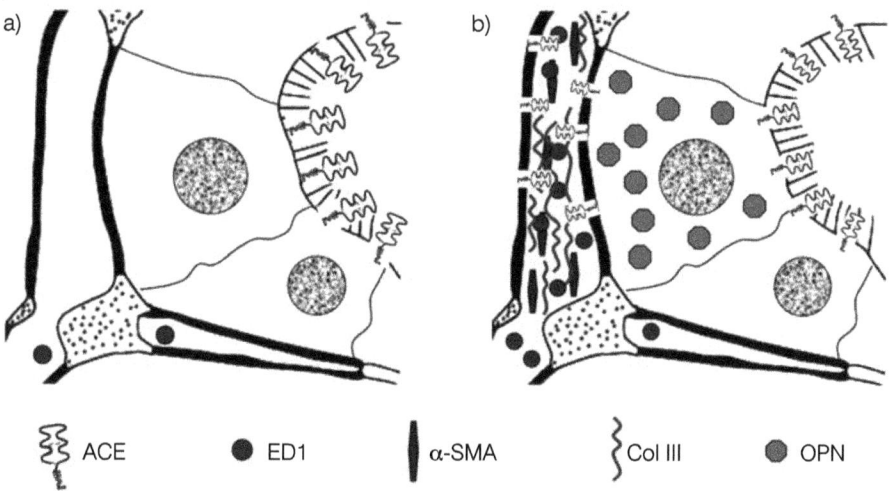

Fig. 12.2 Structure of the tubulointerstitial space in normal and pathological conditions. In (**a**) the structure of the normal tubulointerstitial space is shown where it is a space with few cells and where the reabsorbed liquid is carried into the systemic circulation. (**b**) shows the pathological structure of the tubulointerstitial space. ACE: angiotensin I-converting enzyme, *ED1* macrophages, *α-SMA* myofibroblast smooth muscle actin, *Col III* collagen III, *OPN* osteopontin

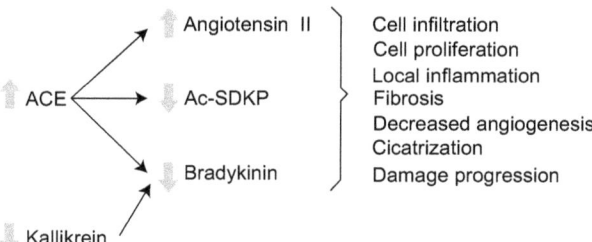

Fig. 12.3 Effects of increased ACE and decreased kallikrein. The local increase in ACE in the kidney increases the levels of angiotensin II and decreases those of bradykinin and those of the regulatory tetrapeptide Ac-SDKP, contributing to tubulointerstitial injury

space gradually loses its physiological capacity, and in fact studies in human kidney biopsies have shown that the greater the interstitial injury, the greater the progressive loss of kidney function. Consequently, therapeutic strategies aimed at halting the progression of kidney disease are geared towards locally inhibiting ACE, antagonizing the effect of angiotensin II, and locally increasing bradykinin and AcSDKP. In the latter case, recent studies have shown that the administration of AcSDKP inhibits macrophage infiltration, cell proliferation, and collagen deposition in the left ventricle of hypertensive animals (the infiltrative, proliferative, and fibrotic process of the left ventricle is similar to that which occurs in the kidney).

12.2.1.2 Is There a Common Mechanism for the Induction of ACE in Kidney Damage?

As we saw previously (Table 12.1), ACE is increased in pathological conditions of diverse nature and seems to correspond to a common mechanism in the progression of kidney damage. The explanation for the local increase in ACE can be local ischemia could correspond to a common mechanism that accounts for their increase, as we have published recently that mild hypoxia induces renal ACE.

Renal ACE originates mainly from segment S_3 of the proximal tubule, a segment located in the outer medulla, an area of the kidney with low oxygen levels (Fig. 12.4). If vasoconstriction occurs during kidney injury, these oxygen levels will decrease and hypoxia has been shown to directly increase ACE levels and indirectly through a VEGF (vascular endothelial growth factor)-mediated effect. Local induction of ACE favors vasoconstriction, which maintains external medullary hypoxia (Fig. 12.5). In at least three models where local induction of ACE has been described (angiotensin II infusion, phenylephrine infusion, and hypokalemia), kidney damage is greater in the outer medulla and there is uptake of pimonidazole (a marker of tissue hypoxia), suggesting a chronic hypoxic effect.

12.2.2 Deficiency of the Kallikrein-Kinin System and Kidney Damage

Consistent with the hypothesis of the balance between RAS and KKS, it can be postulated that a deficiency in KKS also contributes to the genesis and progression of

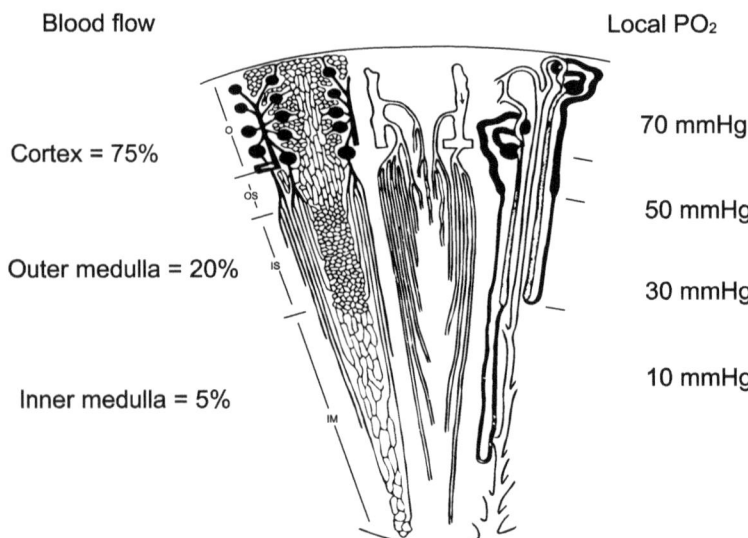

Fig. 12.4 Distribution of renal blood flow and O_2 partial pressure. The scheme of the distribution of the renal vascular architecture in the cortex and medulla is shown. Note that the partial pressure of O_2 has a zonal distribution from 70 mmHg in the cortex to 10 mmHg in the medulla

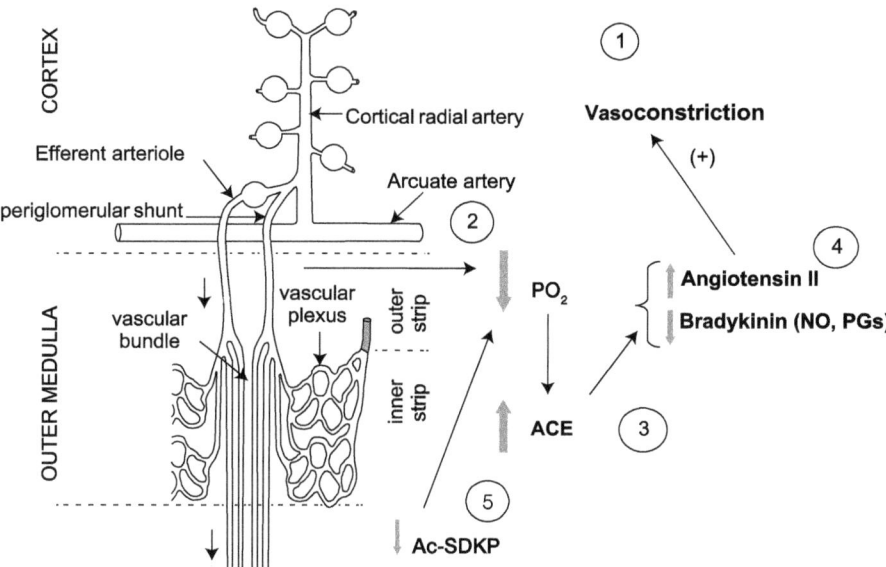

Fig. 12.5 Simplified hypothesis of the local induction mechanism of ACE in kidney damage models. The hypothesis integrates the concepts of relative medullary hypoxia, the effect of vasoconstrictors on medullary blood flow and oxygen partial pressure, and the local induction of ACE by hypoxia. Numbers *1* through *5* describe the possible sequence of events

kidney damage. Thus, a decrease in kallikrein in the presence or absence of an increase in RAS can contribute to tubulointerstitial kidney damage by leaving the RAS to predominate without the counterbalance of KKS. In addition, since renal sodium excretion is a function of KKS, a deficiency of kallikrein, kininogen, or BKR2 receptor can induce sodium-sensitive hypertension by the inability to excrete excess sodium from the diet.

There is now a wealth of information obtained from rodents whose KKS is altered, either in rats that are deficient in kallikrein and its substrate (kininogen) or in rats that are subject to chronic inhibition of the BKR2, as well as in mice knock out for the BKR2. In general, all of these models are found to develop hypertension when faced with a high-sodium diet (salt-sensitive hypertension) and are more susceptible to developing kidney damage. It has been observed in humans that patients with lower kallikrein excretion develop high-sodium diet-sensitive hypertension, and, in fact, it has been postulated that the measurement of low urinary kallikrein levels could be used to predict salt-sensitive hypertension.

In several of the experimental models described above (Table 12.1), such as the short administration of catecholamines (phenylephrine) and infusion of angiotensin II, it is observed that there is a decrease in kallikrein which persists after the initial noxa has been suspended. All these experimental models are of sodium-sensitive hypertensives, since they develop high blood pressure when their sodium intake is high.

In addition to the increase in ACE described above, the decrease in kallikrein contributes to a greater imbalance between the vasoactive systems.

In several experimental models in rats or in human kidney diseases, a decrease in renal kallikrein has been described, leading to a vicious circle, where the imbalance in vasoactive hormones is accentuated. This imbalance occurs in favor of the vasopressor, pro-fibrotic, proliferative, and sodium-retaining effect of angiotensin II and against the vasodilator, anti-fibrotic, and sodium-releasing effect of bradykinin. Finally, the antiproliferative and angiogenic effect of the tetrapeptide AcSDKP must be added (Fig. 12.3).

We have demonstrated, studying nearly 2000 renal genes, that in the stage of normalization after acute renal failure there is a persistent decrease in the expression of the renal kallikrein gene and its protein; when placed in high-sodium diet, these animals develop sodium-sensitive hypertension. When translated or extrapolated to humans, this observation would explain the increased susceptibility to developing kidney damage and high blood pressure observed in patients who have had an episode of acute kidney failure in the past.

Gene therapy with the human kallikrein gene has so far been used quite successfully in rats to prevent or reverse kidney damage and as antihypertensive therapy. However, this therapeutic strategy is far from being able to be administered to humans and even less can it be thought of as a preventive measure considering that approximately 20% of the world's population suffers from essential hypertension and that a significant fraction of them are sensitive to salt. For this reason, knowledge of the factors that stimulate KKS is of great relevance and importance in the prevention and treatment of high blood pressure and chronic kidney disease.

12.3   Regulation of Kallikrein Synthesis, Regulatory effect of Dietary Potassium

Renal kallikrein is regulated by sodium and potassium balance and by steroid hormones, thyroid, insulin, and catecholamines. Among the various factors involved in the regulation of kallikrein, the contribution of potassium has been underestimated, receiving less attention than the others. In humans, kallikrein excretion varies in direct relation to potassium intake, and there is a high correlation between urinary potassium and urinary kallikrein. Therefore, potassium deserves serious consideration, not only because it is an important kallikrein regulator (Fig. 12.6) but also because of the therapeutic implications of a high-potassium diet. Renal kallikrein originates exclusively from connecting tubule cells (CNTc), cells highly specialized in potassium secretion, with abundant Na^+, K^+-ATPase in the basolateral membrane and apical potassium channel (ROMK). In response to a high-potassium diet, hypertrophy of CNTc, increased basolateral surface area, and increased Na^+, K^+-ATPase as a compensatory mechanism for increased potassium intake by increasing its secretion are observed. This mechanism is known as "potassium adaptation" and allows the body to remove this ion by keeping the potassium in the extracellular fluid relatively constant. Together with hypertrophy

Fig. 12.6 Effect of dietary potassium on renal kallikrein. The increase in potassium in the diet increases the production of renal kallikrein and that of the B2 receptor and bradykinin, thereby increasing the excretion of sodium and water and causing vasodilation. This effect of dietary potassium appears to be on the expression of the hK1 kallikrein gene. The increase in kallikrein produces an increase in bradykinin. The scheme details the structure of the hormones of the kallikrein system and their physiological effects

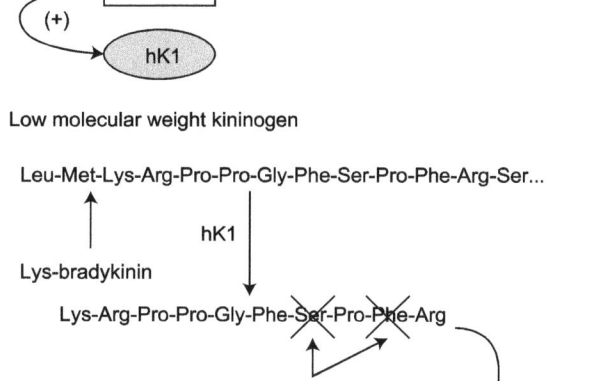

of CNTc there are an increase in renal kallikrein synthesis and increase in kallikrein mRNA and bradykinin B2 receptor. Thus, a high-potassium diet increases KKS activity and the number of bradykinin receptors. It is interesting to note that bradykinin, in physiological concentrations, inhibits the apical entry of sodium into cells of the collecting duct. In addition, recent studies show that by limiting the apical sodium intake, bradykinin produces increased urinary elimination of this cation. Bradykinin, via BKR2 receptor, antagonizes the water-retaining effect of vasopressin arginine mediated by the V_2 receptor and together with nitric oxide participates in the control of papillary blood flow and pressure natriuresis, thus contributing to the long-term control of blood pressure. Thus, it can be postulated that stimulation of KKS by a high-potassium diet increases sodium and water excretion with a net effect of lowering blood pressure. It is well established that increased potassium intake dephosphorylates the thiazide-sensitive sodium/chlorine cotransporter (NCC) located in the distal convoluted tubule, thus inducing natriuresis, a mechanism mediated by Kir4.1 potassium channel. More recently we have described a similar effect of bradykinin; BK dephosphorylates the NCC located in the distal convoluted tubule, thus inducing natriuresis.

Thus, increased dietary potassium increases kallikrein and BKR2 receptors; kallikrein produces BK which dephosphorylates NCC, inducing sodium excretion; this effect is in addition to the direct effect on Kir4.1 and dephosphorylation of NCC and natriuresis.

Several studies in hypertensive humans have shown that dietary potassium supplementation significantly increases kallikrein and urinary sodium excretion and lowers blood pressure. The therapeutic implications in the management of hypertensive patients are of great importance, particularly in those who are sensitive to salt. The partial replacement of sodium salts by potassium salts in the diet has an additive effect with the double benefit of reducing sodium intake and stimulating KKS at the same time. This change in dietary salt content (less sodium and more potassium) has been shown to be effective in reducing blood pressure and deaths from coronary and cerebral vascular accidents in large-scale studies in Finland and recently in large population in China. In addition, the Intersalt study has shown that blood pressure levels are directly related to sodium intake and inversely and independently related to potassium intake.

The high-sodium diet corresponds to a relative value; it is generally accepted that the normal sodium diet corresponds to an intake of 100 mmol/day; however, changes in eating habits have meant that the usual sodium diet is now close to 150–200 mmol/day which corresponds to a high-sodium diet.

On the other hand, an adequate diet in potassium corresponds to our ancestral diet, which as a result of civilization and food preservation was changed to a diet rich in sodium. However, our body is not adequately adapted to handle the high-sodium diet, characteristic of our current diet, resulting in sodium-sensitive hypertension. This occurs in subjects with normal kidney function and in subjects with minimal kidney injury. Since sodium restriction is unrealistic and processed foods are high in sodium, the increased intake of potassium has been promoted in the population which in the long term may constitute the main non-pharmacological measure of prevention of high blood pressure and kidney damage.

12.4   Importance of an Adequate Diet in Potassium and Its Natriuretic Effect

There is consensus in the world that our modern diet is potassium deficient; consumption from a balanced diet with meat, fruits, and vegetables provides approximately 150–200 mmol/day, but data shows that current potassium intake is close to 50 mmol/day. For this reason, the USDA and the FDA have announced the existence of a deficiency in this fundamental component of our diet. This moderate potassium deficiency cannot be detected in blood levels but is associated with the development of high blood pressure, cerebral vascular infarction, osteoporosis, and kidney damage.

Current recommendations for minimum potassium intake are 100–120 mmol/day. Large rural and urban epidemiological studies show that people with low-potassium diets develop hypertension.

There is strong evidence that the beneficial health effects of a low-sodium diet with adequate potassium are synergistic, and both factors are necessary.

Since 1928, studies by Addisson have shown that potassium salts lower blood pressure and eliminate sodium. Today we know that increased potassium intake dephosphorylates the apical thiazide-sensitive sodium/chlorine cotransporter (NCC) expressed in the distal convoluted tubule cells. This potassium-induced dephosphorylation deactivates the NCC cotransporter and sodium is eliminated by urine.

This occurs together with increased KKS where BK also deactivates the NCC cotransporter resulting in BK-induced sodium excretion natriuresis.

12.5 Sodium and Potassium in the Diet and Salt-Sensitive (Sodium) Hypertension

High blood pressure affects 20–30% of the population (>1200 million) and contributes to the leading cause of death in men and women. Depending on the characteristics of the population, 40–70% of hypertensives are sensitive to salt (sodium). Less than 50% of hypertensives know their condition and only 10–20% have their pressure controlled; the Chile National Health Survey 2010 shows similar numbers to those in the literature. Regarding the knowledge of their hypertension, women (60%) are better than men, but the effectiveness of the treatment is equally insufficient (Fig. 12.7).

With similar numbers around the world, one cannot help but wonder if we are missing an important factor in the management of hypertension, especially since we have, as never before in history, a wide variety of effective antihypertensive drugs.

We believe that the forgotten factor is the potassium content of the diet whose importance has been relegated in favor of the importance placed on sodium content, but both sodium and potassium are closely linked in their health effects.

Our proposal is that potassium in the diet is very important for health, that the population is exposed to a diet poor in potassium, and that decreasing sodium is neither sufficient nor feasible, at the recommended levels with contemporary nutrition.

There is abundant evidence that the kidney is responsible for high blood pressure, especially the salt (sodium)-sensitive one that accounts for half of all hypertensives. Sodium-sensitive hypertension was defined as a 10 mmHg increase in sodium intake from 100 to 200 mmol/day. This definition comes from the 1980s when sodium intake was 100 mmol/day, but the current sodium intake is close to 200 mmol/day.

The healthy kidney contributes in a central way to the balance of sodium, potassium, and water; this balance depends on the one hand on our intake and on the other hand on the elimination or conservation by the kidney. The kidney plays a major role in the management of sodium, potassium, and water and in the regulation

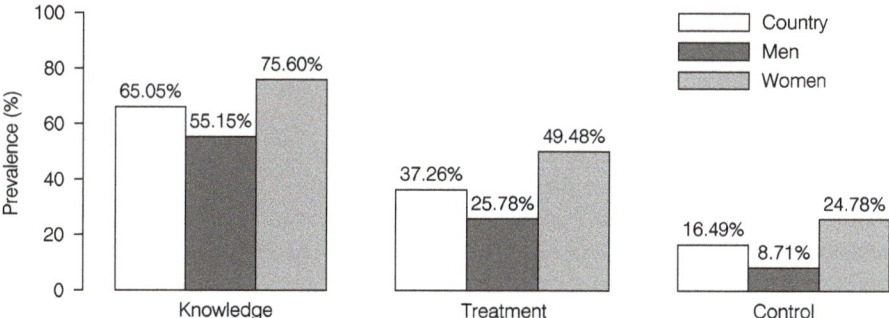

Fig. 12.7 Hypertension in the Chile National Health Survey 2010. The Chile National Health Survey 2010 (NHS) shows similar numbers in arterial hypertension to those in the world literature regarding the knowledge, treatment, and control of blood pressure. Low knowledge of this condition, undertreatment, and insufficient control of blood pressure

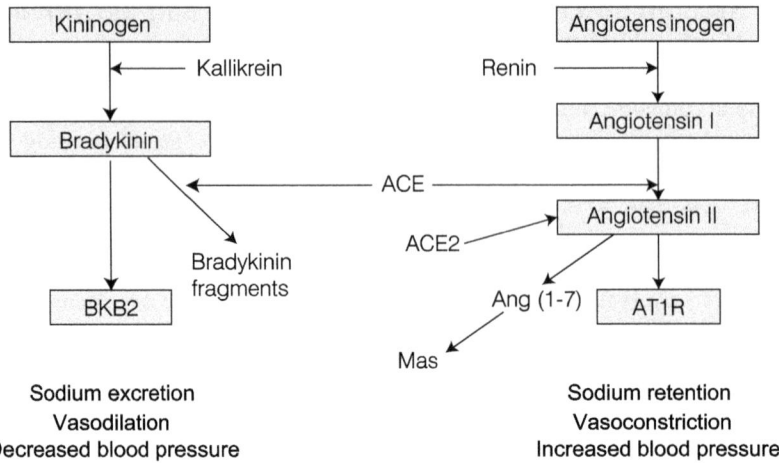

Fig. 12.8 Kallikrein-kinin and renin-angiotensin systems and renal function. Diagram illustrating the components of the main renal vasoactive systems and their contribution to renal excretory function and blood pressure regulation

of blood pressure through the production of hormones (angiotensins, bradykinins, prostaglandins) that eliminate or retain sodium and potassium and that are involved in the control of vascular tone and in the regulation of blood pressure and hypertension (Fig. 12.8).

The alteration of these vasoactive systems produces renal injury, the control mechanisms are altered, and high blood pressure is produced which in turn damages the kidney, contributing additionally to hypertension.

Factors contributing to hypertension include the high-sodium and low-potassium intake characteristic of the modern Western diet.

Animal or vegetable cells from natural, unprocessed foods have an intracellular concentration of 140 mEq of potassium and 10 mEq of sodium (Fig. 12.9). Therefore, natural foods are rich in potassium and contain little sodium. However, processed foods have 5–10 times more sodium than natural foods. Today, it can be stated that in industrialized countries, a large part of the population is exposed to a low daily intake of potassium, a situation correlated with the development of salt-sensitive hypertension.

It is relevant to remember that humans have inhabited the earth for 3.5 million years and that during most of that time they have evolved by adapting to a natural diet that is low in sodium (<0.7 g/day); therefore, to survive has evolved with several mechanisms that reabsorb sodium; >99% of the sodium filtered at the glomerular level is reabsorbed. Currently sodium intake is about 5 g/day, a sevenfold increase in sodium intake. On the other hand, this natural diet of our ancestors has a potassium amount of 280 mmol/day, which is a big difference with the current one of about 60 mmol/day, five times less in potassium intake.

Over the past 40 years, there has been a global effort (by the WHO and other organizations) to reduce sodium consumption through health programs, educational

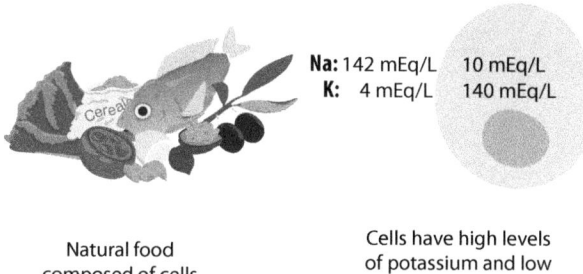

EXTRACELLULAR INTRACELLULAR

Na: 142 mEq/L   10 mEq/L
K: 4 mEq/L   140 mEq/L

Natural food
composed of cells

Cells have high levels
of potassium and low
levels of sodium

Fig. 12.9 Sodium and potassium distribution of intra- and extracellular concentration. The animal or plant cells of unprocessed natural foods have an intracellular concentration of 140 mEq/L of potassium and 10 mEq/L of sodium; therefore, natural foods are rich in potassium and contain little sodium. Industrially processed foods alter this composition by increasing the sodium content and decreasing the potassium content

policies, and food labeling. However, there is little public awareness of the risks of a potassium-deficient diet and what might be the most effective way to prevent cardiovascular disease by increasing potassium intake. Increased potassium intake through unprocessed foods helps to lower blood pressure, prevent strokes, eliminate sodium, and prevent osteoporosis and has vascular protective effects.

The medicinal advice has attempted to reduce dietary sodium intake and has failed because according to the prospective urban-rural study (PURE) being conducted in 18 countries (102,000 participants), sodium intake today exceeds 200 mmol/day (about 5 g of sodium, or 12 g of NaCl), with the recommendation being between 2.3 g/day (3.3% of the population) and 1.5 g/day (0.6% of the population). Potassium consumption should be greater than 100 mmol/day (3.9 g of potassium), a level reached by less than 1% of the population, and the current level is only 50 mmol/day (2 g) (Fig. 12.10).

Additionally, it was thought that antihypertensive drugs could control this disease independent of high-sodium intake, but it has been shown that the "gold standard" of these treatments (ACE inhibitors and angiotensin II receptor antagonist) does not decrease the damage in the face of high-sodium intake and if the sodium in the diet is not lowered.

Public resistance to a low-sodium diet is explained, in part, because we know little about the biomedical basis for why a high-sodium diet produces high blood pressure and cardiovascular-renal damage. In studies in collaboration with Dr. Richard Johnson, we have shown that a low-potassium diet induces an alteration of vasoactive systems, mainly a decrease in kallikrein, nitric oxide, and prostaglandins, and an increase in ACE expression. These vasoactive changes coexist with evidence of tubulointerstitial kidney injury (Fig. 12.2) and the acquisition of salt-sensitive phenotype. Rats become hypertensive on a high-sodium diet despite normalization of dietary potassium; more recently, renal ACE has been shown to be required to generate salt sensitivity in an experimental hypertension model.

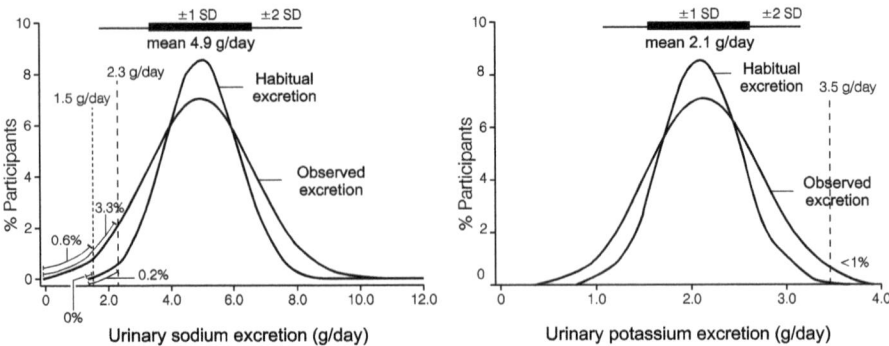

Fig. 12.10 Distribution of sodium and potassium consumption in the population. Distribution of sodium and potassium consumption in the prospective urban and rural study (PURE) carried out in 18 countries (102,000 participants). Sodium intake today exceeds 200 mmol/day (about 5 g of sodium, or 12 g of NaCl), and the recommendation is 2.3 g/day (3.3% of the population). Potassium consumption should be greater than 100 mmol/day (3.9 g of potassium), a level reached by less than 1% of the population. Consumption of sodium and potassium is obtained by measuring your urinary excretion for 24 h

The benefits of an adequate potassium diet have been known for decades, but the biomedical basis for this beneficial effect is unknown. This is why we have studied the biomedical basis that explains the beneficial effect of an adequate potassium diet. The imbalance of sodium and potassium has effects on the heart, arteries, and kidney, but it is important to study its effects on the brain and other organs since such a drastic change in our natural diet must have consequences on the whole body.

In the United States of America, increasing dietary potassium intake is beginning to be seen as a public health issue. Its content in processed foods will be incorporated into the new food labeling of the Food and Drug Administration (FDA) to raise awareness of the benefits of increased consumption of potassium-rich foods. The Agency has stated that "we have evidence that people are not consuming enough of this mineral to protect themselves from cardiovascular disease," adding: "if a person is concerned about their blood pressure, they should pay attention to the levels of sodium and potassium they are consuming."

The recommendations of the Dietary Reference Intake (DRI 2005) and Academy of Sciences USA (2006) along with other recent publications show that low-potassium intake is associated with cardiovascular disease, hypertension, and stroke and recommend its increase in the diet. The problem in much of the world is that the importance of potassium content in the diet is not widely known to the public. The concept that our natural diet is high in potassium and low in sodium is a not well-known concept, although it appears in the contents of the secondary education of our schoolchildren, and it is a basic concept in biology.

As mentioned before after 20 years of validity, the USA FDA decided to change the current food labeling, incorporating the potassium content in view of the evidence of potassium deficiency in the diet and the population's lack of knowledge about its content in processed foods (Fig. 12.11).

Current label	New label

Current label

Nutrition Facts
Serving Size 2/3 cup (55g)
Servings Per Container About 8

Amount Per Serving

Calories 230   Calories from Fat 72

	% Daily Value*
Total Fat 8g	12%
Saturated Fat 1g	5%
Trans Fat 0g	
Cholesterol 0mg	0%
Sodium 160mg	7%
Total Carbohydrate 37g	12%
Dietary Fiber 4g	16%
Sugars 1g	
Protein 3g	

Vitamin A	10%
Vitamin C	8%
Calcium	20%
Iron	45%

* Percent Daily Values are based on a 2,000 calorie diet. Your daily value may be higher or lower depending on your calorie needs.

	Calories:	2,000	2,500
Total Fat	Less than	65g	80g
Sat Fat	Less than	20g	25g
Cholesterol	Less than	300mg	300mg
Sodium	Less than	2,400mg	2,400mg
Total Carbohydrate		300g	375g
Dietary Fiber		25g	30g

New label

Nutrition Facts
8 servings per container
Serving size   2/3 cup (55g)

Amount per serving

Calories   230

	% Daily Value*
Total Fat 8g	10%
Saturated Fat 1g	5%
Trans Fat 0g	
Cholesterol 0mg	0%
Sodium 160mg	7%
Total Carbohydrate 37g	13%
Dietary Fiber 4g	14%
Total Sugars 12g	
Includes 10g Added Sugars	20%
Protein 3g	

Vitamin D 2mcg	10%
Calcium 260mg	20%
Iron 8mg	45%
Potassium 235mg	6%

* The % Daily Value (DV) tells you how much a nutrient in a serving of food contributes to a daily diet. 2,000 calories a day is used for general nutrition advice.

Fig. 12.11 Change in food labeling adopted by the US FDA, 2018–2019. After 20 years in force, the Food and Drug Administration (FDA, USA) decided to change the current food labeling, incorporating the potassium content due to the evidence of its insufficient presence in the diet and the population's ignorance of its content in processed foods. The current label and the new label are displayed in detail for comparison

Potassium eliminates sodium in the short term and quickly. Today we know that this effect is mediated by the inactivation (dephosphorylation) of the NCC (sodium-chlorine cotransporter) or thiazide-sensitive cotransporter. This way, a diet high in potassium eliminates sodium. Conversely, with a low-potassium diet the opposite occurs; there is increased activation (phosphorylation) of the NCC in the distal tubule and therefore increased reabsorption of sodium. Under the current conditions of low-potassium and high-sodium intake, we are in the worst possible condition for cardiovascular health since the higher the sodium intake, the higher the sodium reabsorption and the increase in blood pressure and cardiovascular risk.

The reason why we have reached this situation is determined by our diet which has changed drastically from natural foods to industrial substitutes, which do not preserve the right proportions of sodium and potassium, adding an excess of sodium and eliminating potassium.

The challenge is to increase potassium intake in the population, along with reducing sodium (Dietary Guidelines for Americans 2015–2020). The increase in potassium is achieved with natural foods (meat, vegetables, fruits) without processing, or with processing that maintains the contents of the food.

It is important to educate people about the potassium content of natural and processed foods so that they can make appropriate choices. The current time is appropriate due to the entry into force of Chilean Law No. 20606 on the "nutritional composition of foods and their advertising" (June 27, 2016), which does not consider the potassium content of foods and which omits the new measures being taken by the FDA.

In this way, and along with the above, an adequate diet in potassium may be especially suitable for the sodium-sensitive population by preventing hypertension or lowering the pressure of those already hypertensive.

12.6   Importance of Renal Medullary Circulation in Sodium and Blood Pressure Regulation

As mentioned in Chap. 2, the kidney receives approximately 20–25% of cardiac output; this high output, however, is not evenly distributed throughout the kidney, as shown in Fig. 12.4. While cortical blood flow (75% of total kidney) is generated from the cortical nephrons, which are the majority, medullary (20% of total kidney) and papillary (5% of total kidney) blood flow is almost entirely derived from the efferent arterioles of the juxtamedullary nephrons.

Coinciding with this blood distribution, the PO_2 of the cortex (70 mmHg) falls sharply into the inner medulla, where it can reach 20–30 mmHg. This low oxygen pressure is the result of the balance between supply and demand (consumption) of O_2 that exists in this area of the kidney.

While the main functions of the renal cortex are filtration and reabsorption of most of the filtered substances, the renal medulla has different functions, such as concentration of urine by the countercurrent mechanism and regulation of sodium reabsorption in response to changes in extracellular fluid volume, playing an important role in the long-term regulation of blood pressure. Considering the different functionality of both zones, it is easy to understand that their irrigation is regulated independently.

Part of the differential regulation can be explained by the anatomical differences that exist between the vessels of cortical glomeruli compared to those of juxtamedullary glomeruli. While in the cortical, the afferent artery has a larger diameter than the efferent artery, allowing them to regulate the filtration pressure, in the juxtamedullary, the diameter of the afferent artery is smaller than that of the efferent artery (Fig. 12.5). Because the medullary circulation originates from a few glomeruli (10% of the total), large changes in medullary blood flow can occur without detectable changes in renal blood flow. On the other hand, many vasoactive agents have differential effects on cortical and medullary circulation, e.g., some vasoconstricting agents such as vasopressin mostly affect medullary circulation, while angiotensin II, endothelin-1, and noradrenaline act on cortical circulation. Vasodilators, on the other hand, have a greater effect on the medullary circulation than on the cortex.

On the other hand, although the renal circulation is classically a self-regulating vascular territory (Fig. 12.12), it is now known that this corresponds only to the cortical circulation. The papilla, unlike the cortical circulation, does not exhibit self-regulation, varying its flow proportionally with changes in blood pressure (Fig. 12.13). However, due to its low contribution to total renal blood flow (5%),

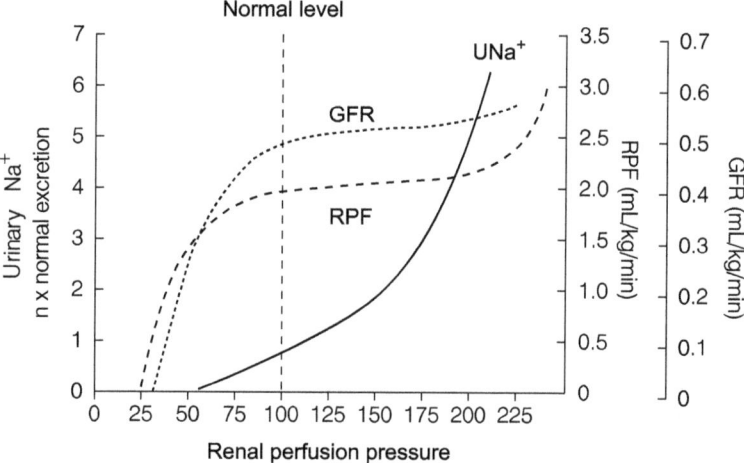

Fig. 12.12 Self-regulation of renal blood flow (RPF) and glomerular filtration (GFR), and pressure natriuresis

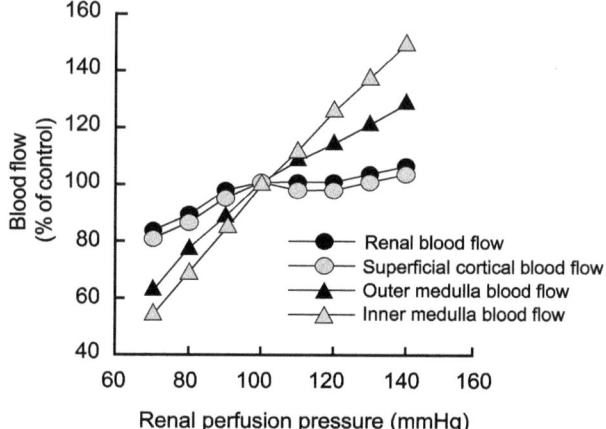

Fig. 12.13 Self-regulation of renal blood flow. Effect of renal perfusion pressure on cortical, external medullary, and internal medullary blood flow and total blood flow. Note that cortical blood flow is self-regulating unlike medullary blood flow which is not self-regulating

Fig. 12.14 Hormonal control of renal medullary circulation. Medullary circulation is hormonally regulated, vasodilator hormones increase the medullary blood flow, and vasoconstrictor hormones decrease it. The excretion of water and sodium is directly proportional to the medullary blood flow

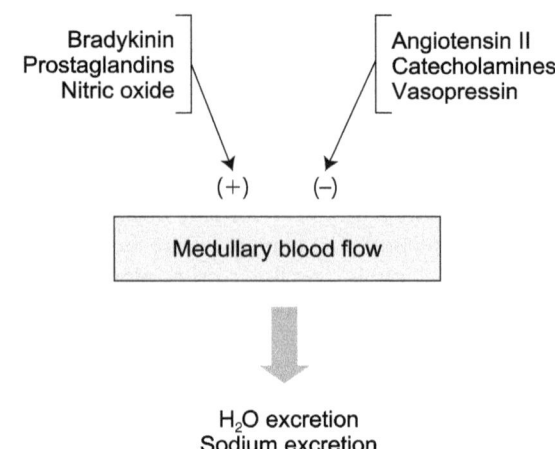

the effect goes unnoticed. The medulla has an intermediate location between the cortex and the papilla (Fig. 12.13).

Medullary circulation is hormonally regulated (Fig. 12.14) and the excretion of H_2O and Na^+ is directly proportional to the magnitude of medullary-papillary blood flow. Since papillary circulation is not self-regulating, this could explain diuresis and pressure natriuresis (Fig. 12.12), which is a renal mechanism for regulating blood volume in the long term and indirectly maintaining constant blood pressure.

12.7   Cyclooxygenase-2

Among the mediators of vasoactive systems, one that has received increasing attention is cyclooxygenase-2 (COX-2). This enzyme was described as an enzyme inducible by pro-inflammatory cytokines whose expression is inhibited by glucocorticoids and therefore responsible for inflammation, pain, and fever. Thus, a paradigm was established that indicated that cyclooxygenase 1 (COX-1 or constitutive) was responsible for the synthesis of PGs that participated in physiological phenomena (gastric protection, renal sodium excretion, regulation of renal hemodynamics, platelet function, etc.), while COX-2 was absent from normal tissues and was induced in tissue damage being responsible for the synthesis of PGs that participated in tissue inflammation (Fig. 12.15). Based on this paradigm, potent selective anti-inflammatory COX-2 inhibitors were developed and were used in the clinic with the promise of no adverse kidney effects. This paradigm was challenged when the existence of COX-2 in renal tubular cells (thick ascending loop of Henle) was discovered under normal conditions. Currently, there is abundant evidence of the physiological regulation of COX-2 and its contribution to physiological processes (Fig. 12.15), such as renal postnatal development, renin secretion, glomerular tubular feedback, and sodium excretion. COX-2 is regulated under physiological conditions by glucocorticoids, angiotensin II, sodium intake, and bradykinin

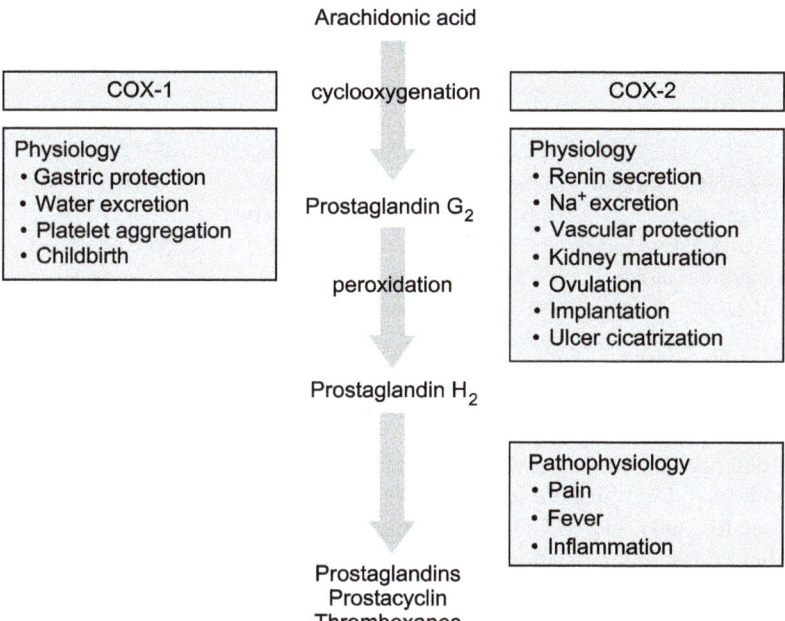

Fig. 12.15 Revised COX-2 paradigm. COX-2 like COX-1 has an important participation in physiological processes

Table 12.2 Contribution of cyclooxygenases to renal function

Function	Site	COX
Renin secretion	JGA	COX-2
Glomerular hemodynamics	JGA	COX-2
Tubuloglomerular feedback	TAL-JGA	COX-2
Na$^+$ excretion	TAL, CCD	COX-2/1
Water excretion	CCD	COX-1
Kidney maturation	Cortex	COX-2

Specific physiological roles of isoforms COX-1 and COX-2 in renal function and their site of action. JGA: juxtaglomerular apparatus; TAL: thick ascending limb of Henle; CCD: cortical collecting duct

(Table 12.2). The expression of COX-2 in normal tissues, its contribution to physiological phenomena, and its regulation by physiological stimuli suggest that its inhibition is not free of adverse effects, as proposed by the pharmaceutical industry. In fact, the mice knock out for the COX-2 gene develop severe kidney damage and ventricular fibrosis and have a high mortality from kidney failure. Postnatal COX-2 inhibition produces altered glomerulogenesis and cortical atrophy. The set of results described suggests that COX-2 has a relevant role in normal kidney function. On the other hand, it is important to note that some work has also postulated beneficial effects of selective COX-2 inhibitors in the treatment of kidney

Table 12.3 Comparison of specific physiological roles of COX-1 and COX-2 enzymes

Physiological process	COX-1	COX-2	Prostaglandin involved
Ovulation	Not essential	Essential	PGE_2
Implantation	Not essential	Essential	PGI_2
Birth	Essential	Compensatory	$PGF_{2\alpha}$
Resolution of inflammation	Not essential	Essential	PGD_2, 15-deoxy-PGJ_2
Platelet aggregation	Essential	Without effect	TXA
Postnatal kidney development	Not essential	Essential	Not determined
Ductus arteriosus closure	Compensatory	Essential	TXA_2/PGH
Gastric ulcer	Inhibition of both isoforms needed		Not determined
Healing	Not essential	Essential	Not determined
Intestinal cancer	Both isoforms play an essential role		Not determined

disease and high blood pressure. The subject, however, is not without controversy as other authors did not observe any beneficial effect. Regulation of COX-2 is still an open issue for study, just as COX-2 inhibition is a controversial issue. It is important to highlight other functions in which COX-2 is essential, i.e., in its absence either by gene deletion (knock-out) or pharmacological inhibition, this physiological function is altered. Among the extrarenal functions in which COX-2 is essential are ovulation, implantation, resolution of inflammation, healing of ulcers, and closure of the ductus arteriosus (Table 12.3). Because of the physiological and therapeutic implications, this is a subject of ongoing study.

An established fact in humans is that COX-2 is the main factor responsible for PGI_2 production without affecting TXA_2 production. When COX-2 is inhibited, PGI_2 and its antithrombotic effect are decreased; with platelet (COX-1-derived) production of TXA_2 predominating, the prothrombotic effect was predictable, and in fact while in use, in the United States alone it is estimated that over 100,000 people suffered from arterial thrombosis. With this evidence, the FDA ordered a recall, and at the time of preparation of this text, all selective COX-2 inhibitors in the world market have been withdrawn from the market because of their serious adverse effects in humans, particularly the induction of arterial thrombosis and kidney damage. Along with this recall, the main manufacturer of these anti-inflammatories has so far (2017) paid US\$ 8.5 billion to compensate for 44,000 people or their relatives who suffered from coronary or cerebral thrombosis.

The data generated in our laboratory and others were available in the medical and scientific literature. Exposed to the fact, during the renal physiology course, students in the early years of their health careers have concluded that a decrease in PGI_2 without changes in TXA_2 would produce arterial thrombosis, which indeed did occur.

This tragic story should remind us of the ethical imperatives of being informed about the medicines we prescribe, and that this knowledge is generated by research and gets into prestigious medical journals to be used responsibly by health professionals.

12.8 Conclusions

In Chap. 11, we described the renal vasoactive hormones and their contribution to kidney excretory function and blood pressure regulation. This chapter addresses the pathophysiological alterations of these vasoactive hormonal systems and their contribution to chronic kidney disease and salt-sensitive high blood pressure.

Renin-angiotensin and kallikrein-kinin systems and their mediators have traditionally been described as having opposite physiological effects on sodium excretion, vascular tone regulation, blood flow distribution, glomerular hemodynamic, and blood pressure regulation. Equally important to these functional effects, RAS and KKS have opposite effects on cardiovascular and renal architecture.

Thus, one of the RAS effectors, angiotensin II (AII), promotes cell proliferation, collagen synthesis, and fibrosis. On the other hand, bradykinin (BK), the hormone that affects KKS, is important in antagonizing the effects of angiotensin II. In this way, BK inhibits cell proliferation, stimulates collagenases, and inhibits ventricular, renal, and vascular fibrosis. A third hormone, the hemoregulatory tetrapeptide N-acetyl-seryl-aspartyl-lysyl-proline (AcSDKP), is involved in the control of hematopoietic cell proliferation and is normally degraded by the angiotensin I-converting enzyme (ACE) or kininase II.

In the determination of local levels of AII, BK, and AcSDKP, ACE and kallikrein, the enzymes that generate angiotensin II and bradykinin, have a critical contribution. Consistent with this, an important part of the beneficial effects of ACE inhibitors are explained by the triple effect of inhibiting the formation of angiotensin II and inhibiting the degradation of BK and AcSDKP, both in rats and humans.

From a morphofunctional perspective are presented the alterations of the vasoactive systems and its anatomical grounds, to the contribution of salt-sensitive hypertension and chronic kidney diseases.

With a focus on the more relevant information such as the importance of the tubulointerstitial space, the induction of angiotensin I-converting enzyme, the deficiency in kallikrein, and the contribution of mild hypoxia to kidney injury and hypertension.

Adequate dietary potassium intake and its contribution to health and renal and cardiovascular disease is underlined as an important factor.

The importance of COX-2 and derived prostaglandins to renal physiology and pathology is currently well established, and the lessons we learned from its pharmacological inhibition are discussed.

Review Questions
1. Which are the mediators for the KKS and how do they influence renal function?
2. What is the mechanism by which angiotensin II exerts a negative feedback on renin secretion?
3. Describe the renal interstitial space. Which is the hallmark of this compartment in the chronic renal disease?
4. Which is the cellular base for the fact that a high-potassium diet has a natriuretic effect?

Bibliography

Buffin-Meyer B, Younes-Ibrahim M, El Mernissi G, Cheval L, Marsy S, Grima M, Girolami JP, Doucet A (2004) Differential regulation of collecting duct Na^+, K^+-ATPase and K^+ excretion by furosemide and piretanide: role of bradykinin. J Am Soc Nephrol 15:876–884

Cervenka L, Harrison-Bernard LM, Dipp S, Primrose G, Imig JD, El-Dahr SS (1999) Early onset salt-sensitive hypertension in bradykinin B_2 receptor null mice. Hypertension 34:176–180

Dinchuk JE, Car BD, Focht RJ, Johnston JJ, Jaffee BD, Covington MB, Contel NR, Eng VM, Collins RJ, Czerniak PM, Gorry SA, Trzaskos JM (1995) Renal abnormalities and an altered inflammatory response in mice lacking cyclooxygenase II. Nature 378:406–409

Dyer AR, Elliott P, Shipley M (1994) Urinary electrolyte excretion in 24 hours and blood pressure in the INTERSALT Study. II estimates of electrolyte-blood pressure associations corrected for regression dilution bias The INTERSALT Cooperative Research Group. Am J Epidemiol 139: 940–951

Eaton B, Konner M (1985) Paleolithic nutrition: a consideration of its nature and current implications. N Engl J Med 312:283–289

Encuesta Nacional de Salud Chile, ENS 2009–2010 [Internet] (2010) Available from: https://www.minsal.cl/portal/url/item/bcb03d7bc28b64dfe040010165012d23.pdf

Gainer JV, Morrow JD, Loveland A, King DJ, Brown NJ (1998) Effect of bradykinin-receptor blockade on the response to angiotensin-converting-enzyme inhibitor in normotensive and hypertensive subjects. N Engl J Med 339:1285–1292

He FJ, MacGregor GA (2001) Beneficial effects of potassium. Br Med J 323:497–501

Kömhoff M, Wang JL, Cheng HF, Langenbach R, McKanna JA, Harris RC, Breyer MD (2000) Cyclooxygenase-2-selective inhibitors impair glomerulogenesis and renal cortical development. Kidney Int 57:414–422

Masferrer JL, Seibert K, Zweifel B, Needleman P (1992) Endogenous glucocorticoids regulate an inducible cyclooxygenase enzyme. Proc Natl Acad Sci U S A 89:3917–3921

Mezzano S, Droguett A, Burgos ME, Ardiles LG, Flores CA, Aros CA, Caorsi I, Vio CP, Ruiz-Ortega M, Egido J (2003) Renin-angiotensin system activation and interstitial inflammation in human diabetic nephropathy. Kidney Int 64:cS64–cS70

Morham SG, Langenbach R, Loftin CD, Tiano HF, Vouloumanos N, Jennette JC, Mahler JF, Kluckman KD, Ledford A, Lee CA, Smithies O (1995) Prostaglandin synthase 2 gene disruption causes severe renal pathology in the mouse. Cell 83:473–482

Vio CP, Figueroa CD (1987) Evidence for a stimulatory effect of high potassium diet on renal kallikrein. Kidney Int 31:1327–1334

Vio CP, Loyola S, Velarde V (1992) Localization of components of the kallikrein-kinin system in the kidney: relation to renal function. Hypertension 19:10–16

Vio CP, Cespedes C, Gallardo P, Masferrer JL (1997) Renal identification of cyclooxygenase-2 in a subset of thick ascending limb cells. Hypertension 30:687–692

Vio CP, An SJ, Cespedes C, McGiff JC, Ferreri NR (2001) Induction of cyclooxygenase-2 in thick ascending limb cells by adrenalectomy. J Am Soc Nephrol 12:649–658

Zhao C, Wang P, Xiao X, Chao J, Chao L, Wang DW, Zeldin DC (2003) Gene therapy with human tissue kallikrein reduces hypertension and hyperinsulinemia in fructose-induced hypertensive rats. Hypertension 42:1026–1033

Genetic Alterations of Tubular Transport of NaCl and Water

Learning Objectives
- To understand the fundamental role of transporters and ion channels in the function of a specific tubular segment.
- To understand how the loss or gain of function of a specific protein alters the function of a tubular segment.
- To understand how the alteration of a tubular segment is translated into alterations such as arterial hypertension and alterations of the acid-base balance, water balance, and K^+ balance.

A number of disorders of hydrosaline balance originate from mutations with loss or gain of function of transporter, ion channel, or enzyme function that are expressed in different segments of the renal tubule. Pathologies such as Bartter, Gitelman, and Liddle syndrome as well as other syndromes were recognized first by their symptomatology. The development of recombinant DNA techniques and others was of great help in elucidating the molecular origin of the disease. From the point of view of renal physiology, these monogenic pathologies have been of great help since they underscore the key physiological role played by transporters, channels, or enzymes in the function of a specific tubular segment.

13.1 Bartter Syndrome

This syndrome affects the thick ascending loop of Henle. This segment reabsorbs about 25% of the filtered NaCl load and plays a key role in the urinary concentration and dilution mechanism. NaCl resorption is mediated by the apical NKCC2 cotransporter. The K^+ entering the cell is recycled through apical membrane K^+ channels, known as ROMK. The basolateral exit of Na^+ occurs through the Na^+, K^+-ATPase. The main basolateral exit route for the Cl^- is the CLC-K1 and CLC-K2 channels. The latter seems to be the most important. The functionality of both channels requires the coexpression of the Barttin protein. Basolateral Cl^- exit and apical K^+ flux generate a lumen-positive transepithelial potential difference that

P. A. Gallardo, C. P. Vio, *Renal Physiology and Hydrosaline Metabolism*, https://doi.org/10.1007/978-3-031-10256-1_13

favors the paracellular reabsorption of monovalent (Na^+, K^+, NH_4^+) and divalent (Ca^{++}, Mg^{++}) cations. The high paracellular cation permeability is related to the expression of claudin 16 in the tight junction of this tubular segment. The absence of apical and basolateral aquaporins make this tubular segment impermeant to water, the net effect is the dilution of the tubular fluid and is therefore crucial to the mechanism of urinary concentration and dilution (*see* Fig. 4.11).

This syndrome was described by Dr. F. Bartter in 1962 in two patients who presented hypokalemia, metabolic alkalosis, hyperaldosteronism with hyperplasia, and hypertrophy of the juxtaglomerular apparatus. This syndrome is currently characterized as an autosomal recessive disease that is expressed in five distinct forms (Bartter 1 to 5), all due to mutations with loss of function of membrane proteins of the thick ascending loop of Henle. The common clinical feature of the syndrome is the urinary loss of NaCl associated with hypokalemia and metabolic alkalosis. The different forms of Bartter syndrome are detailed in Table 13.1.

The different forms of the syndrome occur during fetal life or in newborns. The disease presents with polydipsia, polyuria, growth retardation, low blood pressure, muscle weakness, and joint pain secondary to chondrocalcinosis.

Why the patients with this syndrome develop these clinical features? Any mutation with loss of function in NKCC2, ROMK, CLC-K, or other of the mentioned proteins will affect transcellular NaCl reabsorption as well as monovalent and divalent cation reabsorption through the paracellular way. NaCl reabsorption in the thick ascending limb plays a key role in the urinary-concentrating and diluting mechanism. The loss of function will reduce interstitial medullary hypertonicity and thus the osmotic gradient for AVP-dependent water reabsorption in the collecting duct. This explains polyuria and polydipsia. The same loss of function of this tubular segment will result in an increase in NaCl delivery to distal segments and an increase in urinary NaCl loss; this will decrease the extracellular volume and arterial pressure. The hypovolemia stimulates the renin-angiotensin-aldosterone system. Aldosterone will increase ENaC-dependent NaCl reabsorption; this will increase the lumen-negative transepithelial potential difference which will favor H^+ secretion in type A intercalated cells. The latter will increase urinary buffer

Table 13.1 Protein affected in the different forms of Bartter syndrome

Type of Bartter syndrome	Protein with mutations and loss of function	Comments
1	NKCC2, coding gene SLC12A1	$1Na^+$:$1K^+$:$2Cl^-$ cotransporter, apical membrane
2	KCNJ1, ROMK channel	K^+ channel involved in apical membrane K^+ recycling
3	CLCNKB coding gene for CLC-Kb	Basolateral Cl^- channel. Mediates basolateral Cl^- exit
4	BSND (Barttin protein)	CLC-Ka and CLC-Kb mutations. Sensorineural deafness
5	CaSR (calcium-sensing receptor)	Activating mutations in CaSR. Autosomal dominant hypoparathyroidism

acidification and bicarbonate regeneration leading to metabolic alkalosis. The alteration in Ca^{++} reabsorption in the thick ascending limb increases hypercalciuria. Hypokalemia can at least be explained by the diminished K^+ reabsorption in the thick ascending limb.

13.2   Gitelman Syndrome

Gitelman syndrome is an autosomal recessive disease that affects the distal convoluted tubule. This segment reabsorbs about 7% of the NaCl filtered load. The apical NaCl entry is mediated by the NCC cotransporter carrier (SLC12A3). The basolateral Na^+ exit is mediated by the Na^+, K^+-ATPase and the Cl^- output is mediated by the CLC-Kb chloride channel (*see* Fig. 4.12).

Gitelman syndrome is due to mutations with loss of NCC function. Compared to Bartter syndrome, Gitelman syndrome has milder effects and is diagnosed in adolescents and adults. Patients with this syndrome have urinary loss of NaCl, hypokalemia, and metabolic alkalosis. Unlike patients with Bartter syndrome, patients with Gitelman syndrome do not have polyuria and polydipsia.

The lack of NCC function causes a urinary loss of NaCl, generating hypovolemia and metabolic alkalosis secondary to the activation of the renin axis, renin-angiotensin II-aldosterone. The increased delivery of NaCl to the connecting tubule and collecting duct increases the electrogenic resorption of Na^+, generating a more negative transepithelial potential difference in the tubular lumen. The latter increases the driving force that favors the tubular secretion of K^+, which explains hypokalemia. A more negative tubular lumen also stimulates the secretion of H^+ by the intercalated A cells, increasing the excretion of acid and the regeneration of bicarbonate, which explains the metabolic alkalosis.

13.3   Liddle Syndrome

Liddle syndrome was described in 1963. It is an autosomal dominant form of hypertension accompanied by hypokalemia, metabolic alkalosis, and low plasma renin and aldosterone levels. The characteristics are typical of primary hyperaldosteronism, but patients with such a condition have low levels of aldosterone. This syndrome affects the electrogenic reabsorption of Na^+, mediated by ENaC of the apical membrane of the connecting and principal cells. The channel consists of subunits α, β, and γ, with a stoichiometry $\alpha2\beta\gamma$. The N and C ends of the subunits are cytosolic (*see* Fig. 4.13). The density of ENaC in the membrane depends on the insertion and endocytosis of channels in the apical membrane. Aldosterone, through SGK1, and AVP, through PKA, are two factors that increase the insertion of ENaC into the membrane. The system formed by Nedd4-2 and ubiquitin is responsible for the endocytosis of ENaC.

This syndrome is due to mutations in the terminal C-tail of the subunits β or γ. The mutations alter regions that are rich in proline. A direct consequence of this is the loss of the interaction of the β or γ subunits with the Nedd4-2 protein. This protein is key to the endocytosis of ENaC from the apical membrane. The net result is an increase in the abundance of ENaC on the membrane. The electrogenic resorption of Na^+ generates the negative transepithelial potential difference in the lumen. This favors the tubular secretion of K^+ and H^+, processes responsible for hypokalemia and metabolic alkalosis found in these patients.

## 13.4	Apparent Excess of Mineralocorticoid (AME)

This syndrome manifests itself as an autosomal recessive disease resulting from mutations with loss of function of the enzyme 11β-hydroxysteroid dehydrogenase type 2. This enzyme is expressed in the main and connecting cells of the distal nephron and converts cortisol into its inactive metabolite cortisone. The action prevents the occupation of mineralocorticoid receptors by cortisol, generating availability of receptors to be occupied by aldosterone. The clinical features of this syndrome are similar to those of Liddle syndrome.

## 13.5	Genetic Alterations of the Tubular Transport of Water

The kidney plays a key role in maintaining water balance and regulating plasma osmolality. Vasopressin (AVP) is a peptide hormone that functions as a signal connecting the state of plasma osmolality with renal water excretion. AVP binds to V_2 receptors located on the basolateral membrane of connecting cells, principal cells, and cells of the internal medullary collecting duct. These receptors stimulate adenyl cyclase through the Gs protein. cAMP generated by adenyl cyclase is an allosteric activator of protein kinase A (PKA). PKA phosphorylates aquaporin 2 (AQP-2) molecules, contained in subapical vesicles, resulting in their insertion in the apical membrane.

This mechanism of action of AVP occurs within minutes and is known as acute or short-term effect. The net result is an increase in the osmotic permeability of the apical membrane. Through the same transduction mechanism, AVP also stimulates transcription of the gene that codes for AQP-2. This long-term mechanism of action increases the abundance of AQP-2 protein.

Water entering through AQP-2 exits through the basolateral membrane via AQP-3 and AQP-4. In the internal medullary collecting duct, AVP increases the activity of the urea transporter UTA1, which promotes medullary urea recycling. In the long term, PKA phosphorylates the transcription factor CREBP, which stimulates the transcription of the AQP-2 gene. Therefore, AVP not only acutely controls the intracellular traffic of AQP-2 but also the abundance of the protein.

13.6 Diabetes Insipidus

Alterations in the tubular transport of water have clinical manifestations that include polyuria and polydipsia, which explains the large volumes of urine. There are three known forms of diabetes insipidus of genetic origin: central or neurohypophyseal diabetes insipidus. Also, there are two forms of nephrogenic diabetes insipidus (NDI): an autosomal form and a less common X-linked form.

In neurohypophyseal diabetes insipidus, plasma levels of AVP are insufficient to promote renal reabsorption of water. These patients show a positive response to the administration of V_2 receptor agonists, such as desmopressin (DDAVP). The molecular origin of this form of diabetes insipidus lies in mutations in the sequence of the preprohormone (preproneurophysin II), which contains the sequence of the AVP.

X-linked NDI is due to mutations with loss of vasopressin V_2 receptor function. Males suffering from this disease have clinical features such as hypernatremia, hyperthermia, or mental retardation due to repeated episodes of dehydration in childhood. These patients do not respond to V_2 agonists such as desmopressin, although there are some individuals who do respond partially due to the presence of functional V_2 receptors. Mutations of the V_2 receiver are translated, for example, in the case of the Gs protein, there is a defective coupling of the receptor to the Gs protein, so there is no stimulation of adenylyl cyclase and, therefore, elevation of cAMP levels.

The autosomal form of NDI is due to mutations with loss of function of AQP-2. Multiple mutations in AQP-2 have been identified that result in alterations in the intracellular traffic of the protein or in altered osmotic unit permeability.

Although it does not have a genetic origin like the other forms of diabetes insipidus, acquired NDI manifests itself as an alteration that may be secondary to the administration of some drugs or some other condition. One of the most studied cases is NDI acquired by the use of lithium in the treatment of psychiatric disorders. The treatment of bipolar mental disorders with lithium salts was started around 1949. Recently lithium is a powerful neuroprotector to slow the progression of degenerative diseases of the nervous system such as amniotic lateral sclerosis, Alzheimer's disease, Parkinson's disease, and Huntington's disease. This neuroprotective action is related to an increase in the survival of neurons through the induction of neurotrophic factor, thus stimulating antiapoptotic pathways and the Bcl2 protein.

One of the adverse effects of lithium treatment is the development of nephrogenic diabetes insipidus in 40% of cases. This alteration is almost irreversible and causes a significant reduction in the kidney's ability to concentrate urine. Clinical manifestations include polyuria, polydipsia, and resistance to the action of V_2 agonists. Lithium is a cation that filters freely in the glomerular filtration barrier. About 80% of the filtered load is reabsorbed in the proximal tubule, and a small fraction is reabsorbed through the Na^+ epithelial channel (ENaC) in the connecting and collecting duct. ENaC is a channel that has a high permeability to Li^+, which facilitates apical entry. However, the basolateral membrane does not have an efficient mechanism to remove lithium, since the Na^+, K^+-ATPase has a low affinity for the cation. The net result of this combination of permeabilities is the accumulation of lithium in the cells.

Recent studies have shown that lithium is a reversible inhibitor of glycogen synthase kinase 3β (GSK3β). The best-known action of this enzyme is the phosphorylation and inhibition of glycogen synthase. In the rat, mouse, and human kidney this enzyme was detected to be expressed in the collecting duct. The GSK3β plays an important role in the signal cascade downstream from the V_2 receiver. Recent studies on a line of mice knock out specific to the GSK3β collector tubule showed that these mice have a lower urinary concentration capacity in response to a period of water restriction. Cellular and molecular studies revealed a decrease in cAMP levels in response to vasopressin and forskolin (direct adenylyl cyclase activator). In addition, the collecting ducts of the transgenic rats showed a reduction in mRNA and AQP-2 protein. These findings suggest that GSK3β is a regulator of the activity of adenylyl cyclase, a key enzyme in the mechanism of AVP V_2 receptor transduction. The lack of cAMP secondary to inhibition of GSK3β reduces intracellular cAMP traffic and protein abundance. Both effects may explain resistance to V_2 agonists and the development of acquired nephrogenic diabetes insipidus.

Acquired nephrogenic diabetes insipidus can occur secondary to hypercalcemia. An increase in plasmatic Ca^{++} concentration activates the calcium-sensing receptor (CaSR) located in the basolateral membrane of the cells of the thick loop Henle. The activation of this receptor results in an inhibition of the activity of the apical NKCC2. The latter results in a reduction of NaCl reabsorption in this segment. It is necessary to keep in mind that the reabsorption of NaCl in the thick loop if Henle plays a key role in the generation of hyperosmolality of the medullary interstitium. Therefore, hypercalcemia reduces the osmolality of the medullary interstitium and reduces the osmotic gradient that favors the reabsorption of water in the collecting duct.

13.7 Conclusions

Monogenic diseases that alter NaCl or water transport along the nephron provoke mild to severe disturbances in electrolyte and/or water balance. They are excellent examples that underline the pivotal role of cotransporters (Bartter, Gitelman), ion channels (Liddle), and aquaporin (nephrogenic diabetes insipidus) in the function of a particular nephron segment.

Review Questions
1. Which is the functional basis to the fact that Bartter's patients but not Gitelman's patients develop polyuria?
2. What would you expect for the transepithelial potential difference in connecting tubules of a patient with Liddle syndrome?

Bibliography

Bartter FC, Pronove P, Gill JR, MacCardle RC (1962) Hyperplasia of the juxtaglomerular complex with hyperaldosteronism and hypokalemic alkalosis. A new syndrome. Am J Med 33:811–828

Bonnardeaux A, Bichet DG (2016) Inherited disorders of the renal tubule. In: Brenner BM, Rector FC (eds) The kidney, 10th edn. Elsevier, Philadelphia, pp 1434–1474

Answers to Chapter Questions

14

Chapter 2

2.1. There are two important differences: position of the renal corpuscle and length of the tubule. Cortical nephrons have their corpuscles near the surface. Juxtamedullary nephrons have their corpuscles more deeply located near the corticomedullary junction. Regarding the length of the tubule, cortical nephrons do not possess the thin ascending limb of Henle's loop; they are called short-looped nephrons. Juxtamedullary nephrons have the loop of Henle with three subsegments: thin descending, thin ascending, and thick ascending limb. It is important to keep in mind that cortical nephrons and yuxtamedullary nephrons are extreme examples. Between them, there is a whole variety of nephrons whose corpuscles are located between cortical and yuxtamedullary ones.

2.2. The glomerular filtration barrier is formed by the fenestrated capillary endothelium, the basal lamina, and the pedicels of podocytes, which are joined together by the filtration slit junction. The basal lamina is composed of three layers: inner rare lamina in contact with the endothelium, dense lamina, and outer rare lamina in contact with the basolateral membrane of pedicels. The basal lamina is made of collagen IV sheet and other proteins like laminin and fibronectin. Proteoglycans are abundant in the basal lamina. These proteoglycans are important because they confer negative charges to the basal lamina.

2.3. The juxtaglomerular apparatus (JGA) is formed at the vascular pole of each renal corpuscle. It is composed of the macula densa, extraglomerular mesangial cells, and granular cells of the afferent arteriole. The macula densa is formed by few epithelial cells that are oriented to the extraglomerular mesangium. They are taller than the normal thick ascending limb cells. The extraglomerular mesangial cells are located at the vascular pole between the wall of the afferent arteriole and the basal lamina underneath the macula densa cells. The granular cells are modified smooth muscle cells; they synthesize and accumulate the endopeptidase renin. The JGA is involved in the regulation of renal blood flow and glomerular filtration rate.

© Springer Nature Switzerland AG 2022
P. A. Gallardo, C. P. Vio, *Renal Physiology and Hydrosaline Metabolism*,
https://doi.org/10.1007/978-3-031-10256-1_14

2.4. The protein should be in the apical and basolateral membrane; a plausible alternative can be aquaporin 1; this water channel is very abundant in the proximal tubule.

2.5. The proximal tubule cell has an extensive amplification of the apical membrane due to abundant and tall microvilli that increase the surface of the apical domain for reabsorption of solutes and water. The lateral membranes from adjacent cells interdigitate, forming a tortuous lateral intercellular channel. The basal membrane exhibits infoldings. Both interdigitations and infoldings increase the basolateral membrane area for reabsorption.

Chapter 3
3.1. Data in table

Period	Urinary flow (mL/min)	[Inulin]$_{plasma}$ (mg/mL)	[Inulin]$_{urine}$ (mg/mL)	C_{inulin} (mL/min)
Protocol A: Increasing plasma inulin concentration				
1	1.2	0.9	90	**120**
2	1.3	1.5	136	**117.8**
3	1.0	2.3	282	**122.6**
4	1.4	3.8	336	**123.7**
5	1.2	5.7	570	**120**
Protocol B: Constant plasma inulin concentration and water overload				
6	1.3	0.5	46	**119.6**
7	2.2	0.6	34	**124.6**
8	3.1	0.4	16	**124**
9	6.0	0.5	10	**120**
10	6.6	0.5	9.2	**121.4**

3.1a. Inulin clearance is constant and independent of its plasma concentration.

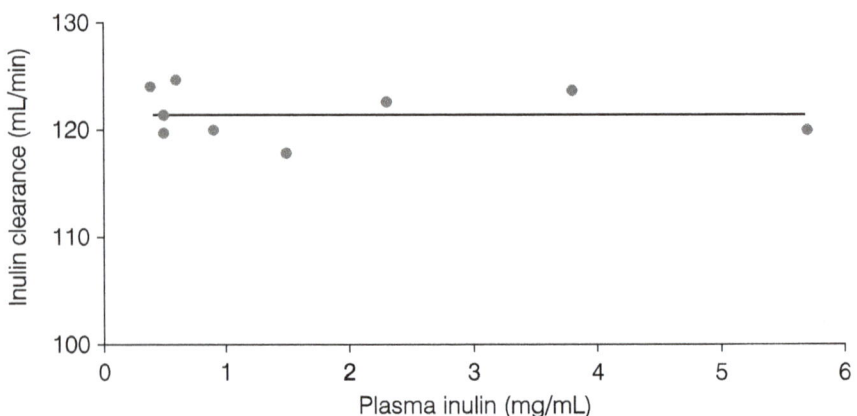

3.1b. All the filtered inulin is not reabsorbed or secreted and therefore is excreted in the urine.

3.1c. Due to tubular inulin handling, the excreted load will be directly proportional to the filtered load.

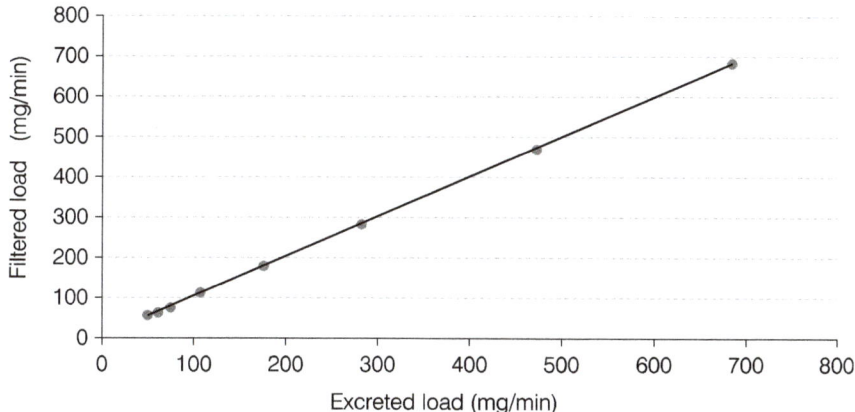

3.2. This can be explained because the filtered inulin is not reabsorbed in the proximal tubule. However, water is reabsorbed abundantly in the proximal tubule raising tubular inulin concentration and thus the ratio TF/P for this solute.

3.3. Within the physiological range of plasma glucose concentration, all the filtered glucose is reabsorbed in the proximal tubule and the excreted load will be 0.

3.4. Data in table.

Period	Creatinine plasma mg/dL	Creatinine urinary mg/dL	Urinary flow mL/d	Clearance creatinine mL/min	Clearance estimated[a]	1/ creatinine
Start study	1.7	146	1280	**76.2**	**71.8**	**0.588**
1 year later	2.1	112	1600	**59.2**	**58.2**	**0.476**
2 years later	3.0	180	1000	**41.6**	**40.7**	**0.33**

[a]Cockcroft's formula

3.4a. The progressive reduction in creatinine clearance might be related to a renal failure with glomerular disease. This will reduce creatinine filtration, elevating plasma creatine concentration.

3.4b. Implies a reduction in the renal functional mass.

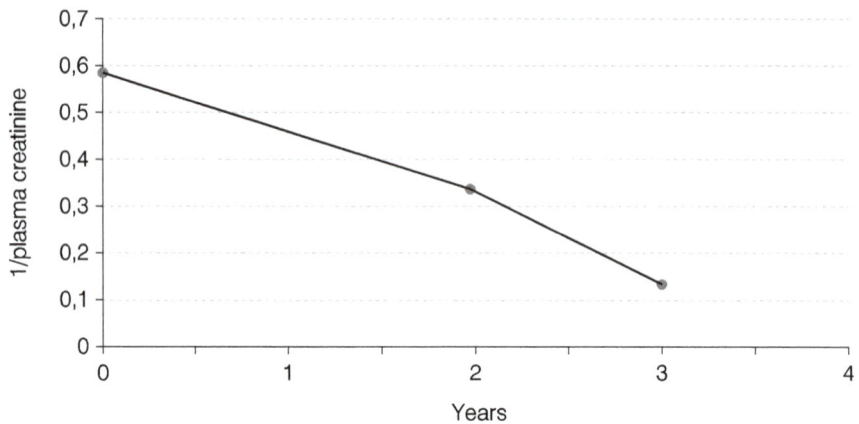

Chapter 4

4.1. Reabsorption is the movement of molecules from the tubular fluid to the blood. Tubular reabsorption occurs through a transcellular and paracellular pathways. In the first one, the molecules should cross the apical and basolateral membrane. In the latter, the molecules move through the channels formed by the proteins forming the tight junction. Tubular secretion is the movement of molecules from blood to tubular fluid.

4.2. Data.

Interval (min)	GFR (mL/min)	[Glucose] pl (mg/mL)	Filtered load (mg/min)	Excreted load (mg/min)	Glucose reabsorbed (mg/min)
0	125	1.0	**125**	0	**125**
Start glucose infusion	125				
26–40	125	1.0	**250**	0	**250**
60–80	125	2.8	**350**	20	**330**
80–100	125	3.5	**437.5**	76	**361.5**
100–110	125	4.0	**500**	125	**375**
130–140	125	5.0	**625**	250	**375**

4.2a. Glucose is freely filtered in the glomerular barrier; therefore, at constant GFR, the filtered load is proportional to the plasma glucose concentration. When all the filtered glucose is reabsorbed, the excreted load will be 0. When glucose reabsorption becomes maximal, excreted glucose increases above 0.

4.2b. The glucose threshold is between 2.0 and 2.8 mg/mL. Maximal transport (Tmax) rate is at 375 mg/min.

4.2c. Glucose is reabsorbed through the transcellular pathway. SGLT2 is the main transporter responsible for glucose reabsorption, SGLT1. The second half of the proximal tubule contributes to the clearance of all glucose from the tubular fluid. Hence, under physiological conditions the tubular fluid leaving the proximal tubule is devoid of glucose. Since both apical glucose transporters are Na^+-dependent, a reduction in Na^+ concentration in the perfusion solution will reduce glucose reabsorption.

4.3. The proximal tubule reabsorbs large amounts of Na^+, Cl^-, HCO_3^-, and organic solutes like glucose and amino acids. Besides, the apical and basolateral membranes have a high osmotic permeability due to the high expression of AQP-1. Transcellular solute reabsorption creates a small osmotic gradient that drives fluid reabsorption. Therefore, both solute reabsorption and high AQP-1 abundance in the same cells and membrane domains explain the coupling between solute and water reabsorption.

4.4. Both segments reabsorb NaCl and are water impermeable due to the lack of expression of aquaporins. Therefore, transcellular NaCl reabsorption is not accompanied by water reabsorption, diluting the tubular fluid.

4.5. The concept of fine regulation of Na^+ and water reabsorption points the fact that the rates of reabsorption of Na^+ and water are adjusted to meet with body needs to maintain Na^+ and water balance. Angiotensin and aldosterone play a key role in adjusting Na^+ reabsorption. Angiotensin II stimulates NaCl reabsorption in the distal convoluted tubule and aldosterone in the connecting tubule and cortical collecting duct, whereas vasopressin is crucial for water reabsorption in the connecting tubule and collecting duct.

Chapter 5

5.1a. This subject has a total body water (TBW) of 42 L, equivalent to 60% of body weight. Roughly, 20% of TBW (14 L) is located in the extracellular compartment (ECW) and 40% (28 L) is in the intracellular compartment (ICW). Since water density is 1, the ingestion of 1 liter is equivalent to 1 kg; thus, the subject will weigh 71 kg. The new TBW will be 42.6 L distributed in 14.2 L of ECW and 28.4 of ICW. Plasma Na^+ concentration before water ingestion was 143 mmol/L. Considering a ECW of 14 L the total amount of extracellular Na^+ will be 2002 mmol. Plasma osmolality is 286 mOsm/kg. After the water ingestion, the same amount of Na^+ will be dissolved in 14.2 L of water; the new Na^+ concentration and plasma osmolality will be 140.9 mmol/L and 281.8 mOsm/kg. The addition 1 L of water increased temporarily the volume of both compartments because the reduction in plasma osmolality will inhibit vasopressin secretion. This will reduce vasopressin-dependent water reabsorption allowing the kidney to excrete an amount of free water roughly equal to the volume ingested.

5.1b. Water balance is characterized because water intake equals water output in a 24 h period. The subject of the previous problem was in water balance and after ingestion of 1 L of water entered in a positive balance. As described in **5.1a**, the homeostatic response to attain the balance state will be the urinary excretion of roughly 1 L of water. The main mechanism to achieving balance is the inhibition of vasopressin secretion, the net result will be an increase in diuresis.

5.2. A vegetarian diet is associated with a low protein intake and therefore with less urea production compared with the ingestion of a typical protein-rich Western diet. Urea plays a key role in the urine-concentrating mechanism in the renal medulla. Almost 50% of the total extracellular solute in the renal medulla is urea. Medullary interstitial hyperosmolality is crucial in establishing the osmotic gradient for vasopressin-dependent water reabsorption along the collecting duct. Thus, the kidneys of a vegetarian subject will have a lower urine-concentrating capacity.

5.3. Animal A is the only animal that is perfused with a hypertonic NaCl solution. Since Na^+ is an effective osmol, it will remain in the extracellular compartment raising plasma osmolality and stimulating vasopressin secretion. Animals B and C are perfused with hypertonic solutions made of ineffective osmols that can permeate the plasma membrane.

5.4. Data (u: urine; pl: plasma).

Subject	Urinary flow (mL/min)	[Inulin]u mg/mL	[Inulin]pl mg/mL	GFR (mL/min)	u/pl for inulin	% of water reabsorbed
A	1.2	15.8	0.151	126	105	99.1
B	0.75	25.2	0.155	122	163	99.4
C	15.0	1.23	0.154	120	8	87.5

Subject	$[Na^+]$pl (mEq/L)	Filtered Na^+ (mEq/min)	$[Na^+]$u (mEq/L)	Excreted Na^+ (mEq/min)	Na^+ reabsorption (%)
A	136	17.1	128	153.6	99.1
B	144	17.6	192	144	99.2
C	134	16.1	10.2	153	99.1

Subject	Uosm mOsm/kg	Posm mOsm/kg	C_{H2O} mL/min	[Urea]u mg/dL	[Urea]pl mg/dL	C_{urea} (mL/min)	Urea reabsorbed (%)
A	663	290	−1.54	480	12	48	62
B	1000	300	−1.75	720	15	36	71
C	100	287	+9.77	48	10	72	40

5.4a. Because inulin is not reabsorbed or secreted along the renal tubule, urine to plasma concentration ratio increases as water is reabsorbed. Subject B, submitted to a water-restriction period, has the greater ratio due to the vasopressin-dependent water reabsorption. Subject C, submitted to a water loading period, has the lower urine to plasma ratio for inulin. His plasma vasopressin concentration is low, associated with a low water permeability in the collecting duct favoring water diuresis.

5.4b. Table 2 underscores the fact that under physiological conditions, water balance and Na^+ balance are handled independently. Subjects B and C were submitted to alterations in their water balance. However, their Na^+ reabsorption is similar. The differences in urinary Na^+ excretion are explained by their different rates of vasopressin-dependent water reabsorption. Thus, the three subjects are roughly excreting the same amount of Na^+ but dissolved in different volumes of water.

5.4c. Free water clearance is a parameter indicative of water vasopressin-dependent water reabsorption. If AVP is present, water is reabsorbed along the collecting duct, then the excreted urine becomes hypertonic and the free water clearance will be negative. If AVP is absent, the urinary solutes are dissolved in a greater volume of water and excreted as a diluted urine; in this case free water clearance will be positive.

5.4d. Urea is freely filtered in the glomerulus; 50% of the filtered load is reabsorbed in the proximal tubule. It is secreted from the medullary interstitium into the thin descending limb of Henle's loop. Except for the inner medullary collecting duct, the rest of the tubular segments are always impermeable to urea. Vasopressin stimulates carrier-mediated urea reabsorption in the inner medullary collecting duct.

Chapter 6

6.1a. Amphibians have an extracellular fluid that is hypertonic to freshwater in the pond; the osmotic gradient thus established favors a water influx to the body. Nephrons with glomeruli allow water filtration and excretion as a diluted urine.

6.1b. NaCl reabsorption at this level allows Na^+ recovery that balances Na^+ outflow due to a gradient with the freshwater. Also, Na^+ reabsorption contributes to the excretion of dilute urine.

6.2. Isoosmolality in this species is given by NaCl and plasma organic osmolytes. However, NaCl concentration is nearly half the concentration in the seawater.

6.3a. The amphibian urinary bladder functions very similarly to the collecting duct of the mammalian nephron. Water reabsorption across the urinary bladder epithelium is stimulated by vasotocin (amphibian antidiuretic hormone) through the insertion of aquaporin water channels in the apical membrane of the bladder epithelium.

6.3b. The amphibian bladder epithelium reabsorbs Na^+ through epithelial Na^+ channels in the apical membrane. Therefore, Na^+ reabsorption is electrogenic, developing a lumen-negative transepithelial potential.

6.4a. Salt overload received by these birds, through an unknown mechanism, stimulates the activity of the Na^+, K^+-ATPase of the basolateral membrane of salt-secreting gland cells. This allows the excretion of the salt excess.

6.4b. The epithelial cells lining the lumen of avian salt glands submitted to a salt overload should have an increased basolateral membrane, consistent with deeper and more abundant basolateral infoldings that contain an increased number of Na^+, K^+-ATPase. Both the morphological and the physiological adaptation would allow a great capacity for NaCl excretion.

6.5. Carbachol is a muscarinic receptor agonist. The avian salt gland receives abundant parasympathetic cholinergic innervation; cholinergic muscarinic receptors are in the basolateral membrane. Carbachol stimulation mimics glandular parasympathetic cholinergic stimulation. Bumetanide is an inhibitor of NKCC cotransporters. The basolateral membrane of the glandular epithelial cells expresses NKCC1 cotransporter that mediates Na^+-dependent Cl^- transport into the cell. The physiological relevance of these data underscores the importance of the parasympathetic innervation for basal and salt-stimulated glandular activity.

6.6. NKCC1 cotransporter uses the Na^+ electrochemical potential, maintained by the basolateral Na^+, K^+-ATPase. The cotransporter has a stoichiometry of $1Na^+$:$1 K^+$:$2Cl^-$. If the Cl^- secondary active transport would be performed by a NCC cotransporter with a stoichiometry of $1Na^+$:$1Cl^-$, oxygen consumption would be greater for the transport of 1 instead of two Cl^-.

6.7a. Salmons in this experiment were transferred from freshwater to seawater with 25% NaCl concentration. Therefore, fish are passing from a hypotonic to a hypertonic aquatic medium. Under this condition, the increase of the Na^+, K^+-ATPase is related with NaCl secretion that occurs in the gill mitochondrial-rich cells (chloride cells).

6.7b. β-actin gene expression is constitutive. Therefore, an increase in the Na^+, K^+-ATPase/actin ratio for mRNA means an increase in α isoform Na^+, K^+-ATPase mRNA. This increase in Na^+, K^+-ATPase mRNA supports the idea of increase pump abundance in gills that support increased NaCl secretion in mitochondrial-rich cells.

Chapter 7

7.1. High Na^+ ingestion will increase Na^+ concentration and osmolality of the extracellular fluid. The thirst mechanism will increase water ingestion that will return plasma osmolality towards normal range. However, the volume of water ingested will increase the extracellular volume, which can be monitored as an increase in body weight. Later, the homeostatic response will increase urinary Na^+ excretion in order to return the total body Na^+ content to normal.

7.2. Plasma vasopressin concentration should be elevated; the mechanism responsible for this increase comes from the medullary cardiovascular regulatory centers. Extracellular volume is reduced due to the loss of gastrointestinal fluids (diarrhea and vomit). Since Na^+ concentration is within its normal range, plasma osmolality is normal. Urine osmolality is increased due to high levels of vasopressin.

Plasma renin activity should be elevated due to the drop in arterial pressure. Renin secretion will increase due to sympathetic stimulation. Norepinephrine secreted by postganglionic sympathetic fibers binds to β1 adrenergic receptor coupled to the Gs-adenylyl cyclase transduction signal cascade; protein kinase A stimulates renin exocytosis. Hypotension also reduces the perfusion pressure in the afferent arteriole, which also stimulates renin secretion. High plasma renin activity will increase angiotensin II formation by the angiotensin I-converting enzyme (ACE).

Plasma aldosterone levels should be also elevated. Binding of angiotensin II to AT_1 receptor in adrenal glomerulosa cells will stimulate aldosterone synthesis. Aldosterone stimulates ENaC-dependent Na^+ reabsorption in the connecting and cortical collecting duct. Angiotensin II increases proximal tubule NaCl and $NaHCO_3$ reabsorption and NaCl reabsorption in distal convoluted tubule. The net result will be a reduction in Na^+ urinary excretion.

7.3. ENaC is located in the apical membrane of connecting cells and principal cells of the connecting tubule and cortical collecting duct. The fine regulation of Na^+ reabsorption to achieve Na^+ balance and extracellular volume is accomplished in the segments of the distal nephron. NCC cotransporter density in the distal convoluted tubule cells is stimulated by angiotensin II, whereas aldosterone increases ENaC abundance in the connecting and principal cells. Therefore, the sympathetic system and the renin-angiotensin II-aldosterone system play key roles in the maintenance of Na^+ balance and extracellular volume, which are crucial for a normal arterial pressure.

Chapter 8

8.1. The value for the concentration ratio is 20; this means that at physiological pH there are 20 times more base than acid. The base predominance helps to neutralize fixed acids derived from metabolism.

8.2. Bicarbonate reabsorption is defined as the recovery of filtered bicarbonate from the filtered fluid; this process occurs mainly in the proximal tubule and second in the thick ascending limb. Bicarbonate regeneration is defined to bicarbonate synthesis to replenish the amount of bicarbonate consumed during neutralization of H^+ derived from fixed acids mainly from protein metabolism.

8.3a. The acid source in this closed system is CO_2. Suppose that "x" is the mmol of HCl that should be added to bring pH from 7.4 to 7.1. The numerator in the Henderson-Hasselbalch equation is (24-x) and the denominator will be equal to the expression (αPCO_2 + x); α is the dissolution constant for CO_2 in plasma. In this setting 1.2 mmol of HCl will be required to make the pH change from 7.4 to 7.1.

8.3b. H^+ derived from HCl will react with HCO_3^- forming CO_2, increasing PCO_2 in the system. In this setting, the CO_2 excess will be eliminated to maintain a constant PCO_2 of 40 mmHg. Under this condition, the numerator in the Henderson-Hasselbalch equation will be (24-x), where "x" is the mmol of HCO_3^- that reacts with H^+ derived from HCl. The denominator will be the expression ($\alpha PCO_2 = 0.03 \times 40$ mmHg), corresponding to a CO_2 partial pressure of 40 mmHg held constant. In this case 12 mmol of HCl is required to make the pH change.

8.3c. In this setting, the system allows that PCO_2 descends from 40 to 20 mmHg. Thus, the numerator in the Henderson-Hasselbalch will be ($\alpha PCO_2 = 0.03 \times 20$ mmHg). The numerator will be (24-x). In this setting, 1 mmol of HCl is required for the pH change.

8.3d. To summarize, when the system is a closed one, only 1.1 mmol of HCl is required for the pH change. If the system becomes regulated allowing PCO_2 to be maintained at 40 mmHg, 12 mmol of HCl is needed for the desired pH change. If the system makes a compensation by lowering PCO_2 from 40 to 20 mmHg, 18 mmol of HCl is required to accomplish the pH change. The first or closed system and the second or regulated system are far from the physiological situation. Instead, the third system or compensated system is closer to the physiological situation. This is because acidosis stimulates ventilation and reduces PCO_2.

8.4. Acid excretion will be reduced due to the increase in bicarbonate urinary excretion.

8.5. Carbonic anhydrase inhibition in the proximal tubule will result in the inhibition of Na^+, Cl^- and water reabsorption, increasing NaCl delivery to the thick ascending limb and distal convoluted tubule. The glomerular-tubular balance mechanism allows that these tubular segments reabsorb the fraction of NaCl that was not reabsorbed in the proximal tubule. The fraction of water not reabsorbed proximally will be reabsorbed in the thin descending limb of juxtamedullary nephrons. The presence of bicarbonate in the tubular fluid in the connecting tubule and cortical collecting duct (at this point, bicarbonates in the tubular fluid behave as a non-reabsorbable anion) will reduce Na^+ reabsorption and K^+ secretion. The net effect of the carbonic anhydrase inhibitor will be an increase of bicarbonate excretion and a reduction in net acid excretion.

8.6. Table with data of patients with acid-base alterations.

Alteration	pH	PCO$_2$ (mmHg)	[HCO$_3^-$] plasma (mEq/L)	Disorder acid-base
Prolonged vomiting	7.55	46	**26.8**	**Metabolic alkalosis**
Intake of NH$_4$Cl	**7.1**	28	10	**Metabolic acidosis**
Uncompensated diabetes mellitus	**7.2**	28	12	**Metabolic acidosis**
Anxiety attack	7.57	**23.7**	21	**Respiratory alkalosis**
Chronic bronchitis	7.33	68	**34.6**	**Respiratory acidosis**

8.6a. See last column in the table.

8.6b. In the first case of metabolic acidosis, the value of the anion gap is 15 and in the normal range. In the uncompensated diabetes mellitus case, the anion gap is 38. This value is abnormally high. In this setting, organic acid (β-hydroxybutyric acid, acetoacetic acid) production is increased. H^+ derived from these acids is neutralized with bicarbonate, thus lowering the plasma bicarbonate concentration. The anions accompanying the H^+ are not included in the anion gap formula. In this case, the anion gap is useful to define the origin of the metabolic acidosis.

8.6c. Bicarbonate concentration in this subject will be elevated. The vomiting episodes have two consequences: acid loss (gastric secretion is rich in HCl) and a reduction of extracellular fluid. Acid loss is the first cause of metabolic alkalosis in this subject. The H^+ secreted through the canalicular membrane of the gastric parietal cell is produced in an equimolar ratio with HCO_3^- by the activity of cytosolic carbonic anhydrase. HCO_3^- is exported to plasma through the basolateral membrane and H^+ is secreted to the canalicular lumen through the gastric H^+, K^+-ATPase. Thus, when vomit occurs H+ is eliminated from the extracellular fluid, but the bicarbonate ions remain in the extracellular fluid and represent a gain of base accounting for part of the metabolic alkalosis. Vomit is a transcellular fluid and thus part of the extracellular fluid. The loss of extracellular fluid activates the sympathetic outflow to the kidney and the renin-angiotensin-aldosterone system. Norepinephrine and angiotensin II stimulate $NaHCO_3$, NaCl, and fluid reabsorption in the proximal tubule. Angiotensin II stimulates NaCl reabsorption in the distal convoluted tubule by increasing NCC cotransporter phosphorylation. The hormone also stimulates aldosterone synthesis which in turn increases ENaC-mediated Na^+ reabsorption in the connecting and principal cells of the distal nephron. This effect will increase the lumen-negative transepithelial potential difference. The latter will favor H^+ secretion from adjacent type A intercalated cells. This H^+ acidifies urinary buffers, while bicarbonate is transported to plasma as new bicarbonate molecule that increases the metabolic acidosis.

8.6d. In the first case, net acid excretion is reduced and increased in the cases of metabolic acidosis.

Chapter 9

9.1. Insulin plays a key role in the translocation of K^+ from the extracellular to intracellular compartment. This effect is mediated by Na^+, K^+-ATPase stimulation. The insulinic effect lowers plasma K^+ concentration; this effect is very important in the postprandial period. In a subject with insulin deficiency, an increase in plasma K^+ is expected. The first is the reduction in GLUT-4 dependent glucose transport in skeletal muscle and adipose tissue, generating hyperglycemia. Second, hyperglycemia generates plasma hyperosmolality which moves water out of the cell, increasing cell K^+ concentration. The latter will increase the electrochemical gradient that favors K^+ efflux from the cell.

9.2. In this setting, the ENaC-mediated Na^+ reabsorption will be increased; the process is electrogenic, increasing the lumen-negative transepithelial potential difference. The increase in electrogenic Na^+ reabsorption also stimulates Na^+, K^+-ATPase activity that favors tubular K^+ secretion in the connecting and principal cells. The increase in K^+ secretion favors the development of hypokalemia.

9.3a. Blocking electrogenic Na^+ reabsorption will reduce the electrochemical gradient that favors K^+ secretion. The latter favors the development of hyperkalemia.

9.3b. Aldosterone deficiency will reduce the electrogenic Na^+ reabsorption in the connecting tubule and cortical collecting duct. The hormonal deficiency will reduce the abundance of two important proteins for transepithelial electrogenic Na^+ reabsorption: ENaC and Na^+, K^+-ATPase. The net effect will be the development of hyperkalemia.

9.3c. The loss of extracellular fluid will activate the renin-angiotensin-aldosterone system. Aldosterone action tends to decrease plasma K^+ concentration.

9.4. The phenomenon can be explained by the effect of hyperkalemia affecting the resting membrane potential in all cells, but especially in axons and muscle fibers. An increase in extracellular K^+ will decrease the electrochemical potential difference, depolarizing the resting membrane potential and inactivating voltage-dependent Na^+ channels; these channels participate in the depolarization phase of the action potential. Second, hyperkalemia also affects the repolarization of cells. Both effects affect the electrical activity of cardiac muscle cells, which is manifested as alterations in the electrocardiogram.

Chapter 10

10.1. NaCl reabsorption in the thick ascending limb is mediated by apical NKCC2. The Na+ is transported across the basolateral membrane by the Na^+,K^+-ATPase. K^+ that entered the cell through NKCC2 is recycled to the lumen via ROMK channel. Cl^- exits across the basolateral through CLCk channels. The movement of K^+ to the lumen and Cl^- through the basolateral membrane generates a lumen-positive transepithelial potential of about +15 mVolt. The tight junctions in the thick ascending limb contain claudin 16. This claudin is essential for creating cation-permeable paracellular channels that allow the passive reabsorption of divalent cations like Ca^{++} and also Mg^{++}.

10.2. PTH receptors are located in the basolateral membrane of the thick ascending limb cells. Activation of PTH receptors results in increased levels of cAMP and PKA activation which phosphorylates and increases the activity of NKCC2 cotransporter. The mechanism described in answer 10.1 becomes enhanced. The net effect is an increase in paracellular calcium reabsorption. The same kind of receptors is located in the basolateral membrane of proximal tubule cells. In this case, the cAMP-PKA signal transduction cascade results in the internalization of apical Na^+-phosphate (NPT) cotransporters. The net result is an inhibition of phosphate reabsorption in the proximal tubule.

10.3. The proximal tubule is very important for calcium homeostasis. First, proximal tubule reabsorbs passively about 70% of the calcium filtered load. Therefore, this segment plays an important role in renal calcium handling. Second, in the proximal tubule cells the last step in the synthesis of 1,25(OH)2 cholecalciferol occurs. This is the only hormone that stimulates intestinal (duodenal) calcium absorption. Therefore, the importance of the proximal tubule can be stated in two points: first, major role in tubular calcium reabsorption and second, hormonal synthesis for intestinal calcium absorption.

10.4. A reduction in plasma free Ca^{++} concentration will stimulate PTH secretion through a calcium-sensing receptor expressed in the plasma membrane of parathyroid chief cells. In the bone, PTH will stimulate osteoclast formation and hence bone resorption, releasing Ca^{++} and phosphate to plasma. In the kidney, PTH inhibits proximal tubule phosphate reabsorption; this effect is mediated by PTH receptors that couple through Gs to adenylyl cyclase. The net effect is the endocytosis of apical NPT cotransporters. Thus, PTH increases phosphate excretion. Also, in the proximal tubule, PTH stimulates calcitriol synthesis by increasing the activity of the 1α hydroxylase. In turn, calcitriol stimulates transepithelial duodenal Ca^{++} absorption. PTH also stimulates Ca^{++} reabsorption through ECaC in the distal convoluted tubule. The net effect is the recovery of free plasma Ca^{++} concentration.

Chapter 11

11.1. Angiotensin-converting enzyme (ACE) has a dual action: converting inactive Ang I to active Ang II and degrading active bradykinin. This exopeptidase digests the angiotensin I peptide (10 amino acids) by cutting the last two amino acids generating the peptide hormone angiotensin II (8 amino acids). Another substrate for ACE is bradykinin (9 amino acids); in this case the enzyme cuts the last two amino acids and generates the inactive bradykinin 1–7. Thus, ACE activity generates angiotensin II, an active peptide hormone that increases sodium reabsorption, stimulates aldosterone synthesis, and increases total peripheral resistance and inactivates bradykinin, a vasodilator hormone that increases sodium excretion. ACE expression is very abundant in the pulmonary endothelium and the brush border of S2 proximal tubule cells.

11.2. The kidney is a rich source of ACE and contains 5–10 more than the lung. ACE is expressed in the brush border of the proximal tubule cells where it can generate angiotensin II, since the proximal tubule contains angiotensinogen, the substrate of renin. This is considered the proximal tubule or intrarenal renal-angiotensin system. Renin can filter across the glomerular barrier and enter the tubular fluid. Therefore, the proximal tubule contains all elements of the intrarenal system.

11.3. The connecting tubules begin after the distal convoluted tubule. It is composed of connecting cells, whose ultrastructure is similar to that of the distal convoluted tubule cells. Connecting cells are the only kallikrein-secreting cells in the kidney. Kallikrein is secreted through the apical and basolateral membrane. Its activity originates in the peptide hormone bradykinin, which increases sodium and water excretion.

11.4. Immunoreactive cells for EPO should be located in the peritubular space of cortical labyrinths. In situ hybridization and immunocytochemistry studies demonstrated that immunoreactive cells are fibroblasts.

11.5. COX-2 has been mentioned as an inducible enzyme, involved in inflammatory reactions. However, Vio et al. demonstrated that COX-2 is constitutively expressed in normal conditions in cells of the thick ascending limb of Henle of the rat adult kidney. This strongly suggests that COX-2 has a physiological role in adult renal function and kidney maturation. Furthermore, COX-2 KO transgenic mice develop several renal abnormalities, demonstrating that the enzyme is required for normal kidney development.

Chapter 12

12.1. The mediators of KKS are prostaglandins (PGs) and nitric oxide (NO). PGs, synthesized by COX-1 and COX-2 enzymes, are important regulators of renal blood flow, renin secretion, and NaCl reabsorption. NO is synthesized by three NOS enzymes, all of which are expressed in different renal sites including the macula densa and distal collecting tubules. NO is a potent vasodilator through activation of soluble guanylyl cyclase in smooth muscle cells in the vascular wall. In general, NO and PGs increase NaCl excretion.

12.2. Angiotensin II activates AT1 receptors in the granular cell plasma membrane. These receptors are coupled to Gq and phospholipase C signal transduction system. The net effect is the inhibition of renin secretion and accumulation in the afferent arteriole. Prolonged high levels of angiotensin II inhibit renin synthesis.

12.3. The renal interstitium is the intertubular, extraglomerular, and extravascular space of the renal tissue. It is bounded on all sides by tubular and vascular basement membranes and contains scarce cells, extracellular matrix, and interstitial fluid. Its distribution varies within the kidney as it represents approximately 8% of the total volume of the cortex parenchyma and up to 40% in the internal medulla. The term "renal interstitium" is often used to refer to the peritubular interstitium (the cortical labyrinth space between tubules, glomeruli, and capillaries); periarterial connective tissue and extraglomerular mesangium are considered specialized interstitium. The tubular interstitium in the cortex and medulla differ with respect to cell content, composition of the extracellular matrix, relative volume, and endocrine function, which justifies considering the cortical and medullary interstitium as separate or different structures. A central component of chronic kidney injury is cell infiltration and fibrosis that occurs in the tubulointerstitial space. Under normal conditions this space is composed of the capillaries and the interstitium surrounding the tubules and with the occasional presence of few macrophages, fibroblasts, and dendritic cells.

12.4. First, a high-potassium diet decreases the expression of renin, tubular ACE, and angiotensinogen. This reduces intrarenal angiotensin II and Na+ reabsorption. Second, a high-potassium diet increases plasma potassium. This inhibits the basolateral channel Kir4.1/5.1 of distal convoluted cells. This event sets an intracellular cascade that ends with NCC reduced activity, contributing to increased NaCl excretion. Third, the high-potassium diet increases the synthesis of kallikrein and the bradykinin B2 receptor, a natriuretic system.

Chapter 13

13.1. Bartter syndrome affects the function of the thick ascending limb. This segment reabsorbs NaCl without water and contributes significantly to the hypertonicity of the medullary interstitium which is the driving force for vasopressin-dependent water reabsorption in the medullary collecting duct.

13.2. The transepithelial potential difference should be increased and more negative in the lumen. In this syndrome, there is an increase in the abundance of ENaC channels in connecting and principal cells. This channel is responsible for electrogenic Na^+ reabsorption in these cells of the distal nephron.

Index

A

ACE, *see* Angiotensin converting enzyme
    (ACE)
Acid
    nonvolatile, 172
    titratable
        in formation of new HCO_3^-, 166
    volatile, 155
Acid base balance
    formation of new HCO_3^-, 158
    HCO_3^- buffer system and, 158, 159
    HCO_3^- reabsorption along the nephron in
        at the cellular level, 164
        by the connecting tubule and collecting
            duct, 162
        by the proximal tubule, 160–162
        segmental, 162
        by the thick ascending limb of the loop
            of Henle, 162
    overview of, 160
    and plasma [K^+], 162
    regulation of H^+ secretion in, 160, 162
    renal net acid excretion, 163, 165
Acid-base disorders
    metabolic, 172
    respiratory, 173
    response to
        extracellular and intracellular buffers in,
            173
        renal, 173
        respiratory, 173
Acid excretion, renal net, 2, 156, 157, 163, 165,
    166, 169, 170, 173, 262, 263
Acidosis
    metabolic
        arterial blood values of, 170
        compensatory response of, 170, 173
        and H^+ secretion, 170

        in insulin-dependent diabetes, 155
        in patients with insulin-dependent
            diabetes, 170
        and plasma [K^+], 171
        with renal compensation, 170
        respiratory acidosis and, 170
        in regulation of P_i excretion, 166, 169, 170,
            173
    respiratory, 173, 262
    response of the nephron to, 168, 169
Active transport
    secondary, 118
Adenosine
    in regulation of renal blood flow and
        glomerular filtration rate, 15, 40, 41
    renal blood flow and, 15, 39–41
Adenosine triphosphate (ATP)
    in regulation of renal blood flow and
        glomerular filtration rate, 39, 41
    in tubuloglomerular feedback, 39, 40
Afferent arteriolar resistance, in regulation of
        renal blood flow and glomerular
        filtration rate, 40, 49
Afferent arteriole, 9, 10, 12, 15, 25, 27, 33, 34,
        37–41, 44, 48, 140, 141, 143, 149,
        208, 210, 211, 217, 220, 253, 260,
        266
Albumin
    in proximal tubule, 49, 65
Aldosterone
    increases reabsorption of sodium, 76
    in K^+ secretion, 22, 23, 79, 80, 150, 188,
        189, 247
    in regulation of NaCl and water
        reabsorption, 74, 143, 169
    in regulation of plasma [K^+], 2, 137
Aldosterone-sensitive distal nephron (ASDN),
    151

P. A. Gallardo, C. P. Vio, *Renal Physiology and Hydrosaline Metabolism*,
https://doi.org/10.1007/978-3-031-10256-1

Milton Keynes UK
Ingram Content Group UK Ltd.
UKHW020112280923
429512UK00001B/14